Geomembranes
Identification and Performance Testing

Other RILEM Reports available from Chapman and Hall

1 Soiling and Cleaning of Building Façades
  Report of Technical Committee 62-SCF
  Edited by L. G. W. Verhoef

2 Corrosion of Steel in Concrete
  Report of Technical Committee 60-CSC
  Edited by P. Schiessl

3 Fracture Mechanics of Concrete Structures:
  From Theory to Applications
  Report of Technical Committee 90-FMA
  Edited by L. Elfgren

4 Geomembranes – Identification and Performance Testing
  Report of Technical Committee 103-MGH
  Edited by A. Rollin and J. M. Rigo

Publisher's Noter

This RILEM Report has been produced from the typed chapters provided by the members of RILEM Technical Committee 103-MGH, whose cooperation is gratefully acknowledged. This has facilitated rapid publication of the Report.

Geomembranes

Identification and Performance Testing

**Report of Technical Committee 103-MGH
Mechanical and Hydraulic Testing of Geomembranes
RILEM**

(The International Union of Testing and Research Laboratories
for Materials and Structures)

EDITED BY

A. Rollin and J-M. Rigo

CRC Press
Taylor & Francis Group
Boca Raton London New York

CRC Press is an imprint of the
Taylor & Francis Group, an **informa** business
A TAYLOR & FRANCIS BOOK

CRC Press
Taylor & Francis Group
6000 Broken Sound Parkway NW, Suite 300
Boca Raton, FL 33487-2742

First issued in paperback 2019

ISBN-13: 978-0-412-38530-8 (hbk)
ISBN-13: 978-0-367-86401-9 (pbk)

British Library Cataloguing in Publication Data
RILEM. Technical Committee 103-MGH Mechanical and Hydraulic
  Testing of Geomembranes
  Geomembranes – identification and performance testing.
  1. Synthetic Materials
  I. Title.   II. Rollin, A.   III. Rigo, J.M.
  620.11

  ISBN 0-412-38530-9

Library of Congress Cataloging-in-Publication Data
Available

Visit the Taylor & Francis Web site at
http://www.taylorandfrancis.com

and the CRC Press Web site at
http://www.crcpress.com

Contents

Preface

RILEM Technical Committee TC-103 on Geomembranes was formed in 1986 under the chairmanship of Professor K. Gamski. It was believed at that time that the use of synthetic components as construction materials was such that technical committees should be set up to study these materials, to identify the problems related to their nature and their uses, to investigate the different installation techniques, and to assess their ageing and durability. Since then, four meetings were held consecutively in Paris (September 1987), in Montreal (June 1988 and June 1989), and in Porto Conte, Sardinia (October 1989). This book is a result of the work achieved by the experts and originating from 12 countries.

A geomembrane is defined as a very low permeability synthetic or re-inforced bituminous sheet placed in contact with soil. They are used extensively in transportation applications (tunnels), environmental applications (landfills, sand reservoirs) and geotechnical related applications (dams) in substitution or in addition to compacted clay, bentonite, concrete, steel and bituminous materials. They are used to contain or restrain gas and liquid movement. Geomembranes are manufactured in industrial plant and are polymeric and/or bituminous products shipped on site in rolls prior to being installed and seamed to obtain a continuous impermeable layer. Welding or adhesive techniques are used to bond the sheets together. In this book, *in situ* geomembranes are excluded.

Historically, geomembranes were used in the 1930s while widespread adoption started in the 1940s. In 1987 more than 50 million m^2 were installed worldwide and future growth of 20 to 25% per year is forecast for the next five years.

Due to the extremely large quantities of synthetic materials used in earth structures and of the potential risks involved in environmental applications, it was a RILEM concern to ensure that knowledge was available about the installation techniques, the seaming methods and the testing procedures, once the ageing and durability testing of these products was established. Since the use of geomembranes is relatively recent, since many new products have been produced specifically for these applications and since new seaming techniques are being used, information is not readily available to engineers, installers, contractors and others.

This book has been written to provide information on all important facets of geomembrane engineering and is divided into two parts: general and selected topics. Each chapter was written by one or more committee members and revised by all committee members in the hope that all necessary information was collected on each topic.

The active members of the RILEM TC-103 committee comprise:

J. P. Benneton*, Ministère de l'équipement, CETE (France)

C. Bernard*, CEMAGREF (France)

D. A. Cazzuffi*, ENEL/CRIS (Italy)

E. Dembiski, Gdansk Technical University (Poland)

D. Fayoux*, Plavina, Solvay Group (Belgium)

R. Frobel*, R. K. Frobel and Associates (USA)

J. P. Gourc*, Université Joseph Fourier (France)

F. Gousse*, CEMAGREF (France)

H. E. Haxo*, Matrecon (USA)

G. Hemond*, SOLEM–UTC/CEB (Switzerland)

S. E. Hoekstra*, Akso Industrial Systems (The Netherlands)

T. S. Ingold*, Consultant (United Kingdom)

R. H. Koerner*, Drexel University, GRI (USA)

I. D. Peggs*, Geosyntec (USA)

P. Pierson*, Université Joseph Fourier (France)

P. de Porcellinis, RODIO (Spain)

J-M. Rigo*, secretary, Université de Liège, GRC (Belgium)

A. L. Rollin*, chairman, Ecole Polytechnique de Montréal (Canada)

H. Schneider*, Polyfelt (Austria).

P. M. Sembenelli, Consultant (Italy)

H. Steffan, Consultant (Germany)

M. Van Wijk, AMOCO Fabrics (The Netherlands)

*Contributing authors

The authors wish to thank RILEM secretariat for their support, Ecole Polytechnique, University of Liege (GRC), CISA (Environmental Sanitary Engineering Center, Caligari) and CEMAGREF for their administrative and financial support, and all participating organizations and producers for allowing their experts to attend the meetings. We are grateful to the following organizations for the participation of their members as authors of this book.

AKZO   Industrial Systems bv, Arnhem, The Netherlands.
CEMAGREF   (Centre national du Machinisme agricole, du Génie Rural, de Eaux de Forêts), Lyon, France.
ENEL-CRIS   (Research Centre on Hydraulics and Structures–Italian National Electricity Board), Milan, Italy.
FROBEL and Associates (Geosynthetics Consultants), Colorado, United States of America.

GEOSEP	(Geosynthetics Research Group–Ecole Polytechnique), Montreal, Canada.
GEOSYNTECS	(Testing and Consulting Laboratory), Florida United States of America).
GRC	(Geotextiles and Geomembranes Research Centre – Université de Liège), Liege, Belgium.
GRI	(Geosynthetic Research Institute – Drexel University), Philadelphia, United States of America.
IRIGM	(Institut de recherches interdisciplaines en géotechnique et mécanique – Université Joseph Fourier), Grenoble, France.
MATRECON	(Geosynthetics Consulting Firm), California, United States of America.
PLAVINA	(Solvay Group), Brussels, Belgium.
SOLEM	(Société et Laboratoire d'Etudes, Maintenance et d'Expertises – UTC/CEB), Geneva, Switzerland.

A. L. Rollin and J-M. Rigo
Editors

DEDICATION

The authors of this book wish to express their deep gratitude to the late Professor Dr K. Gamski. Professor Gamski was the first Chairman of RILEM Technical Committee 10C and proposed the basis of this manual. His sudden death in September 1988 created a void.

The authors commemorate Professor Gamski as a charming and wise man, a great pioneer in the field of geosynthetics.

PART ONE

GENERAL TOPICS

1 GEOMEMBRANE OVERVIEW – SIGNIFICANCE AND BACKGROUND

R. M. KOERNER
Geosynthetic Research Institute, Drexel University,
Philadelphia, PA, USA

1-1 Introduction

The containment, storage and movement of liquids, solids and gases within geomembrane enclosures represents a distinct and unique challenge facing the engineering and scientific community. Depending upon the type of materials being retained, one must be concerned about health, environment, aging, durability and safety against uncertainties; all within the framework of an appropriate cost-to-benefit ratio for the situation being considered. Clearly it is a demanding task, yet one that is necessary and worthwhile. Within the various types of materials to be contained, stored and sometimes transported are the following:

Liquids — potable water, industrial water, safe-shutdown water, industrial waste water, municipal waste water, liquid process chemicals, liquid waste chemicals, solid waste leachates, etc.

Solids — radioactive waste, hazardous waste, industrial waste, municipal waste, hospital waste, incinerator ash, power plant ash, mining waste, heap leach ores, building demolition, solar ponds, etc.

Gases — industrial product gases, industrial waste gases, landfill gases, radon gases, radioactive waste gases, water vapor, ground released gases, etc.

Indeed, the range of materials to be contained, stored or transported is as wide as technology is itself… a formidable, yet imminently worthwhile task.

1-2 Historic solutions

Many of the challenges listed in the previous section are certainly not new. Ancient history often mentions the burden of bringing potable water from streams and wells to villages and towns for use and consumption. Early crude attempts in Egypt and neighboring lands were brought into a superb art-form by the Romans, whose aqueducts represent Herculean feats still marveled today. Their use of various stone block paving schemes for leakage control, while undoubtedly troublesome and difficult to maintain, were successfully implemented.

Not until the advent of man-made materials, however, have the choices been so numerous and capable of being adapted to a specific substance being contained, stored or transported. The following list, modified from Kays (1987), gives some idea as to the many different types of liner systems available to a designer or owner.

- Steel
- Concrete
- Soil cement
- Gunite
- Waterborne treatment additives
- Asphalt concrete
- Compacted clay
- Bentonite modified soil
- Bituminous panels and rolls
- Thermoset polymers
- Thermoplastic amorphous and semicrystalline polymers

Within the last three liner categories comes the area of geomembranes to which this book is directed. Also called liners, seals, or flexible membrane liners, geomembranes are polymeric or bituminous sheets which are factory manufactured in a single thickness or in multiple built-up layers. This latter possibility makes possible a number of variations. Figure 1 illustrates the possible variations in different manufacturing methods. These manufacturing techniques result in products varying in thickness from 0.5 mm to 5.0 mm. Note that this is significantly less in thickness than the traditional civil engineering liner materials made from concrete or soil. This, in itself, underscores the necessity of proper design and testing, of flawless construction, and of adequate post-installation maintenance.

Polymer + Additives

Fabric

1. Extrusion   2. Calendering   3. Spread Coating

Nonreinforced   Reinforced

Fabrication

Panel

Installation

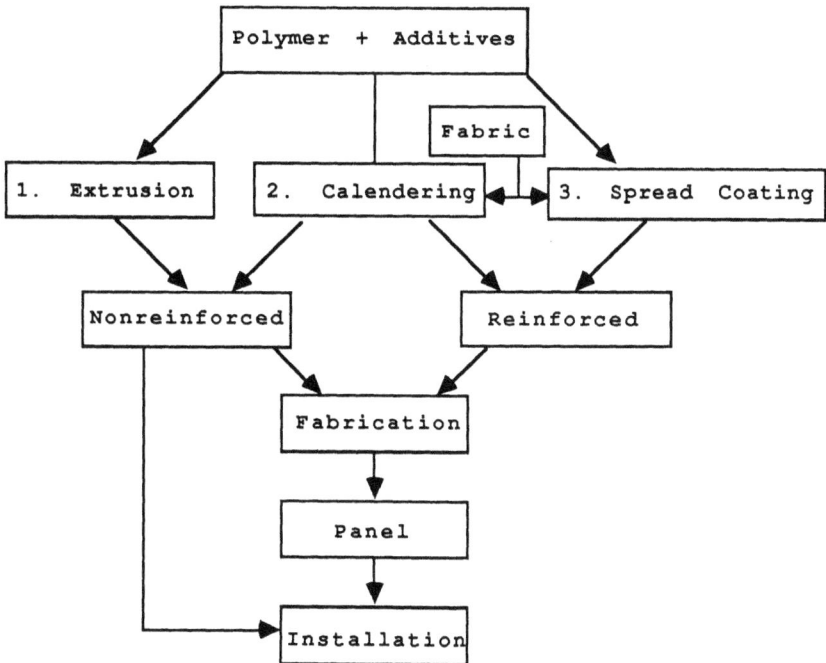

Fig. 1   Major methods for manufacturing geomembranes, after Haxo (1986)

1-3 Background of modern liners

The genesis of modern liners, which in the context of geo-synthetics are called geomembranes, is intrinsically tied to the development and growth of the polymer industry. Perhaps the earliest type of thin prefabricated sheets, placed on a prepared soil base as currently practiced, was the use of polyvinyl chloride as swimming pool liners in the early 1930's, see Staff (1984). Transportation related trials were practiced by the Bavarian Highway Department using low density polyethylene seepage barriers in the late 1930's, see Bell and Yoder (1957). For large fabricated sheets, however, it seems as though the synthetic thermoset materials made the first significant impact. Potable water reservoirs sealed with butyl rubber liners are reported in the late 1940's, see Staff (1969). A research oriented approach toward the lining of water canals was undertaken by the U.S. Bureau of Reclamation in the 1950's. The research was performed by Lauritzen who worked close with the manufacturer and resin suppliers. A bevy of worldwide activity using a wide range of polymeric materials ensued. As seen in Figure 2, polyvinyl chloride (PVC) canal liner installations were made in Canada, Russia, Taiwan, and in Europe throughout the 1960's and 1970's. Chlorosulfonated polyethylene (CSPE) was developed in the USA and Europe and created a major impact. Polyethylene (PE) liners were developed in West Germany and spread to all of Europe, Africa, Australia and North America. South Africa also was involved in PE liner development which it exported and then further developed in North America. Bituminous (asphalt) panels were used by agricultural groups for canal and reservoir liners since the 1940's. Their origins were probably worldwide and by numerous groups of both private and public agencies.

Today, geomembranes are indeed worldwide in both their availability and applicability. They are truly a subset of geosynthetics and a major category in their own right. In some applications, such as solid waste landfill liners, they are the primary focus of attention. In such cases the other geosynthetics such as geotextiles, geonets, geogrids and geocomposites are auxiliary to the proper and long-term functioning of the geomembrane.

1-4 Various geomembrane types

It should be recognized from the outset that a particular polymer type is actually a generic classification wherein the compounder has a wide variety of choices insofar as additives, fillers, extenders, antidegradients, etc., is concerned. Table 1 lists the three common classes of polymeric geomembranes where the basic resin (polymer or alloy) is seen to be only one part, albeit the major part, of the final product.

Table 1 Major components in various types of polymeric geomembranes, after Haxo (1986)

COMPONENT	COMPOSITION IN PARTS BY WEIGHT		
	Thermoset (Elastomers)	Thermoplastic (Amorphous)	Thermoplastic (Semicrystalline)
Polymer or Alloy	100	100	100
Oil or Plasticizers	5-40	5-55	0-10
Fillers			
carbon black	5-40	5-40	2-5
inorganics	5-40	5-40	--
Antidegradants	1-2	1-2	1
Crosslinking			
inorganic	5-9	0-5	--
sulfur	5-9	--	--

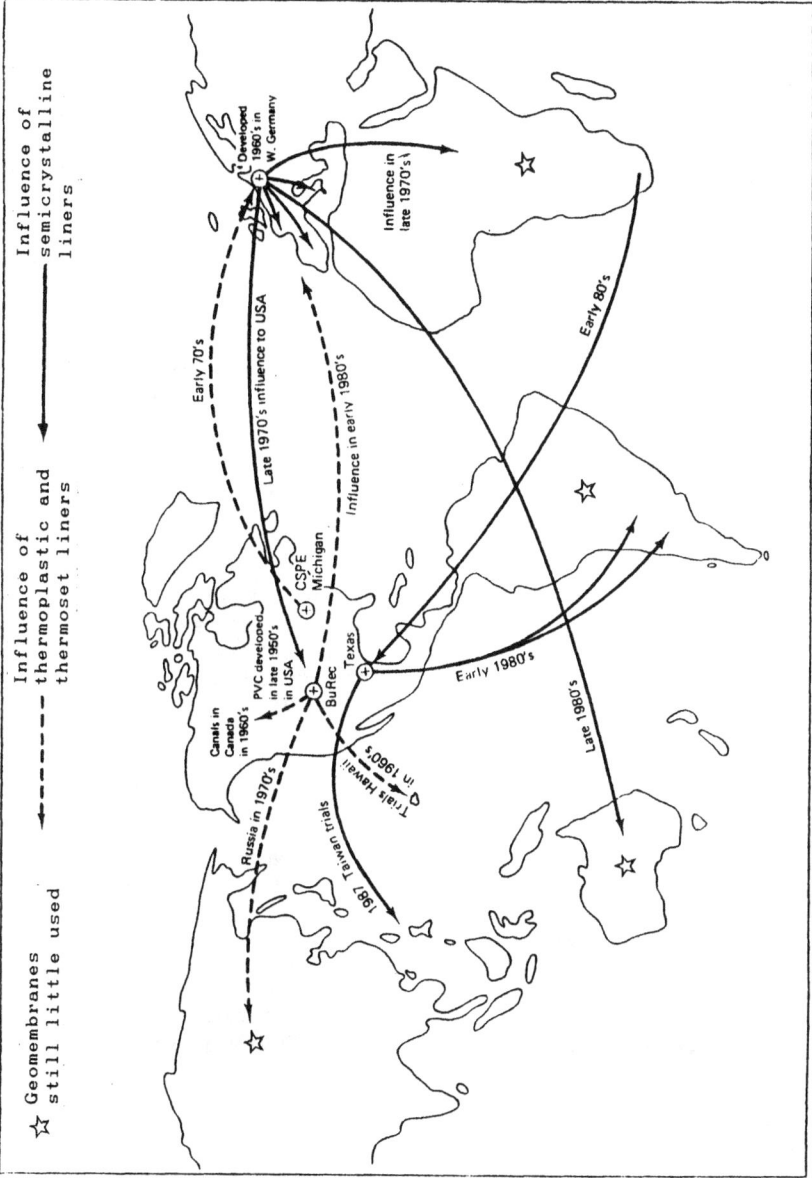

Fig. 2   Initial development and movement of geomembranes on a worldwide basis,
ref, Koerner (1990)

Thus it is seen that the thermoplastic semicrystalline geomembranes have the least amount of additives, while the elastomeric category has the largest amount. The thermoplastic amorphous are intermediate between the other two classes. Listed by generic name within each of the above classifications are the following geomembrane types which are in common use today. They are given in Table 2 with the addition of bituminous-based geomembranes. Also included in the table are some generalized comments as to advantages and disadvantages of each type. These products make up for the vast majority of geomembranes currently being used.

Table 2 Commonly used geomembranes, based on an original by Woodley (1978)

ADVANTAGES	DISADVANTAGES
(a) ELASTOMERIC GEOMEMBRANES	
EPDM Thermoplastic Rubber	
Good weathering	Fair in high temperatures
Cold crack resistance below 60°C	Blocking possible
Good seams - heat-bonded	Fair chemical resistance
No adhesives required	
Butyl, Butyl/EPDM, EPDM-Cured Rubbers	
Fair to good weathering	Poor field seams
Low permeability to gases	Small panels
High temperature resistance is good	Fair chemical resistance
Nonblocking	
Chloroprene (Neoprene) Cured Rubber	
Good weathering	Fair field seams - solvent and tape
Good high temperature	Fair seams to foreign surface
Good chemical resistance	
(b) THERMOPLASTIC AMORPHOUS GEOMEMBRANES	
Polyvinyl Chloride (PVC) Thermoplastics	
Tough without reinforcement	Plasticizer leaches over time
Lightweight as single ply	Cold crack must be verified
Good seams - dielectric, solvent, and heat	High-temperature performance must be verified
Large variation in thickness	Blocking (sticking) possible
Large variation in blends	Chemical resistance must be verified
Chlorinated Polyethylene (CPE) Thermoplastic	
Good weathering	Seam Reliability
Easy seams - dielectric and solvent	Delamination possible
	Blocking (sticking) possible

Table 2 Continued

ADVANTAGES	DISADVANTAGES
Cold crack resistance is good	
Chemical resistance is good	

Chlorosulfonated Polyethylene CSPE Thermoplastic Rubber

Excellent weathering	Fair in high temperatures
Cold crack resistance is good	Blocking (sticking) possi-
Chemical resistance is good	ble
Good seams - heat and adhesive	

(c) THERMOPLASTIC SEMICRYSTALLINE GEOMEMBRANES

High-Density and Medium-Density Polyethylene (HDPE, MDPE)
Thermoplastic

Chemical resistance is excellent	Low-friction surfaces
Good seams - thermal and	Stress crack sensitive
extrusion	Seam workmanship critical
Large variation in thickness	High thermal expansion/
	contraction

Low-Density and Very-Low-Density Polyethylene (LDPE, VLDPE)
Thermoplastic

Chemical resistance is good	Moderate thermal expansion/
Good seams - thermal and	contraction
extrusion	LDPE rarely used
Large variation in thickness	VLDPE recently introduced
No stress crack	Chemical resistance must be
	verified

Linear-Low-Density Polyethylene (LLDPE) Thermoplastic

Good seams - thermal and	LLDPE newly introduced
extrusion	Chemical resistance must be
Large variation in thickness	verified
High-friction surface	
No stress crack	

(d) POLYMER/BITUMEN AND BITUMINOUS GEOMEMBRANES

Large variation in manufac-	Chemical resistance must be
turing	verified
Good conformability to subgrade	Seam reliability must be
Easily reinforced	verified
Self healing possible	Blocking (sticking) possi-
	ble

1-5 Geomembrane Seam Types

Certainly the quality control of the manufacturing of geomembranes is considerably better than traditional civil
engineering liners which must be necessarily thick, placed
in lifts and placed under variable field conditions. These
same considerations, however, place added incentive to
prepare flawless geomembrane seams which are required to

make the factory panels or rolls into a continuous liner system. Field seams are therefore of paramount importance. Field seaming of geomembranes (as well as factory seams when they are required) is material specific insofar as the procedure is concerned. Table 3 summarizes the current seaming methods together with the type of geomembrane for which they are best suited. As the text to follow indicates, there is considerable choice.

Table 3 Commonly used geomembrane seaming methods, modified after an original by Frobel (1984)

Seaming Method	Elastomeric Polymers			Thermoplastic Amorphous Polymers			Thermoplastic Semi-crystal Polymer	Polymer/ Bitumen and Bituminous
	EDPM	Butyl	Neo-prene	PVC	CPE	CSPE	Various PE's	
Solvent				√	√	√		
Bodied solvent				√	√	√		
Solvent adhesive			√	√	√			
Contact adhesive						√		
Vulcanizing tape	√	√						
Tapes	√				√	√		
Hot air				√	√	√	√	
Hot wedge				√	√	√	√	
Dielectric				√	√	√		
Ultrasonic							√	
Electric				√			√	
Extrusion							√	
Mechanical	√		√	√	√	√		
Flame								√
Liquid								√

Solvent Seams use a brushed liquid solvent between the two geomembrane sheets to be joined, followed by pressure to make complete contact. As with any of the solvent-seaming processes to be described, a portion of the two adjacent geomembranes is truly dissolved, resulting in both liquid and gaseous phases. Too much solvent will weaken the adjoining geomembrane, and too little solvent will result in a weak seam. Therefore great care is required in providing the proper amount of solvent with respect to the particular type and thickness of geomembrane. Care must also be exercised in allowing the proper amount of time to elapse before contacting the two surfaces, and in applying the proper pressure and duration of rolling. These seams are used primarily on thermoplastic amorphous materials. *Bodied solvent seams* are similar except that 8 to 12% of the parent lining material is dissolved in the solvent and then the seam is made. The purpose being to compensate for the lost material while the seam is in a liquid state. Pressure is necessary, and the use of heat guns or radiant heaters aids the process. A *solvent adhesive* uses a adherent left after dissipation of the solvent. The adhesive thus becomes an additional element in the system. Sufficient pressure must be used. Most thermoplastic materials can be seamed in this manner. *Contact adhesives* have the widest applicability to all geomembrane types. The solution is applied to both mating surfaces by brush or roller. After reaching the proper degree of tackiness, the two sheets are placed on top of one another, followed by roller pressure. The adhesive forms the bond and is an additional element in the system. *Vulcanizing tapes* are used on very dense thermoset elastomer materials such as butyl and EPDM. Here an uncured tape or adhesive containing the polymer base of the geomembrane and cross-linking agents are placed between the two sheets. Upon application of heat and pressure, crosslinking occurs, which gives the necessary bond. Factory seams are made in large vulcanizing presses or autoclaves, while field seams require a portable machine to provide the necessary heat and pressure. *Single-sided tapes* can be applied over the edge of the seam, while double-sided tapes can be used between sheets to be joined.

There are a number of thermal methods that can be used on thermoplastic amorphous and semi-crystalline geomembrane materials. In all of them, the opposing geomembrane surfaces are truly melted into a liquid state. This being the case, temperature, time, and pressure all play important roles, in that too much melting weakens the geomembrane and too little melting results in a weak seam. The same care as was described for solvent seams holds for thermal seams as well. *Hot air* uses a machine consisting of a resistance heater, a blower, and temperature controls to blow air between two sheets to actually melt the opposing surfaces. Usually, temperatures greater than 260°C are required. Immediately following the melting of the surfaces, pressure by rollers is applied. For some devices this last aspect is automated in that pressure is applied by counter-rotating knurled rollers. The *hot wedge or hot knife* method consists of an electrically heated resistance element in the shape of a blade that is passed between the two sheets to be sealed. As it melts the surfaces, roller pressure is applied. Most of these units are automated as far as temperature, speed of travel, and amount of pressure applied is concerned. An interesting variation is the dual-hot-wedge method, which forms two parallel seams with an unbonded space between them. This space is subsequently pressurized with air and any lowering of pressure signifies a leak in the seam. Lengths of 100 meters, and more, can be field-tested in one step. *Dielectric* bonding is used only for factory seams. An alternating current is used at a frequency of approxi-

mately 27 MHz, which excites the polymer molecules, creating
friction and thereby generating heat. This melts the
polymer, and when followed by pressure, a seam results. A
variation of this method has recently been introduced for
the manufacture of field seams on polyethylene liners.
Ultrasonic bonding utilizes a generated wave form of 40 kHz
transmitted via a blade passed between the geomembrane
sheets which produces a mechanical agitation of the opposing
geomembrane surfaces. Mixing and pressure is applied via a
set of knurled wheels which follows the melting process.
Electric welding is yet another new technique for polyethylene seaming. In this technique a stainless steel wire
is placed between overlapping geomembranes and is energized
with approximately 36 volts and 10 to 25 amps current. The
hot wire radially melts the entire region within about 60
seconds thereby creating a bond. It is later used with a
low current in it and a questioning probe outside of the
seamed region, thereby becoming a nondestructive testing
method.

   Extrusion (or fusion) welding is used exclusively on
polyethylene geomembranes. It is a direct parallel of
metallurgical welding in that a ribbon of molten polymer is
extruded between the two slightly buffed surfaces to be
joined. The electrode causes some of the sheet material to
be liquefied and the entire mass then fuses together. One
patented system has a mixer in the molten zone that aids in
homogenizing the extrudate and the molten surfaces. The
technique is called *flat welding* when the extrudate is
placed between the two sheets to be joined and *fillet
welding* when the extrudate is placed over the leading edge
of the seam.

   Mechanical seams might suffice in certain situations
where complete watertightness is not necessary. Various
types of clamping systems that are effective in certain
circumstances have been devised. Sewn seams subsequently
waterproofed by a suitable liquid that solidifies or a cap
strip have also been used. In all of these mechanically
joined seams, however, some leakage must be tolerated.

   Bituminous panels and rolls are seamed by either a
controlled *flame* or torch method or by pouring *liquid*
bitumen directly over the joint to be sealed.

1-6 Major application categories

Although quite subjective, one can view current geomembrane
application areas in three categories: transportation
related, environmental related, and geotechnical related.
Each category has a number of specific applications which
have been reported in the literature, see Koerner and Hwu
(1989). These are illustrated in Figures 3, 4 and 5
respectively.

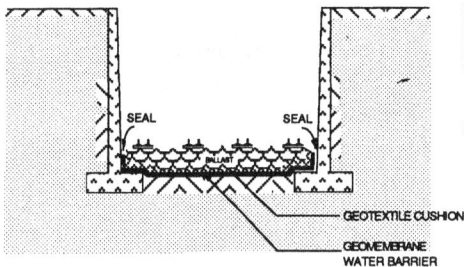

UPWARD MOVEMENT OF GROUNDWATER IN RAILROAD CUT
(Relief of Pore Water May Be Necessary)

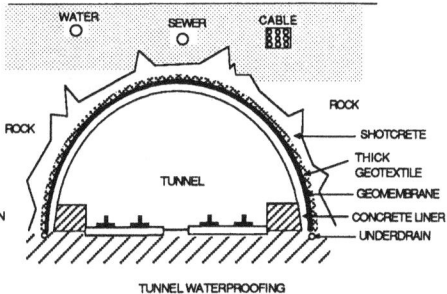

SEAL   SEAL
BALLAST
GEOTEXTILE CUSHION
GEOMEMBRANE
WATER BARRIER

WATER   SEWER   CABLE

ROCK
ROCK
TUNNEL
SHOTCRETE
THICK GEOTEXTILE
GEOMEMBRANE
CONCRETE LINER
UNDERDRAIN

TUNNEL WATERPROOFING

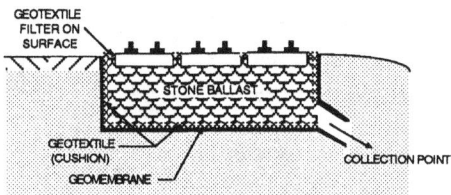

GEOTEXTILE FILTER ON SURFACE
STONE BALLAST
GEOTEXTILE (CUSHION)
GEOMEMBRANE
COLLECTION POINT

RAILROAD REFUELING AREAS

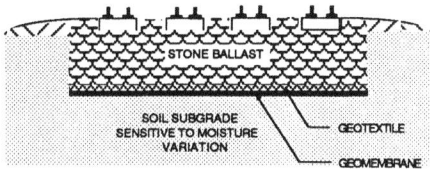

STONE BALLAST
SOIL SUBGRADE SENSITIVE TO MOISTURE VARIATION
GEOTEXTILE
GEOMEMBRANE

SOIL SUBGRADE MOISTURE PROOFING
(Relief of Pore Water May Be Necessary)

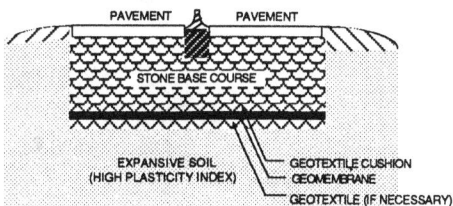

PAVEMENT   PAVEMENT
STONE BASE COURSE
EXPANSIVE SOIL (HIGH PLASTICITY INDEX)
GEOTEXTILE CUSHION
GEOMEMBRANE
GEOTEXTILE (IF NECESSARY)

CONTROL OF EXPANSIVE SOILS

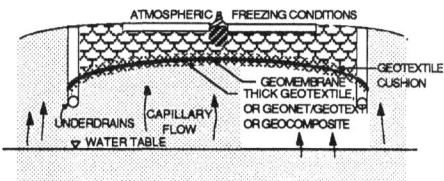

ATMOSPHERIC & FREEZING CONDITIONS
GEOTEXTILE CUSHION
GEOMEMBRANE
THICK GEOTEXTILE OR GEONET/GEOTEXTILE OR GEOCOMPOSITE
UNDERDRAINS
CAPILLARY FLOW
WATER TABLE

PREVENTION OF FROST HEAVE

Fig. 3   Sketches of geomembranes used in transportation
         applications

13

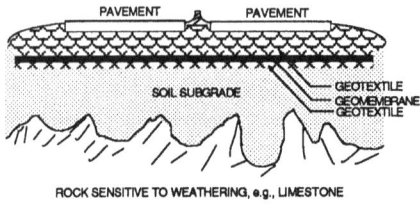

ROCK SENSITIVE TO WEATHERING, e.g., LIMESTONE

PREVENT FORMATION OF KARST-TYPE SINKHOLES

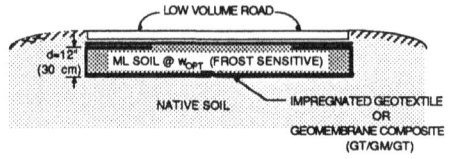

MAINTAIN OPTIMUM WATER CONTENT

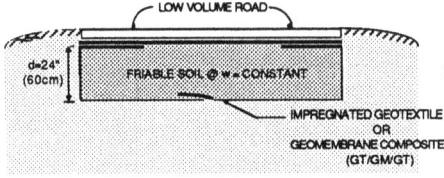

MAINTAIN CONSTANT WATER CONTENT

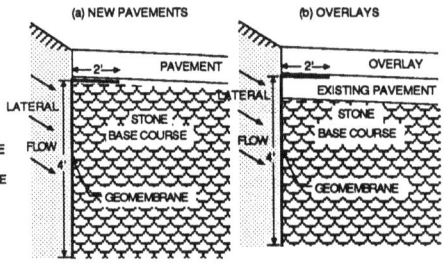

PREVENT STONE SATURATION FROM SIDES

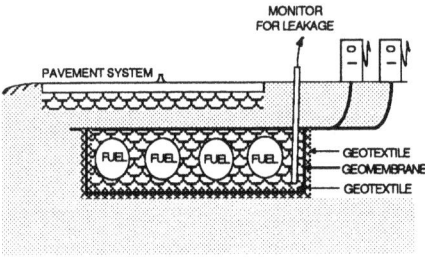

SECONDARY CONTAINMENT OF UNDERGROUND STORAGE TANKS

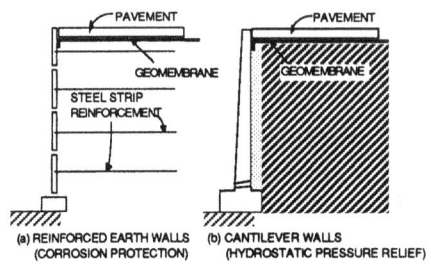

WALL WATERPROOFING SYSTEMS

Fig. 3 (continued)

14

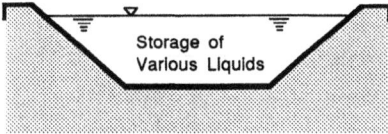

Reservoir Liners

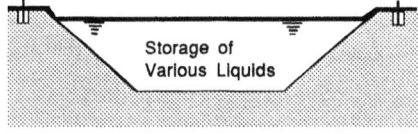

Reservoir Covers

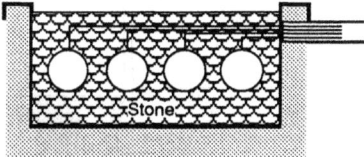

Storage of Various Liquids

Storage of Various Liquids

Underground Storage Tank Liners and Piping Systems

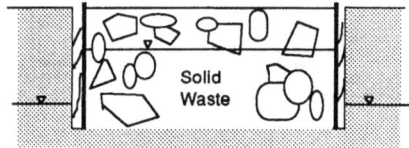

Cutoff Wall Barriers

Stone

Solid Waste

Landfill Liners

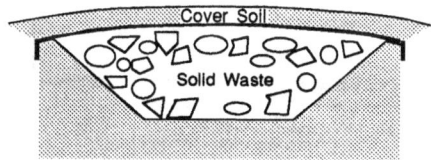

Landfill Covers

Solid Waste

Cover Soil

Solid Waste

Fig. 4   Sketches of geomembranes used in environmental applications

15

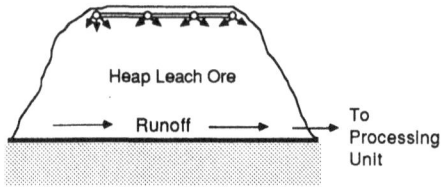

Heap Leach Ore

Runoff

To Processing Unit

Heap Leach Pads

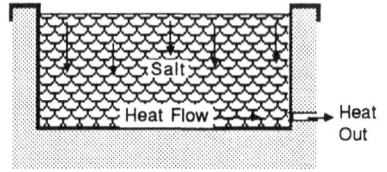

Salt

Heat Flow

Heat Out

Solar Pond Liners

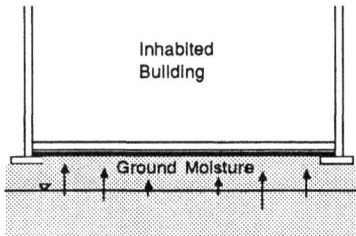

Inhabited Building

Ground Moisture

Moisture Barriers

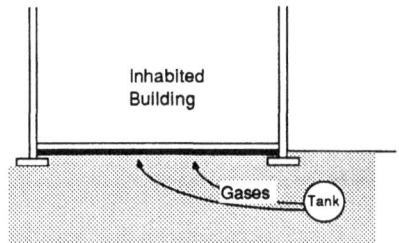

Inhabited Building

Gases

Tank

Hydrocarbon Barriers

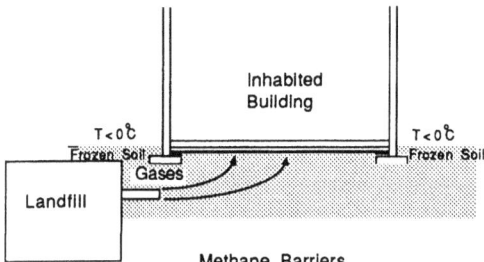

Inhabited Building

$T<0°C$
Frozen Soil

$T<0°C$
Frozen Soil

Gases

Landfill

Methane Barriers

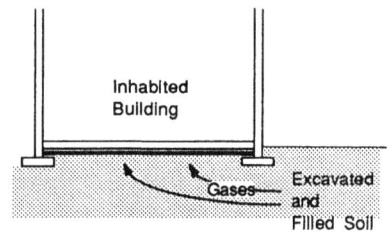

Inhabited Building

Gases

Excavated and Filled Soil

Radon Barriers

Fig. 4 (Continued)

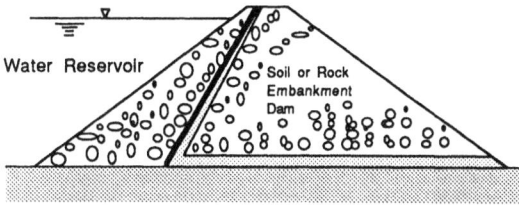

Earth and Earth/Rock Dams

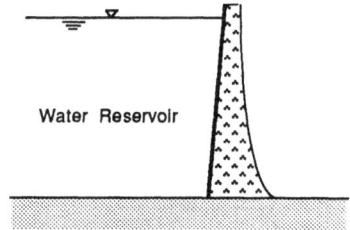

Concrete Dams

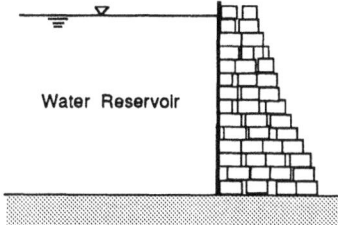

Stone Masonry Dams

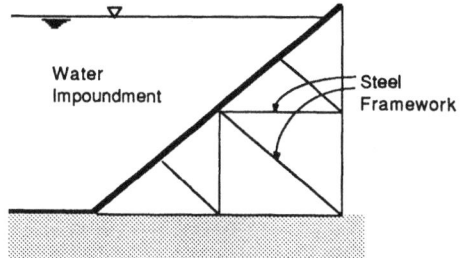

Portable (Temporary) Dams

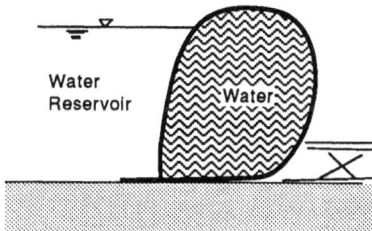

Inflatable Dams

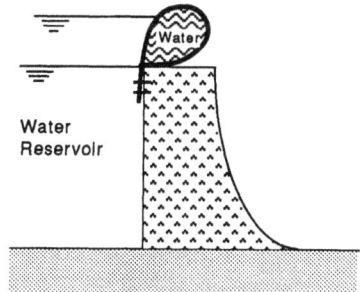

Increase Spillway Capacity

Fig. 5   Sketches of geomembranes used in geotechnical
applications

17

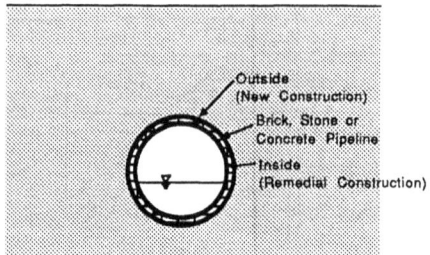

Seepage Control Inside or Outside
of Pipelines and Utilities

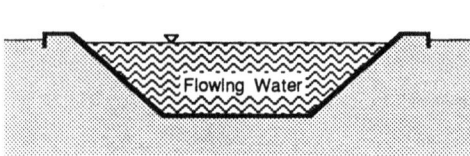

Canal Liners

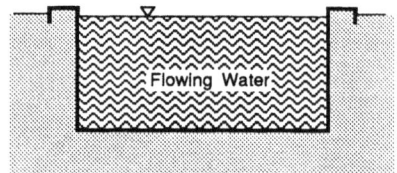

Water Diversion Systems

Fig. 5   (Continued)

18

Transportation Related (see Figure 3)
 • railroad cut beneath the water table
 • tunnel moisture - proofing (NATM)
 • railroad refueling areas
 • subgrade moisture - proofing
 • prevention of soil expansion or contraction
 • prevention of frost heave
 • prevention of karst sinkhole formation
 • maintain water content for friable soils
 • maintain water content for expansive soils
 • prevent stone saturation
 • secondary containment systems
 • wall waterproofing systems
Environmental Related (see Figure 4)
 • Liquid Containment for:
   • reservoir liners
   • reservoir covers
   • underground storage tank liners
   • cutoff wall barriers
 • Solid Containment for:
   • landfill liners
   • landfill covers
   • heap leach pads
   • solar pond liners
 • Vapor Containment for:
   • moisture barriers
   • hydrocarbon barriers
   • methane barriers
   • radon barriers
Geotechnical Related (see Figure 5)
 • liners for earth and earth/rock dams
 • liners for concrete dams
 • liners for masonry dams
 • portable dams
 • inflatable dams
 • increase spillway capacity
 • liners for seepage control
 • canal liners
 • water diversion systems

1-7 Future trends and growth

Projections into the future growth of geomembranes are
difficult at best. Furthermore, they are usually biased on
the part of the person making the prediction by reason of
personal knowledge of certain specific situations. The
position from the author's point of view is that of a very
strong environmental awareness in the area of groundwater
pollution. Solid waste containment is clearly powering the
geomembrane market in North America and every indication is
that this "mind-set" will continue and be spread worldwide.
Thus environmental applications will continue to be strong
with a long-term extended growth pattern.

With an ever increasing worldwide population growth, water supplies will become more scarce and require storage reservoir and surface impoundment liners. Similarly, canal liners for the transportation of this water to areas of primary use is an important applications. Recent concerns of potable water reservoir pollution from the atmosphere has ushered in the concept of floating geomembrane covers.

With industrialization on a worldwide basis comes the necessity of canal liners to transport chemicals and waste waters. The application here is among the most demanding of all geomembrane applications.

With continued applications of geomembranes in the above environmentally related categories will come an awareness on the part of many transportation and geotechnical engineers for the use of geomembranes in many of the applications listed previously. Certainly rehabilitation of our infra-structure such as remediation of concrete, masonry and soil/rock dams is an area which can nicely use geomembranes of a wide variety of types, see Cazzuffi (1987). With familiarity, comes additional use and even forays into areas where geomembranes have not previously been used. The flow chart of Figure 6 of potential use and opportunities (along with education) should bring about a strong growth for geomembranes in the near term, e.g. 20 to 25% over the next 5 years, and, quite possibly, will be continued even beyond that time frame.

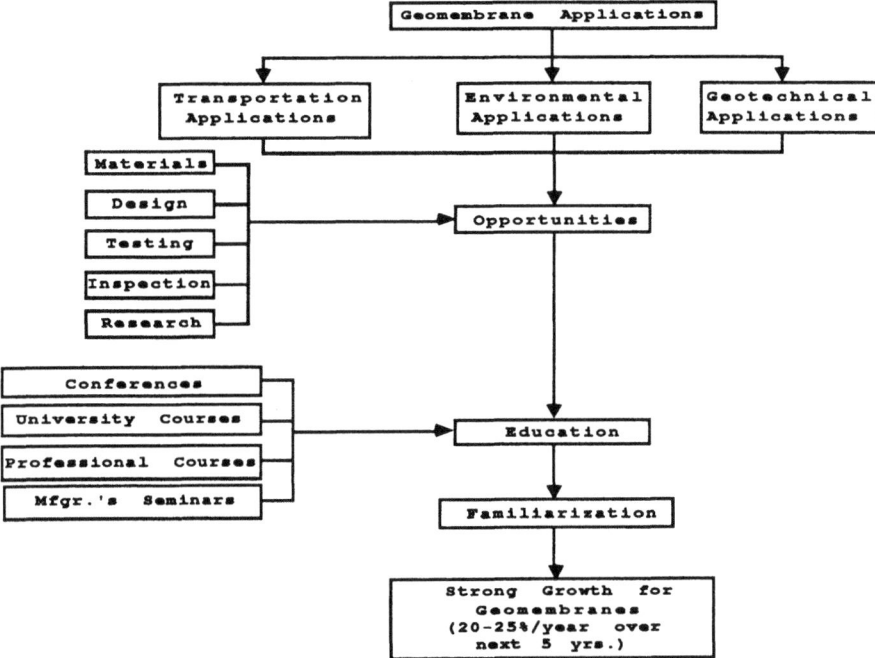

Fig. 6 Future growth and trends in the use of geomembranes

References

Bell, J. R. and Yoder, E. J., "Plastic Moisture Barrier for Highway Subgrade Protection," Proc. Highway Research Board, Washington, DC, Vo. 36, 1957, pp. 713-735.

Cazzuffi, D., "The Use of Geomembranes in Italian Dams," Intl Jour. Water Power and Dam Construction, March, 1987, pp. 44-52.

Frobel, R. K., "Methods of Constructing and Evaluating Geomembrane Seams," Proc. Intl. Conf. on Geomembranes, Denver, CO, 1984, IFAI, pp. 359-364.

Haxo, H. E., Jr., "Quality Assurance of Geomembranes Used as Linings for Hazardous Waste Contaiment," Jour. Geotextiles and Geomembranes, Elsevier Appl. Sci. Publ. Ltd., England 1986, pp. 225-247.

Kays, W. B., Construction of Linings for Resources, Tanks, and Pollution Control Facilities, Second Edition, J. Wiley & Sons, Inc., NY, 1987.

Koerner, R. M. and Hwu, B.-L., "Geomembrane Use in Transportation Systems," Proc. 1989 Transportation Research Board Mtg., Washington, DC, 1989, Public Works, Sept. 1989, pp. 135-138.

Koerner, R. M. Designing with Geosynthetics: Second Edition, Prentice Hall, Englewood Cliffs, NJ, 1990.

Staff, C. E., "Seepage Prevention with Impermeable Membranes," Civil Engineering, Vol. 37, No. 2, 1969.

Staff, C. E., "The Foundation and Growth of the Geomembrane Industry in the United States," Proc. Intl. Conf. on Geomembranes, Denver, CO, 1984, IFAI, pp. 5-8.

Woodley, R. M., "Pond Liners Anyone?", Military Engineer, Vol. 70, No. 458, Nov.-Dec., 1978, pp. 392-395.

2 TEST STANDARDS AND THEIR CLASSIFICATION

J-M. RIGO
Geotextiles and Geomembranes Research Centre,
University of Liege, Belgium
D. A. CAZZUFFI
ENEL Research Centre on Hydraulics and Structures,
Milan, Italy

2.1 Introduction

The characterization of a geomembrane requires two major steps:
  identification tests;
  performance tests, that can be divided as following:
    on the geomembrane;
    on the joints;
    for durability studies.
In order to evaluate geomembrane characteristics, test procedures have been developed, but generally according to a very specific approach, that takes into account the different types of products and sometimes the different types of applications. For example, for applications in hydraulic works (ponds, canals and reservoirs), ASTM prescribes different specifications, according to the different types of products, and respectively for polyethylene and ethylene copolymer geomembranes (ASTM D 3020-89), for polyvinyl chloride geomembranes (ASTM D 3083-89), for elastomeric unreinforced geomembranes (ASTM D 3253-81) and for elastomeric reinforced geomembranes (ASTM D 3254-81).
At the present stage of testing development, the type of geomembrane has a particular influence on the choice of procedures for the identification tests, while performance tests are more influenced by the particular application.
If the geomembrane is included into a "system", the testing programme should also include analysis on each component of the system.
If a complete information on the characteristics of the products is needed for a technical agreement, a reduced programme of tests is necessary for the reception of the products on site.
Certain test procedures on geomembranes can be applied to all the types of geomembranes, no matter the nature of the constitutive materials. On the contrary, other procedures are closely associated to the physical and chemical nature of a particular type of geomembrane. The different test procedures are analysed hereafter and classified in function of their first objective (identification or performance). The test procedures applicable specifically to certain types of products are clearly identified.
The descriptions of the tests presented in this chapter 2 are rather simple and short. More details can be obtained in the standards listed in the references. On the other hand, detailed descriptions and evaluations on some specific aspects of geomembrane testing are presented by other Authors in the following chapters of this book.

2.2 Identification tests

In order to identify the constitutive material or the elements of the system, some simple tests can be performed: they permit to ensure that the products delivered on site are in conformity with the specifications. They also permit a good manufacture quality control. Table 2.1 hereafter gives the list of the identification tests.

Table 2.1 Identification tests on geomembranes

Thickness
Density
Mass per unit area
Tensile test
Flexibility at low temperature
Infra-red spectroscopy (IR)
Differential scanning calorimetry (DSC)
Thermogravimetric analysis (TGA)
Thermomechanical analysis (TMA)
Chromatography (GC and HPLC)
Melt index (MI)
Extractables
Ash
Hardness
Ring-ball (for bitumen)
Penetration (for bitumen)
Dehydrochlorination (for PVC)
Carbon black content (for HDPE)
Carbon dispersion (for HDPE)

2.2.1 Thickness
The geomembrane thickness is measured in various points of its plane. It corresponds to the distance of two plates of a given area applied with a certain load on the geomembrane sheet. The plates area and the pressure are depending from the selected standards.

Thickness can be measured as follows:
    ASTM D 1593-81 for thermoplastic geomembranes;
    ASTM D 3767-83 for elastomeric geomembranes.

2.2.2 Density or specific gravity
The density or specific gravity express the ratio between the unit weight of the tested product versus the unit weight of water.
This characteristic can be measured as follows:
    ASTM D 792-79 (method by displacement);
    ASTM D 1505-79 (method by density - gradient technique);

ASTM D 297-81 (methods by use of a pycnometer and by hydrostatic weighings).

ISO R 1183-87 and DIN 53479 can also be used for the measure of density.
More details about density determination can be found in this book in Chapter 15 (Halse et al., 1990).

2.2.3 Mass per unit area
The mass per unit area is a physical property commonly used to characterize the weight of a geomembrane.
This measurement can be performed according to ASTM D 3776-79.

2.2.4 Tensile tests
Tensile properties of polymeric materials are generally measured by a stress-strain test. The specific properties that are measured depend on the type of geomembrane. They may include:
    tensile stress at yield (if the geomembrane is semicrystalline);
    strain at yield (if the geomembrane is semicrystalline);
    tensile stress at fabric break (if fabric reinforced);
    stress at specified strains (for the modulus evaluation, e.g. at 100% and 200%);
    tensile stress at failure of geomembrane;
    strain at failure of geomembrane.

Figure 2.1 illustrates, from a qualitative point of view, the tensile behavior of different types of geomembranes (Cazzuffi, 1988, modified from Venesia, 1983).

In case of unreinforced geomembranes, dumbbell shape specimens are commonly used (see Figure 2.2): the exact specimen shape and the rate of strain are depending from the selected standard and from the polymer.
It is presently suggested to perform the tensile test, in case of dumbbell shape specimens, as follows:
    ASTM D 638-84 and ASTM D 882-83 for thermoplastic geomembranes;
    ASTM D 412-83 for elastomeric - unreinforced geomembranes;
ISO R 527-66, DIN 53455 and NF T 54102 may also be used.

In case of reinforced geomembranes (bituminous, PVC reinforced, elastomeric reinforced), rectangular specimens (50 mm width and 200 mm gage length) are commonly adopted. The stresses are usually related to the width of the sample and not to its cross section.
Tensile test can be used as identification test and also performance test, as control after ageing.
Tensile tests are probably the most widely used tests in the rubber and plastics industries also for evaluating polymeric compositions and products, because tensile properties give a good indication of the quality of the compound typical of a geomembrane.

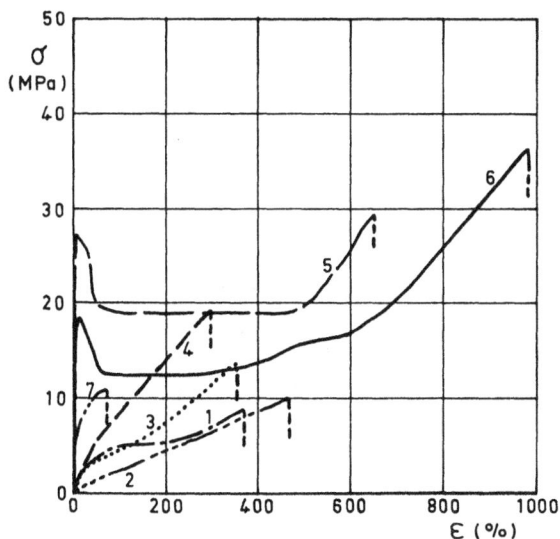

Fig. 2.1 Stress-strain behavior of different types of geomembranes
(from Cazzuffi, 1988).

1. Chlorosulfonated polyethylene (CSPE) geomembrane
2. Butyl rubber (IIR) geomembrane
3. Chlorinated polyethylene (CPE) geomembrane
4. Polyvinil chloride (PVC) geomembrane
5. Polypropylene (PP) geomembrane
6. High density polyethylene (HDPE) geomembrane
7. Bituminous geomembrane, reinforced with polyester fabric

Some research programs are in progress, in order to select only one
specimen size and one methodology for all the types of geomembranes,
both bituminous and polymeric: the results of the more recent
developments of this research are presented in this book in Chapter 5
(Gourc and Perrier, 1990).

A standard for testing all the types of polymeric geomembranes with a
wide strip method (specimen 200 mm width and 100 mm gage length) was
recently adopted in United States (ASTM D 4885-88). According to this
standard the geomembrane strength is given in kN/m.

TYPES M-I, M-III

TYPE M-II

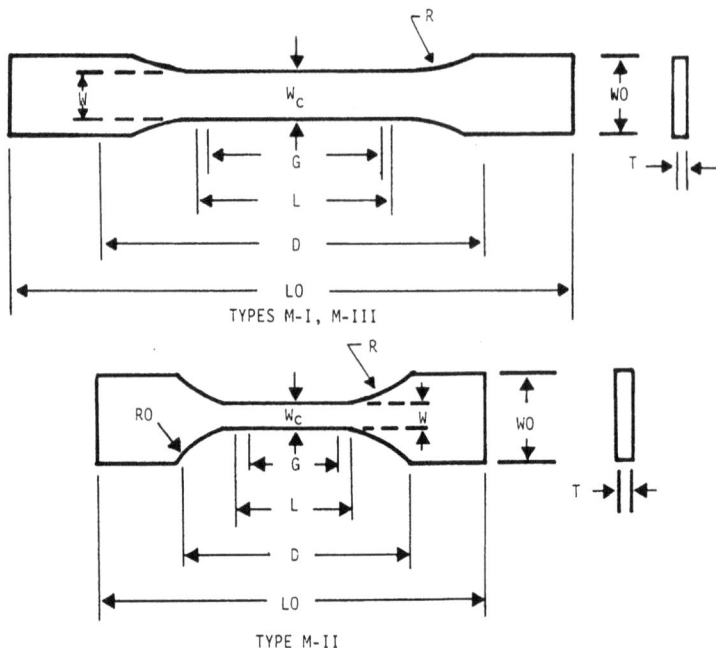

Figure 2.2   Dumbbell shape specimens (according to ASTM D 638-84):
              LO is ranging from 60 to 150 mm and W from 2,5 to 10 mm

2.2.5 Flexibility at low temperature
A particular procedure for the evaluation of flexibility at low
temperatures, especially conceived for bituminous geomembranes, is
proposed in the recommendations published by the European Union for
Technical Agreement in Construction (UEAtc, 1982).
The specimen is bended at 180° around a cylindrical mandrel (Figure
2.3), with the following characteristics:
   20 mm diameter, if the specimen thickness < 5 mm;
   30 mm diameter, if the specimen thickness > 5 mm.
The bending time is 5 seconds. The test is conducted in a conditioned
room, where the temperature is decreased by steps of 5°C. The
temperature at which cracks appear on the specimen, once it is bended,
is registered.
This test is used as:
   identification test;
   test to appreciate the possible rate of ageing of the material.
Information regarding the flexibility behavior of geomembranes at low
temperatures can also be obtained by means of the procedures described
in the following standards: NF T 54110, DIN 53372 (only for PVC
geomembranes), ASTM D 1790-83 and ASTM D 746-79 (impact methods) and
ASTM D 2136-84 (bending method).

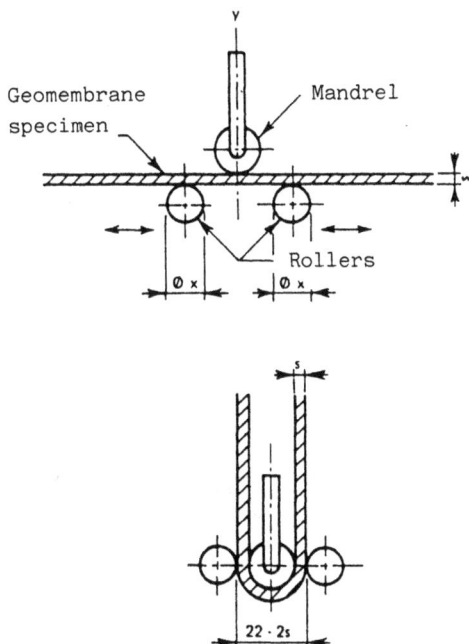

Figure 2.3   Flexibility at low temperature (according to UEAtc, 1982)

2.2.6. Infra-red analysis (IR)

A polychromatic electromagnetic ray is directed on the geomembrane specimen. The wave lengths correspond to the infra-red ("lambda" ranging from 2,5 to 25 microns).

The specimen is a thin film sampled from the geomembrane sheet. The transmitted or reflected ray is analysed. Some wave lengths correspond to the frequencies of vibration typical of the functional groups of the product: this induces ray absorption.

A diagram presenting ray absorption vs. wave length gives a very precise identification card of the product.

This test is used for:

    identification, both for the pure polymer and for the additives (for example determination of plasticizer in PVC geomembranes);
    analysis of the degree of ageing, for instance for the determination of the oxydation in a HDPE, or of the percentage of polymer in a modified bitumen.

More details about IR analysis are illustrated in this book in Chapter 15 (Halse et al., 1990).

2.2.7 Differential scanning calorimetry (DSC)

A small geomembrane specimen (few miligrams) is subjected to a thermal flow. All the phenomena, consisting in a production or an absorption of heat, give endo- or exo-thermal peaks on a diagram (Figure 2.4).

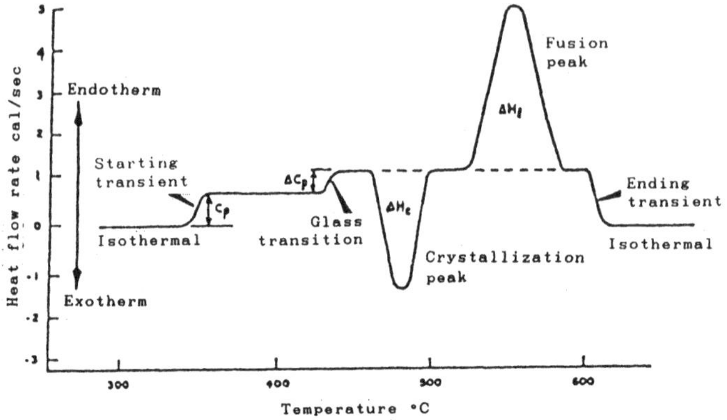

Figure 2.4   Qualitative DSC diagram

A DSC analysis gives:
   percentage of crystallinity (ASTM D3417-82), if the geomembrane is semicrystalline;
   melt temperature (ASTM D3417-82);
   melt enthalphy (ASTM D3417-82);
   glass transition temperature (ASTM D3418-82);
   post-polymerisation;
   decomposition (or oxydation) of the product;
   oxydation stability.

This test is used for:
   identification;
   durability control.

More details about DSC analysis are illustrated in this book in Chapter 15 (Halse et al., 1990).

2.2.8 Thermogravimetric analysis (TGA)

Thermogravimetric analysis (TGA) is a thermal technique for assessing the composition of a material by means of its loss in weight during heating at a controlled rate in an inert or oxidizing atmosphere.

When a geomembrane is heated in an inert atmosphere from room temperature to 600°C at a controlled rate, it will volatilize at different temperatures until only carbon black, carbonaceous polymer residue and ash remain. The introduction of oxygen into the system will burn off the carbonaceous polymer residue and the carbon black.

The weight vs. time curve, which can be expressed also vs. temperature, can be used to calculate the polymer and additives content, the carbon black content and the ash content.
The TGA curve and the derivative of the TGA curve can be used as part of the fingerprinting of a polymeric material.
Figure 2.5 gives a TGA diagramme obtained on a HDPE geomembrane.
More details about TGA analysis are described in this book in Chapter 15 (Halse et al., 1990).

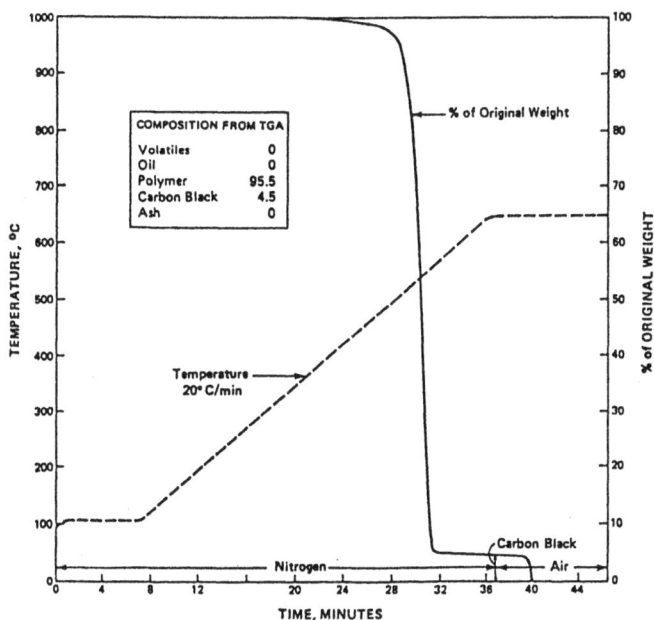

Figure 2.5   TGA curves obtained on a HDPE geomembrane (from EPA, 1988)
               continuous line: thermogram weight vs. time
               dashed line: temperature vs. time

2.2.9 Thermomechanical analysis (TMA)
Thermomechanical analysis (TMA) is a thermal technique in which the physical properties of a material are measured during a process of controlled rate heating.
Many test methods were developed to follow the thermomechanical behavior of polymers. ASTM D 648-82 is to be pointed out, because it was developed long time ago. By measuring the deflection temperature of a specimen progressively heated and maintened under a flexural load, this method permits to determine a temperature very close to glass transition. The standard is applicable only to rigid materials, like HDPE geomembranes.

The techniques of thermal analysis, recently developed, are able to characterize the thermomechanical behavior of geomembranes very quickly and for a very wide range of temperatures. According to these techniques a TMA analysis of a material, measured under a static load, gives the following characteristics:

linear coefficient of thermal expansion (ASTM E 831-81);
glass transition temperature;
melting temperature.

It's possible to perform this test also under a cyclic load; in this way it's furthermore possible to obtain the Young's modulus of the material (vs. temperature).

More details about TMA analysis are illustrated in this book in Chapter 15 (Halse et al., 1990).

2.2.10 Chromatography (GC and HPLC)

Chromatography is a technique that permits to separate and to determine the different components of the complex polymeric mixtures typical of geomembranes, as volatiles and also decomposition products.

Cromatography can be performed, following two different techniques: gas chromatography (GC), is which the mobile phase is gas and high performance liquid chromatography (HPLC), in which the mobile phase is liquid and the stationary phase is liquid or solid.

More details about gas chromatography (GC) and high performance liquid chromatography (HPLC) techniques can be found in this book in Chapter 15 (Halse et al., 1990).

2.2.11 Melt Index (MI)

Melt index is the flow rate of a thermoplastic material as determined by means of an extrusion in a plastometer apparatus, as specified in ASTM D 1238-82. Values are reported in terms of rate of extrusion, in grams per 10 minutes, vs. temperature and vs. load.

This test is used above all for the quality control of PE resins.

More details about melt index (MI) determination are described in this book in Chapter 15 (Halse et al., 1990).

2.2.12 Extractables

The extractable content of a polymeric geomembrane is the fraction of the compound that can be extracted from a geomembrane specimen by means of a suitably selected solvent, that neither decomposes nor disolves the base polymer.

Extractables consist of plasticizers, oils, etc.

The measure of the extractable content is important in geomembrane fingerprinting. The extract and the residual specimen obtained by this procedure can be used for further analytical testing, as TGA, IR, GC, etc.

The determination of the extractables generally follows the procedures described in ASTM D 3421-75 for PVC geomembranes and in ASTM D 297-81 for elastomeric geomembranes.

2.2.13 Ash

The ash content of a geomembrane is the inorganic fraction that remains after a process of burning at 550 ± 25°C in a muffle furnace, to which a geomembrane specimen is subjected after removal of volatiles. The ash consists of the inorganic ingredients that have been used as fillers and components of the liner compound, and of the ash residue from the polymer. Different geomembrane manufacturers formulate their compounds differently, and the ash content is part of the fingerprinting of a polymeric geomembrane compound. The residue obtained by ashing can be retained for other analyses (such as trace metals analyses) needed for further identification.

The test method described in ASTM D 297-81 is generally followed in performing this analysis. Ash content can also be determined by TGA.

2.2.14 Hardness

Hardness is the ability of a material to resist to the indentation by a small probe of specified shape and dimensions. Hardness test is usually performed according to ASTM D 2240-86 by means of an apparatus that can use the two different types of indentors (depending from the material deformability) illustrated in Figure 2.6.

Although no simple relationship exists between hardness and other measured properties, hardness can be related to the modulus of elasticity, or Young's modulus, according to the formula reported in ASTM D 1415-81.

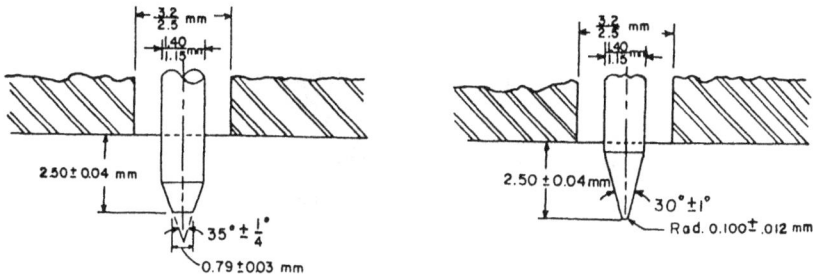

A. indentor for type "A" durometer   B. indentor for type "D" durometer

Figure 2.6 Indentors for hardness test, according to ASTM D 2240-86

2.2.15 Ring-ball temperature (for bitumen)

The ring-ball temperature is the temperature at which a standard specimen of bitumen, loaded by a steel ball (Figure 2.7), reaches a given deformation (2,54 mm). The temperature is increased at a constant rate.

ASTM E 28-82 describes this test method.

This test is used as:
    identification test;
    performance   test (durability studies).

31

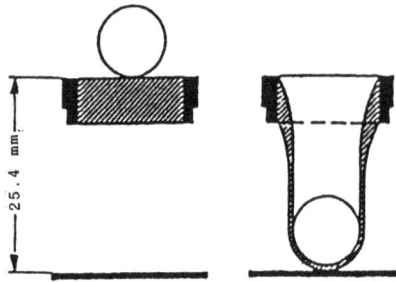

Figure 2.7   The ring-ball temperature device

2.2.16   Penetration (for bitumen)
Bitumen is characterized by the penetration. This characteristic
includes 2 numbers (e.g. 80/100), according to the range of
penetration.
The result of the test corresponds to the penetration of a standard
needle, loaded with a standard weight (100 grams) during 5 seconds.
The penetration is expressed in 1/10th of mm (Figure 2.8). The test is
performed at 25°C. This test is described in ASTM D 5-83.

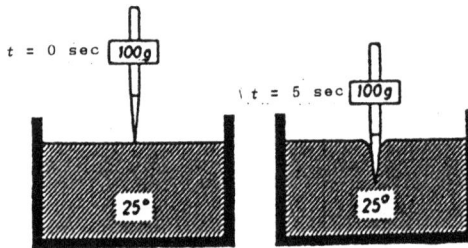

Figure 2.8   Standard penetration test on bitumen

2.2.17   Dihydrochlorination (for PVC)
This test is performed only on PVC geomembranes. It gives information
on the quality of the PVC stabilizers vs. temperature.
The specimen is maintained under a given temperature for the period of
time necessary to obtain the decomposition (i.e. the production
of HCl).

2.2.18   Carbon black content (mostly for HDPE)
ASTM 1603-76 describes this type of determination: the method covers
the definition of the carbon black content in polyethylene,
polypropylene and polybutylene materials. Determinations of carbon
black are made gravimetrically after pyrolysis of the specimen under
nitrogen. This method is not applicable to compositions that contain
non volatile pigments or fillers other than carbon black (for which
TGA analysis can be performed).

32

2.2.19 Carbon black dispersion (mostly for HDPE)

ASTM D 3015-78 describes microscopical examination of plastic compounds to check quality of pigment dispersion. Thin sections of polymers are prepared for observation with transmitted light. Microscopical examination is normally made for grading or classification by comparison vs. observational standards.
The method may be employed to measure the degree of dispersion of pigments, the presence of foreign matter, the spots of unpigmented resin, and the resin degradation.
Serviceability of a compound plastic can be quite sensitive to the quality of pigment dispersion. Black weather-resistant polyethylene serves as a good example where quality of the dispersion is indicative of weatherability which is achieved by effective screening of ultraviolet light by the carbon black.

2.3 Performance tests

Performance testing can cover three different aspects:
    performance tests on the geomembrane;
    performance tests on the joints;
    durability tests.

Performance tests on the geomembrane must demonstrate the ability of the membrane to play its waterproofing function, taking the site stresses and the environmental conditions into account (Table 2.2).

Performance tests on the joints refers to the ability of the geomembrane to be well placed and welded on site or in the factory conditions, if pre-welded (Table 2.3).

The objectives of the durability tests are to control the ability of the geomembrane or of the system including also the geomembrane, to resist to the long-term stresses.In general, these stresses are of mechanical, chemical, physical or biological origin. These stresses can be applied single or in combination.
Therefore, in order to accelerate the ageing process, concentrations, temperature or stresses are increased (for laboratory studies). This is not without danger because generally the ageing process is influenced by the environmental parameters. In spite of that, the accelerated tests presented in Table 2.4 give precious informations on the sensibility of the geomembranes versus the ageing factors.
After ageing, the characteristics of the materials are checked, according to the identification or performance tests on geomembranes already described (Tables 2.1 and 2.2).

Table 2.2   Performance tests on the geomembrane

```
        Tear
        Bursting
        Puncture
        Cyclic movements
        Friction
        Creep
        Thermal expansion
        Dimensional stability
        Water permeability
        Vapor permeability
```

Table 2.3   Performance tests on joints

```
        Shear test
        Peel test
        Non destructive tests:
          mechanical point stressing
          pressurized dual seam
          air lance
          vacuum box
          ultrasonic impedance plane
          ultrasonic pulse echo
```

Table 2.4   Performance tests for durability studies

```
        Volatiles
        Abrasion
        Thermal ageing
        Light exposure
        Chemical resistance
        Biological resistance
        SO₂  ageing (for PVC geomembranes)
        Ozone ageing (for elastomeric geomembranes)
        Stress-cracking (for HDPE geomembranes)
```

2.3.1 Performance tests on the geomembrane

2.3.1.1 Tear test

The tear test procedures and specimen sizes depend from the nature of the constitutive material and from the type of tearing considered.

For bituminous geomembranes, the tear test consists in a nail propagation (Figure 2.9), according to the procedure described by UEAtc, 1982. The displacement rate of the clamp is 100 mm/min.

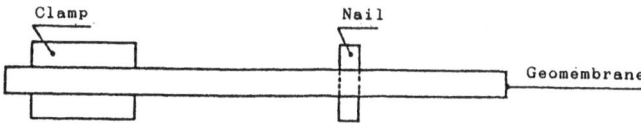

Figure 2.9   Tear test on bituminous geomembranes

For polymeric geomembranes, tear resistance can be evaluated according to two different approaches: tear initiation and tear propagation.

As far as concerns tear initiation, the test can be performed using the specimes illustrated in Figure 2.10. The test procedure is standardized by ASTM D 1004-81 for thermoplastic geomembranes and by ASTM D 624-86 (die C) for elastomeric geomembranes.

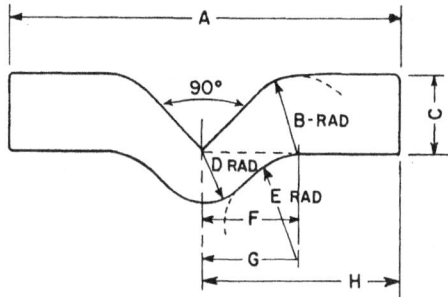

Fig. 2.10   Specimen used for the evaluation of tear initiation (according to ASTM D 1004-81 and ASTM D 624-86): the dimension A is 102 mm

As far as concerns tear propagation, different types of specimens are used.

For unreinforced geomembranes, the specimen presented by ASTM D 624-86 (die B) and illustrated in Figure 2.11 is recommended.

For reinforced geomembranes, the tear propagation is obtained submitting a "trousers" specimen to the type of sollicitation illustrated in Fig. 2.12. The specimen is rectangular; the dimensions vary from 25x75 mm (ASTM D 1938-78, for geomembranes with thickness < 1 mm) to 75x200 mm (ASTM D 751-79).

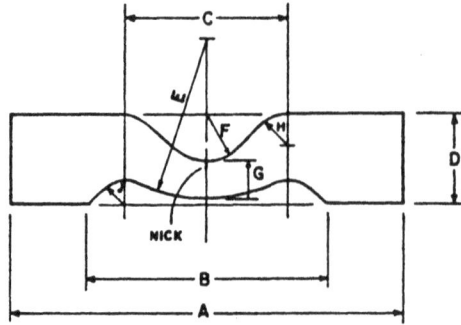

Fig. 2.11   Specimen used for the evaluation of tear propagation for unreinforced geomembranes (according to ASTM D 624-86): the dimension A is 110 mm

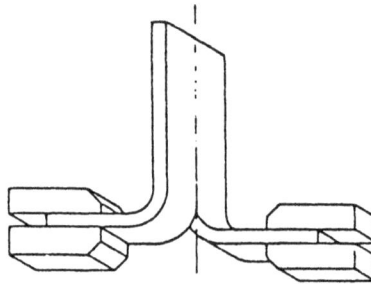

Fig. 2.12   Tear propagation test with "trousers" specimen

2.3.1.2   Bursting test

Bursting test is useful to reproduce in the laboratory the stress-strain condition of a geomembrane when it's subjected to a three-dimensional axi-symmetric tension (due for example to a localized deformation under the liner).

The usual procedure of bursting test is as following: a circular geomembrane specimen is installed between the flanges of the top section of a metallic tank.

In order to submit the geomembrane to bursting, water pressure or air pressure can be used: the pressure is applied by steps until bursting occured. The strain is measured between each step.

Figure 2.13 illustrated one of the first bursting strength apparatus conceived for geomembranes (Rollin et al., 1984). The present tendence is to use greater apparatus (with specimen having diameter > 500 mm), as reported in Chapter 6 of this book (Hoekstra, 1990).

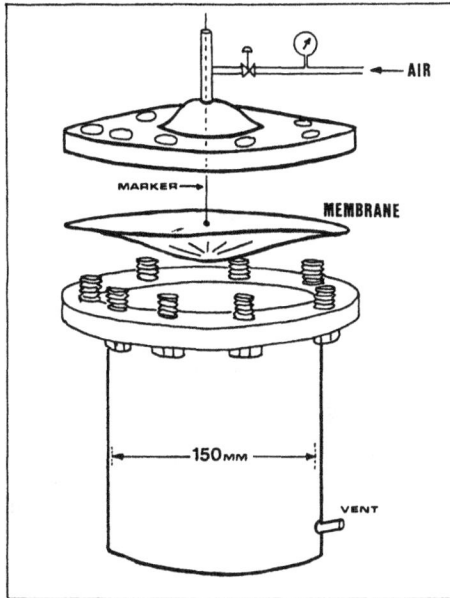

Fig. 2.13 Bursting strength apparatus used at the Ecole Polytéchnique in Montreal (from Rollin et al., 1984)

2.3.1.3 Puncture test

The puncture resistance is the pressure required to puncture a geomembrane: this value gives an indication of the ability of the material to withstand puncture from above (due to a standard probe) or from below (due to a supporting granular soil).

Testing devices have been developed to assess both these aspects. A review of the different test procedures available for puncture is presented in this book in Chapter 7 (Frobel and Rigo, 1990).

The apparatus used to measure the bursting strength of geomembranes is lightly modified to measure the puncture resistance over a granular support. Sub-angular material, gravel, filled up the bottom section of the tank. The geomembrane specimens are installed between the top and bottom sections of the apparatus, with or without a geotextile cushion. Pressure on the specimen is gradually increased by steps until puncture occured. An example of this test device is showed in Figure 2.14.

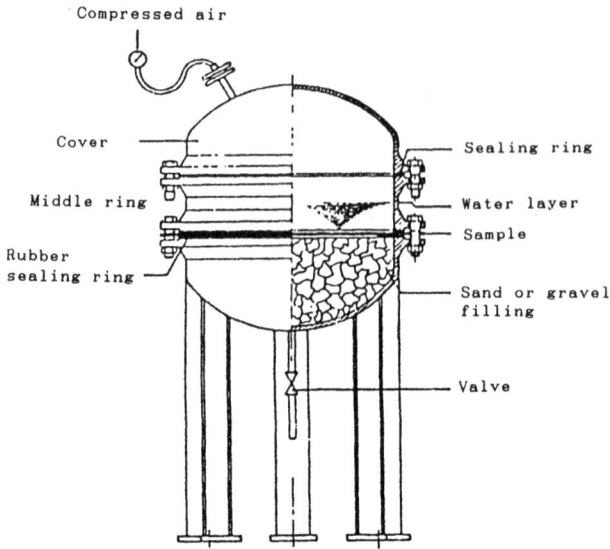

Figure 2.14   Puncture test apparatus for geomembranes, available at
              GRI - Drexel University in Philadelphia

2.3.1.4 Cyclic movements
It is a test to control the ability of the geomembrane to bridge
cracks.
Geomembranes are glued on 2 concrete supports distant of 1 mm. The two
concrete substrates have a relative alternate movement, in order to
simulate the opening and closing of the crack. The deformation rate
and the amplitude are depending from the different constitutive
materials.
The number of cycles is:
    500 for nonaged specimens;
    200 for aged specimens.
The test is usually performed at 0°C.
The UEAtc directives describe the procedure of this test (UETtc,
1982).

2.3.1.5 Friction
The adherence between different materials constituting a waterproofing
system (soil, geomembrane, geotextile, geocomposite, etc.) can be
studied using the apparatus schematically presented in Figure 2.15.
Particularly, the geomembrane specimen can be installed at the
interface between the two boxes subjected to a shear strength. A load
is applied normally on the top box, while a mechanism applied a
horizontal shear force on the bottom box. The horizontal load is
slowly increased as the friction developed through the interface
reaches its maximum value.

More details about friction test are described in this book in Chapter 8 (Ingold, 1990).

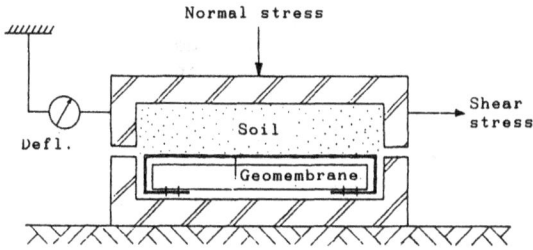

Figure 2.15 Direct shear test device to evaluate the geomembrane - soil friction (from Van Zanten, 1986)

2.3.1.6 Creep

Polymeric materials have a relatively high tendency to creep under constant load, above all if compared to traditional construction materials. Viscoelastic properties of polymer base materials for geomembranes are reviewed by Gamski (1984).

Creep behavior generally results in one of the three different deformation vs. time curves. These qualitative curves are shown in Figure 2.16.

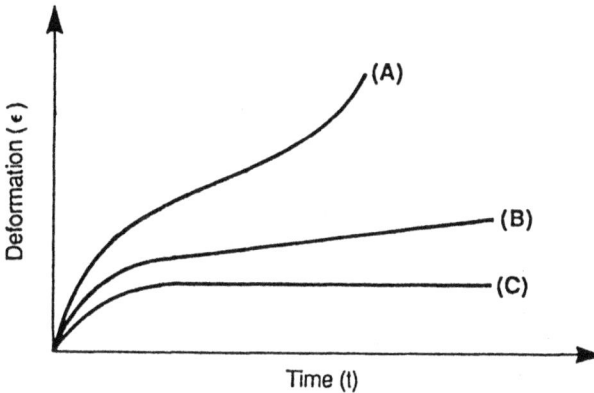

Figure 2.16 Types of creep behavior
   Curve A describes creep failure
   Curve B shows a constant creep after an initial deformation
   Curve C shows no creep after an initial deformation

In order study the rheological behavior of polymeric geomembranes, the failure state described in curve A is beyond consideration and only curves like B or C have to be considered. The empirical relationship defined by these two curves is represented by the following equation:

$$\varepsilon_t = \varepsilon_o + b \log t$$

39

where:

ε_t = strain at a generic service time "t"
ε_o = initial, or elastic, strain
b = constant experimentally obtained
t = service time under consideration.

To simulate the creep behavior of a geomembrane in soil, test specimens need to be evaluated under some type of confinement. Until now little work has been done on the creep testing of confined geomembranes, but work has progressed in assessing polymer behavior under constant stress or constant strain in both geotextile and geogrid areas.

2.3.1.7 Thermal expansion
Some types of materials (as HDPE) present an important thermal expansion.
The coefficient of linear thermal expansion is determined by the recording of the dimensions of a specimen when is subjected to an increasing or a decreasing of temperature.
Figure 2.17 shows the test device, made of quartz tube, recommended by ASTM D 696-79 for the evaluation of the linear thermal expansion.

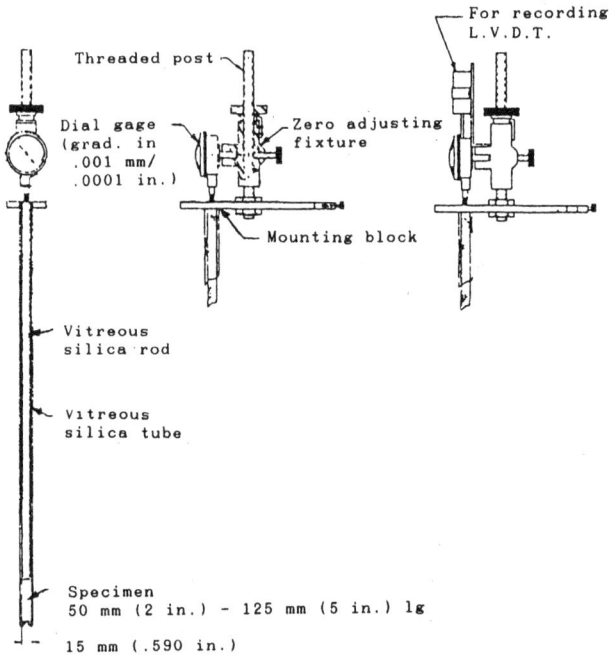

Figure 2.17 Thermal expansion test device: quartz-tube dilatometer (according to ASTM D 696-79)

2.3.1.8 Dimensional stability

High temperatures can cause shrinkage and distorsion due to internal stress relaxation.
ASTM D 1204-84 measures changes in the linear dimensions of specimens (250x250 mm) when subjected for a certain period of time to a process of heating up to 100 °C.

2.3.1.9 Water permeability

Water permeability of geomembranes can be characterized in comparison with a porous medium, by using Darcy's equation.
A specimen of the geomembrane of a given area A (m^2) is submitted to a water flow with an hydraulic gradient i(-). Once the steady state is reached, the measure of the flow rate Q (m^3/sec) permits to calculate the permeability coefficient k (m/s) by the Darcy's equation:

$$k = \frac{Q}{A \cdot i}$$

where: i = H/t
H = hydraulic head (m)
t = geomembrane thickness (m)

Figure 2.18 shows a test device presently used for the evaluation of water permeability on geomembranes at CEMAGREF in Antony (France).

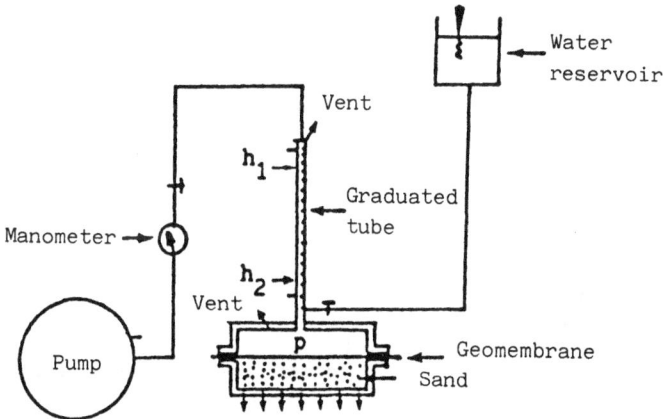

Figure 2.18 Permeability test device for geomembranes, available at CEMAGREF in Antony

The geomembrane is layed down on fine sand. The specimen is circular (diameter of 100 mm).
A water pressure of about 1300 kPa is applied on the top surface of the geomembrane.

41

The geomembrane is first in contact with water during 24 hours.
The flow rate is evaluated afterwards, by measuring the water level
changes in the top vertical graduated tube.
More details about water permeability test are described in this book
in Chapter 10 (Haxo and Pierson, 1990).

2.3.1.10 Vapor permeability
Some methods for the evaluation of the vapor permeability are based on
the diffusion law.
Among different procedures, the method presented by ASTM E 96-80 can
be pointed out.
The principle is to place a circular specimen of geomembrane at the
top of a box. The joint is insured by paraffine; the box is partially
filled with water (Figure 2.19).
Afterwards the box is placed upside down, therefore the water is in
contact with the geomembrane.
The test is performed at 23°C and 50% relative humidity. The weight of
the box is regulary controlled.
The required test time is about 30 days.

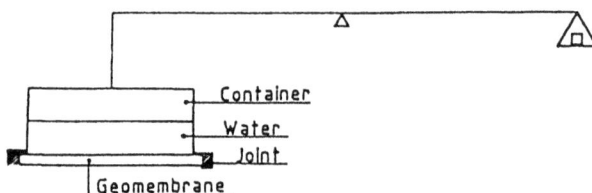

Figure 2.19   Scheme of the vapor permeability apparatus

2.3.2 Performance test on joints

2.3.2.1 Shear test
Shear strength testing is performed by applying a force across the
seam in a direction parallel to the plane of the bond, subjecting the
bond interface to a shearing force (Figure 2.20).
The shear test is standardized by ASTM D 4437-84 for field seams and
by ASTM D 4545-86 for factory seams: the specimens are rectangular,
having a width of 25 mm (for unreinforced geomembranes) and of 50 mm
(for reinforced geomembranes). The gage length for all the types of
geomembranes is fixed to 50 mm, plus the width of the seam. The
displacement rate usually adopted is 50 mm/min.
The test results are related to the width of the specimen and not to
the cross-section.

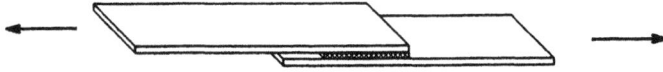

Figure 2.20   Shear test on a joint

2.3.2.2 Peel test
Peel test is performed by applying a load in such a way that the bonded interface is subjected to a peeling force: this peeling force attempts to separate the two geomembranes that have been seamed together (Figure 2.21).
The peel test is standardized by ASTM D 4437-84 for field seams and by ASTM D 4545-86 for factory seams: the specimens are rectangular, having a width of 25 mm and a gage length of 25 mm for all the types of geomembranes. The displacement rate usually adopted is 50 mm/min.
The test results are related to the width of the specimen and not to the cross-section.

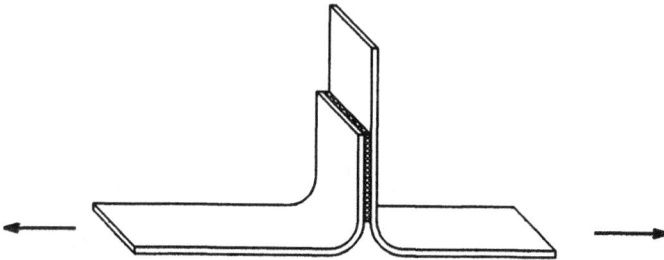

Figure 2.21   Peel test on a joint

2.3.2.3   Non destructive tests
According to Rollin (1984), six non-destructive test methods are used for qualitatively assessing geomembrane seams. These methods are controlling the continuity of a seam, but cannot be used to measure the relative strength of a joint.
These methods are:
    the mechanical point stressing;
    the pressurized dual seam test;

43

the air lance test;
the vacuum box test;
the ultrasonic impedance plane (UIP) test;
the ultrasonic pulse echo (UPE) test.

Mechanical point stressing

The detection of an unbonded area of a seam can be found by sliding a dull tool along the edge of a bonded seam (ASTM D 4437-84 and ASTM D 4545-86).

The tool can easily lift the unbobded areas as long as en edge unbond exists, but cannot detect unbond areas within the width of the seam.

Pressurized dual seam test

This method is used on a double hot wedge bond, as shown in Figure 2.22.

Both ends of a dual wedge seam are sealed and a needle is inserted between the welded areas. A pressure of about 200 kPa is applied inside the air channel and pressure drop is monitored vs. time, in order to detect leaks.

Although this method permits the testing of long lengths at the same time, it must be rechecked with a vacuum box test or other method, if a leak is detected.

Figure 2.22   Pressurized dual seam test

Air lance test

This method, described by ASTM D 4437-84 and ASTM D 4545-86, is using air (at about 300 kPa) that is directed through a nozzle and that is applied at the edge of an overlap seam in order to detect any unbonded areas: in this case, the air stream inflates the unbonded area, as shown in Figure 2.23.

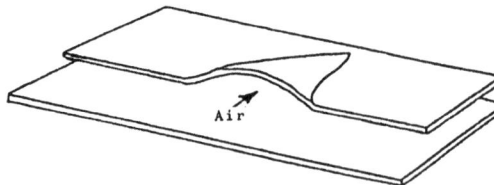

Figure 2.23   Air lance effect on a seam

Vacuum box test

The equipment needed for this test is a vacuum box (with a glass viewport and a portable vacuum pump) that can be sealed to a seam area.

The rubber seal located at the bottom of the box is first soaped to provide a vacuum seal and the seam edge is covered with a soap solution.

A vacuum ranging from 15 to 30 kPa is applied to the box that contain water and unbonded areas are detected by the presence of bubbles (ASTM D 4437-84 and ASTM D 4545-86).

A schematic view of the apparatus is presented in Figure 2.24.

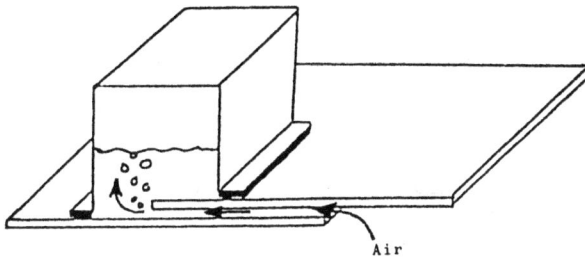

Figure 2.24 Vacuum box test scheme

Ultrasonic impedance plane (UIP) test

The UIP technique works on the characteristic acoustic impedance. A well-bonded seam possesses a certain acoustic impedance and any other area that is unbonded is different. A continuous wave of 160-185 kHz is transmitted through the seam by means of a transducer in contact with the geomembrane and a characteristic dot pattern is registered on a monitor. The location of the dot pattern indicates the quality of the seam.

Ultrasonic pulse echo (UPE) test

This technique, described by ASTM D 4437-84 and ASTM D 4545-86, is a thickness measurement using a high frequency pulse (5-15 MHz) of energy transmitted via a surface transducer through both sheet thickness and returned to a receiving oscilloscope. If the seam is homogeneous, a characteristic curve will be obtained, while a different signal will represent a faulty bond.

The signal patterns of both good and faulty bonds are presented in Figure 2.25.

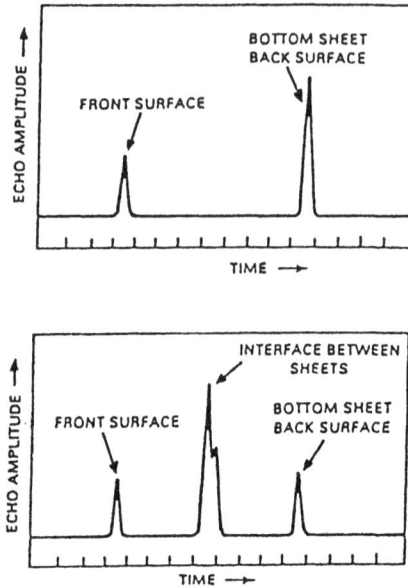

Figure 2.25 Ultrasonic pulse echo test: good seam (above) - faulty
seam (below)

2.3.3 Performance test on durability

2.3.3.1 Volatiles
Determination of volatiles is generally the first test performed on an
exposed geomembrane specimen, therefore it needs to be run as soon as
possible after the specimen has been removed from the site.
This test indicates the amount of volatile constituents that has been
absorbed by the geomembrane during exposure.
The volatile fraction is defined as the weight lost by a geomembrane
specimen when heating in a circulating air oven at 105°C for 2 hours.
The recommended test specimen size is a 50 mm diameter disk.
Volatiles should be removed before determining ash, extractables and
specific gravity. Ash and extractables are reported on a dry basis.

2.3.3.2 Abrasion
In some applications (hydroelectric tunnels, canals,..), geomembranes
can be subjected to sand or fine gravel abrasion, in presence of
water.

Geomembrane specimens can be placed in a rotating drum (Figure 2.26) partially filled with water and sand, as proposed by Cazzuffi et al. (1983). The weight loss is controlled afterwards.
A similar apparatus for the evaluation of the abrasion behavior of geomembranes in presence of water and granular materials is available at BAM in Berlin.

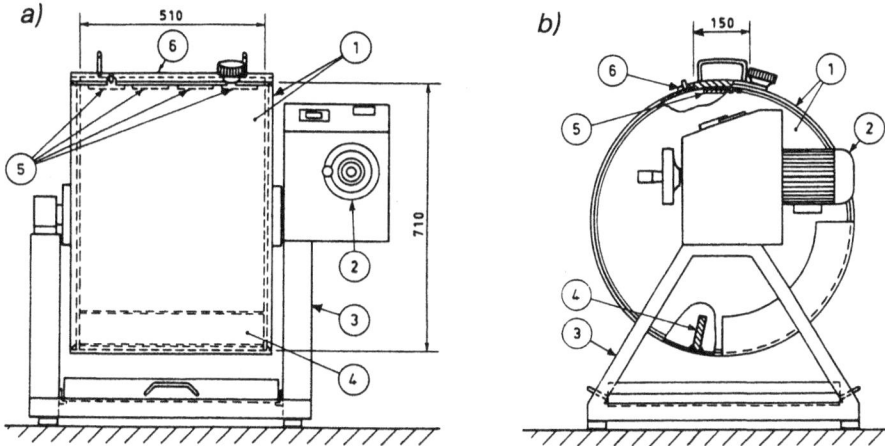

Figure 2.26   Abrasion test on geomembranes, available at ENEL-CRIS in Milano (from Cazzuffi et al., 1983): measures are in mm

    a) Front view
    b) Lateral view
    1. Steel cylinder (12 mm thick)
    2. Engine
    3. Bearing
    4. Steel device for soil particles draging
    5. Geomembrane specimens
    6. Cover

More traditional tests (such as Taber abrasion apparatus for thermoplastic geomembranes and Pico abrader for elastomeric geomembranes) can also be used to reproduce dry conditions of abrasion. The Taber abrasion tester is described by ASTM D 4060-84, while the Pico abrader is described by ASTM D 2228-83.

2.3.3.3 Thermal ageing
A geomembrane may encounter temperatures higher than normal prior to installation, during installation, and during service.
Thermoplastic geomembranes, if allowed to be exposed to heat as rolled or folded panels prior to installation, such as being left in the sun, can block or stick together; afterwards, when unfolded, a coated geomembrane may split or an unreinforced geomembrane may tear and become unserviceable.

Tensile test specimens can be placed for 15 days, 1,2,3,4 and 6 months in an oven at 80°C. The tensile resistance is checked afterwards, according to UEAtc specifications (UEAtc, 1982).
A similar procedure of thermal ageing is also presented by ASTM D 573-81.

2.3.3.4 Light exposure
Plastic materials are naturally sensible to light and specifically to U.V. exposure. This is the reason why admixtures like carbon black are added to the basic compounds during the manufacturing process of geomembranes.
This sensitivity can be determined according to the following procedures, that are different above all because of the type of light source:

Carbon-arc type apparatus (described by ASTM G 23-81), in which the radiation source is represented by one or two carbon-arc lamps;

Xenon-arc type apparatus (described by ASTM G 26-77), in which the radiation source is a watercooled xenon-arc lamp;

Fluorescent UV condensation type apparatus (described by ASTM G 53-77 and illustrated in Figure 2.27), in which the radiation source is an array of fluorescent lamps with emissions concentrated in the UV range.

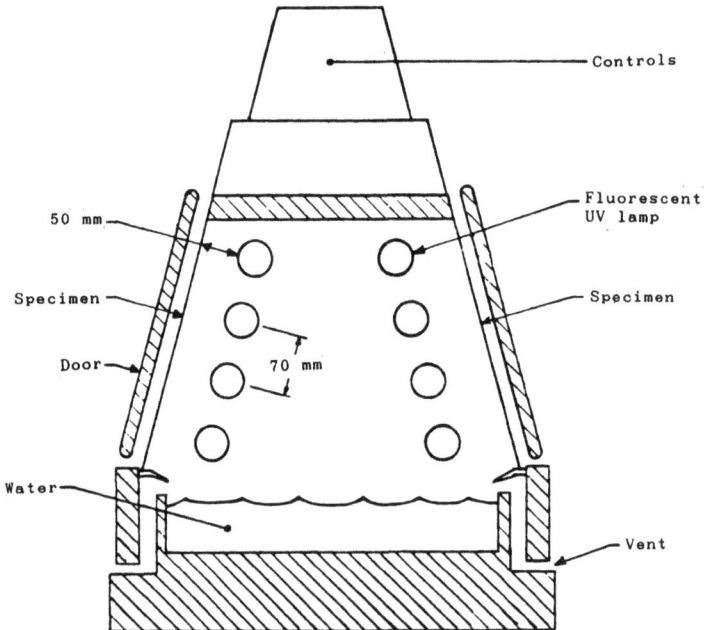

Fig. 2.27 Cross-section of the fluorescent UV condensation type apparatus (according to ASTM D 53-77)

2.3.3.5 Chemical resistance
When in service, geomembranes are subjected to various chemical attacks.
The general chemical resistance of different geomembranes to well defined chemicals is usually known. The problem arises when the chemical is not a single component material: in this case possible synergistic effects, that are not completely known, can be originated (Koerner, 1990).
Therefore the development of a specific compatibility test between chemicals and geomembranes has been requested.
Test specimens can be immersed in special tanks (Figure 2.28) in vertical position, without contact between specimens, for a certain period of time and at fixed temperature conditions (EPA 9090, 1986).
More details about chemical compatibility testing of geomembranes are given in this book in Chapter 13 (Peggs and Tisinger, 1990).

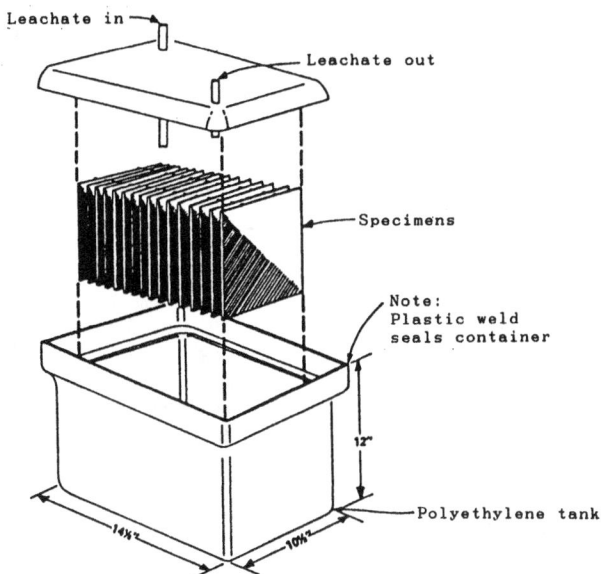

Figure 2.28   Immersion tank for the evaluation of chemical resistance of geomembranes (from Haxo et al., 1982)

The tests which are recommended on the lining materials before and after exposure are:
    mass per unit area;
    thickness;
    hardness;
    tensile;
    tear;
    puncture;
    specific gravity;
    volatiles;
    extractables.

2.3.3.6 Biological resistance

Geomembranes can be attacked by a number of living organisms, particularly when applied in waste facilities: the general term used to describe the ability of a geomembrane to resist to this type of attack is biological resistance.

Biological resistance of geomembranes can be examined, dividing the macrobiological effect (due to animals and plants) from the microbiological effect (due to fungi and bacteria).

As far as concerns resistance to animals (like rats), there are no well-established test procedures available: only intuitively it's possible to say that the stronger, harder and thicker the geomembrane, the better its resistance to animal attack.

As far as concerns resistance to plants, some geomembranes may be crossed by roots: Lupinus Albus plant is well known for the great aggressivity of its roots.

A geomembrane specimen is placed in a container between two layers of vegetal earth. Lupinus Albus grains are sowed in the upper mould layer. After a period of 6 months, the germination and the growth of the plant is achieved and the geomembrane can be removed: no root has to cross the geomembrane, as illustrated in Figure 2.29.

A reference permits to control the Lupinus vivacity when the geomembrane is replaced by 10 mm of bitumen. The roots must cross these 10 mm of bitumen.

As far as concerns micro-biological resistance, geomembranes may be subjected to the action of fungi and bacteria.

The polymer portion of a geomembrane is usually resistant to fungi and bacteria, in that it does not serve as a carbon source for the microbiological growth. It is generally the other components, such as plasticizers, lubrificants, stabilizers and colorants that can be responsible for such attack. It is important to establish the resistance of geomembranes to microbiological attack when are used under conditions of high temperature and humidity, favorable for such attack.

The geomembrane specimens are submitted to bacteria attack for a minimum period of 21 days at a temperature of $36 \pm 1°C$ and more than 85% relative humidity. The complete test procedure for bacteria attack on plastics is described by ASTM G 22-80 (part A).

For fungi attack, geomembrane specimens are submitted for a minimum period of 21 days to a temperature of $29 \pm 1°C$ and to a relative humidity $> 85\%$. The complete test procedure for fungi attack is described by ASTM G 21-80.

Anyway, the great concern regarding the microbiological attack due to fungi and bacteria is not really related to geomembranes, but to the possible clogging and blinding of geotextiles, geonets and drainage geocomposites that are often associated with geomembranes, particularly in waste facilities.

More details about biological resistance testing can be found in this book in Chapter 14 (Rigo et al., 1990).

Figure 2.29   Roots resistance test on geomembranes (after 6 months)

2.3.3.7 SO$_2$ ageing (for PVC geomembranes)
PVC geomembranes may be attacked by SO$_2$ present in the atmosphere.
The test usually performed for this evaluation is described by DIN
50017. The specimen is exposed in an atmosphere, at a temperature of
about 40°C and relative humidity of 100%, in which the content of SO$_2$
is fixed. Tensile test specimens are exposed in these conditions and
tensile resistance is checked afterwards.

2.3.3.8 Ozone ageing (for elastomeric geomembranes)
Elastomeric geomembranes, if containing double liaisons (like:-CH=CH-)
are attacked by ozone present in the atmosphere.

ASTM 1149-81 describes a test procedure, in which geomembrane specimens, under certain conditions of strain, are exposed - for different amounts of time - in an atmosphere containing fixed amounts of ozone. The performance of the geomembrane is satisfactory if no cracks appear on their surfaces. ISO R 1431-72 may also be used.

2.3.3.9 Stress-cracking (for HDPE geomembranes)
A stress-crack is defined as either an external or internal crack in a semicrystalline geomembrane that is caused by tensile stress lower than its mechanical strength.
In fact, under conditions of simultaneous stress and exposure to chemicals (e.g. soaps, soils, detergents, or other surface-active agents), some plastics, such as PE, can fail mechanically by cracking. A test can be run that indicates the susceptibility of a PE sheeting to stress-cracking by exposing bent geomembrane specimens (with standardized imperfections) to a designated surface-active agent.
ASTM D 1693-80, although commonly used to measure susceptibility to stress-cracking, has limitations for assessing the long-term resistance in service of geomembranes to cracking.
In this test 10 notched and bent strip specimens are immersed in a detergent solution, and the time it takes before 5 of the 10 specimens break is determined. The test apparatus is shown schematically in Figure 2.30. This method is not suitable for testing PE seams.

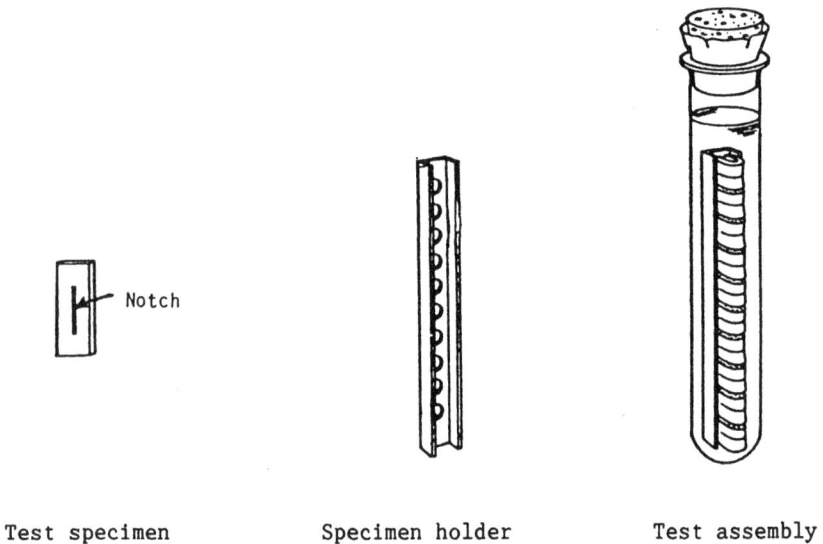

Test specimen   Specimen holder   Test assembly

Figure 2.30   Stress-cracking specimen and equipment of for bent-strip test (according to ASTM D 1693-80)

Another method used to measure the tendency of a semicrystalline product to break, when exposed simultaneously to stress and to a detergent solution, is ASTM D 2552-80. In this test, 20 dumbbell specimens are placed under constant load, and the time it takes before 10 of the 20 specimens break is determined.

The test apparatus is shown schematically in Figure 2.31. This method has been used to test seams by selecting a dumbbell with a neck section of sufficient length to test the full width of the seam and by modifying the specimen holders accordingly.

More details about stress-cracking testing on geomembranes can be found in this book in Chapter 11 (Koerner at al., 1990).

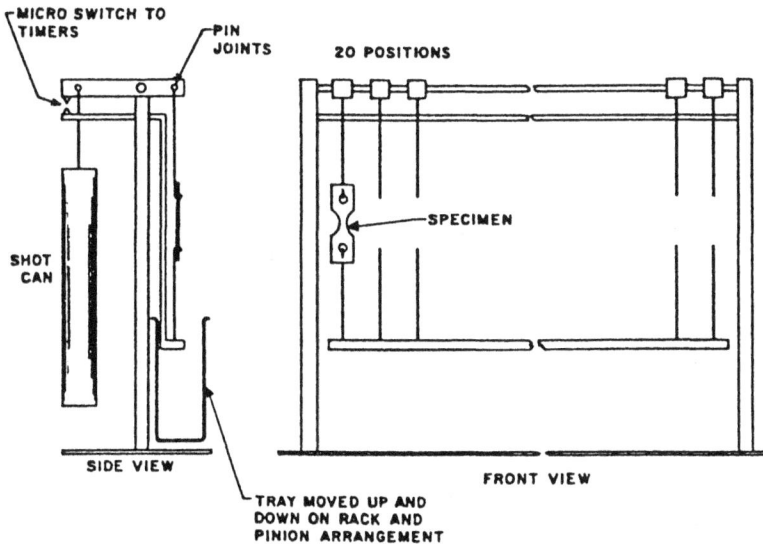

Figure 2.31   Schematic view of constant-load stress test apparatus
              (according to ASTM D 2552-80)

2.4 Conclusions

The success of the use of geomembranes in geotechnical engineering is depending from:

      the conception of the project;
      the quality of the materials;
      the quality of the placement system;
      the final use of the construction.

This presentation is mostly dealing with the second and the third above mentioned points.

Not all the described tests are necessary to give to the designer a good idea of the actual behavior of the geomembrane in the specific application.

Two important points have to be considered in the choice of a geomembrane:
    chemical nature;
    mechanical properties.

The chemical nature of the geomembrane influences its:
    ageing resistance;
    thermal stability;
    thermal expansion;
    chemical resistance;
    mechanical resistance when the membrane is not reinforced;
    ability to be welded on site.

The selection of a geomembrane on the basis of its mechanical properties depends from the type of problem to be solved.
According to Giroud (1983), "the best membrane is not necessarly the strongest one".
In fact, sollicitations on the geomembranes can be applied in stresses (than a reinforced geomembrane shall be used) or in strains (than an unreinforced geomembrane with high elasticity is to be used).

The quality control of a geomembrane must include the checking of its:
    manufacture;
    transport;
    placing;
    welding.

Indeed, after a correct design and after a good manufacture quality control, an on site survey of the placement permits to achieve the work in a correct way.
As far as concerns the construction phase, the causes that seem to be of importance in the waterproofing structure failures are:
    bad weather conditions for the on site welding;
    the lack in the joints control;
    bad junctions of the geomembrane with the earth or concrete constructions.
In fact no single phase of the preparation for the use of a geomembrane can be neglected, each phase is important.

The objective of this Chapter 2 was to present an overview of the test methods presently available for testing geomembranes.
Several test methods will be deeply examined and reviewed in the following chapters.
On the other hand, some available test methods were perhaps omitted in this overview, that didn't in fact pretend to be exhaustive, but only to try to present a comprehensive approach to the complex matter.

The reader is also invited to have a glance at the reference list hereafter and at the next chapters in order to complete his information.
The Authors will appreciate any information regarding omissions or mistakes.

References

ASTM D 5 (1983) Standard Test Method for of Penetration of Bituminous Materials.

ASTM D 297 (1981) Standard Test Methods for Rubber Products - Chemical Analysis.

ASTM D 412 (1983) Standard Test Methods for Rubber Properties in Tension.

ASTM D 573 (1981) Standard Test Method for Rubber - Deterioration in an Air Oven.

ASTM D 624 (1986) Standard Test Method for Rubber Property - Test Resistance.

ASTM D 638 (1984) Standard Test Methods for Tensile Properties of Plastics.

ASTM D 648 (1982) Standard Test Method for Deflection Temperature of Plastics under Flexural Load.

ASTM D 696 (1979) Standard Test Method for Coefficient of Linear Thermal Expansion of Plastics.

ASTM D 746 (1979) Standard Test Method for Brittleness Temperature of Plastics and Elastomers by Impact.

ASTM D 751 (1979) Standard Methods of Testing Coated Fabrics.

ASTM D 792 (1979) Standard Test Methods for Specific Gravity and Density of Plastics by Displacement.

ASTM D 882 (1983) Standard Test Methods for Tensile Properties of Thin Plastic Sheeting.

ASTM D 1004 (1981) Standard Test Method for Initial Tear Resistance of Plastic Film amd Sheeting.

ASTM D 1149 (1981) Standard Test Method for Rubber Deterioration - Surface Ozone Cracking in a Chamber (Flat Specimens).

ASTM D 1204 (1984) Standard Test Method for Linear Dimensional Changes of Non-rigid Thermoplastic Sheeting or Film at Elevated Temperature.

ASTM D 1238 (1982) Standard Test Method for Flow Rates of Thermoplastics by Extrusion Plastometer.

ASTM D 1415 (1981) Standard Test Method for Rubber Property - International Hardness.

ASTM D 1505 (1979) Standard Test Method for Density of Plastics by the Density-Gradient Technique.

ASTM D 1593 (1981) Standard Specification for Non-rigid Vinylchloride Plastic Sheeting.

ASTM D 1603 (1976) Standard Test Method for Carbon Black in Olefin Plastics.

ASTM D 1693 (1980) Standard Test Method for Environmental Stress-Cracking of Ethylene Plastics.

ASTM D 1790 (1983) Standard Test Methods for Brittleness Temperature of Plastic Sheeting by Impact.

ASTM D 1938 (1978) Standard Test Method for Tear Propagation Resistance of Plastic Film and Thin Sheeting by a Single Tear Method.

ASTM D 2136 (1984) Standard Methods of Testing Coated Fabrics - Low Temperature Bend Test.

ASTM D 2228 (1983) Standard Test Method for Rubber Property - Abrasion Resistance (Pico Abrader)

ASTM D 2240 (1986) Standard Test Methods for Rubber Property - Durometer Hardness.

ASTM D 2552 (1980) Standard Test Method for Environmental Stress Rupture of Type III Polyethylenes under Constant Tensile Load.

ASTM D 3015 (1978) Standard Recommended Practice for Microscopical Examination of Pigment Dispersion in Plastic Compounds.

ASTM D 3020 (1989) Standard Specification for Polyethilene and Ethylene Copolymer Plastic Sheeting for Pond, Canal and Reservoir Lining.

ASTM D 3083 (1989) Standard Specification for Flexible Polyvinyl Chloride Plastic Sheeting for Pont, Canal and Reservoir Lining.

ASTM D 3253 (1981) Standard Specification for Vulcanized Rubber Sheeting for Pond, Canal and Reservoir Lining.

ASTM D 3254 (1981) Standard Specification for Fabric-Reinforced, Vulcanized Rubber Sheeting for Pond, Canal and Reservoir Lining.

ASTM D 3417 (1982) Standard Test Method for Heats of Fusion and Cystallization of Polymers by Thermal Analysis.

ASTM D 3418 (1982) Standard Test Method for Transition Temperatures of Polymers by Thermal Analysis.

ASTM D 3421 (1975) Standard Recommended Practice for Extraction and Analysis of Plasticizer Mixtures from Vinyl Chloride Plastics.

ASTM D 3767 (1983) Standard Practice for Rubber - Measurement of Dimensions.

ASTM D 3776 (1979) Standard Test Method for Weight (Mass per Unit Area) of Woven Fabric.

ASTM D 4060 (1984) Standard Test Method for Abrasion Resistance of Organic Coatings by the Taber Abraser.

ASTM D 4437 (1984) Standard Practice for Determining the Integrity of Field Seams Used in Joining Flexible Polymeric Sheet Geomembranes.

ASTM D 4545 (1986) Standard Practice for Determining the Integrity of Factory Seams Used in Joining Manufactured Flexible Sheet Geomembranes.

ASTM D 4885 (1988) Standard Test Method for Determining Performance Strength of Geomembranes by the Wide Strip Tensile Method.

ASTM E 28 (1982) Standard Test Method for Softening Point by Ring and Ball Apparatus.

ASTM E 96 (1980) Standard Test Methods for Water Vapor Transmission of Materials.

ASTM E 831 (1981) Standard Test Method for Linear Thermal Expansion of Solid Materials by Thermodilatometry.

ASTM G 21 (1980) Standard Practice for Determining Resistance of Synthetic Polymeric Materials to Fungi.

ASTM G 22 (1980) Standard Practice for Determining Resistance of Plastic to Bacteria.

ASTM G 23 (1981) Recommended Practice for Light and Water Exposure Apparatus (Carbon Arc Type) for Exposure of Non-Metallic Materials.

ASTM G 26 (1977) Recommended Practice for Operating Light Exposure Apparatus (Xenon Arc Type) with and without water for Exposure of Non-Metallic Materials.

ASTM G 53 (1977) Recommended Practice for Operating Light and Water Exposure Apparatus (Fluorescent UV Condensation Type) for Exposure of Non-Metallic Materials.

Cazzuffi, D. (1988) Geotessili e Geomembrane, Modulo, October.

Cazzuffi, D., Puccio, M. and Venesia, S. (1983) Essai de résistance à l'abrasion de membranes, feuilles et revetements pour des ouvrages hydrauliques, C.R. Colloque sur l'Etanchéité des Ouvrages Hydrauliques, Paris, February.

DIN 50017 (1982) Method of Test in Damp Heat Atmosphere.

DIN 53372 (1970) Testing of Plastic Films - Determination of the Low - Temperature Breakage of Non-Rigid PVC Films.

DIN 53455 (1981) Testing on Plastics.

DIN 53479 (1976) Testing of Plastics and Elastomers - Determination of Density.

EPA 9090 (1986) Compatibility Test for Wastes and Membrane Liners, September.

EPA (1988) Lining of Waste Containment and Other Impoundment Facilities. EPA/600/2-88/052.

Frobel, R. and Rigo, J.M. (1990) Puncture Testing (Chapter 7), Geomembranes: Identification and Performance Testing, RILEM TC-103, Chapman and Hall.

Gamski, K. (1984) Geomembranes: Classification, Use and Performance, Geotextiles and Geomembranes, Vol. 1, No. 2, 85-117.

Giroud, J.P. (1983) General Report on Geomembranes-Lined Reservoirs, C.R. Colloque sur l'Etanchéité des Ouvrages Hydrauliques, Paris, February.

Gourc, J.P. and Perrier, H. (1990) Tensile Test (Chapter 5), Geomembranes: Identification and Performance Testing, RILEM TC-103, Chapman and Hall.

Halse, Y., Wiertz, J., Rigo, J.M. and Cazzuffi, D.A. (1990) Chemical Identification Methods used to Characterize Polymeric Geomembranes (Chapter 15), Geomembranes: Identification and Performance Testing, RILEM TC-103, Chapman and Hall.

Haxo, H.E. et al., (1982) Liner materials exposed to municipal solid waste leachate. EPA/600/2-82/097.

Haxo, H.E. and Pierson, P. (1990) Permeability Testing (Chapter 10), Geomembranes: Identification and Performance Testing, RILEM TC-103, Chapman and Hall.

Hoekstra, S.E. (1990) Burst Testing (Chapter 6). Geomembranes: Identification and Performance Testing, RILEM TC-103, Chapman and Hall.

Ingold, T.S. (1990) Friction Testing (Chapter 8), Geomembranes: Identification and Performance Testing, RILEM TC-103, Chapman and Hall.

ISO R 527 (1966) Plastics: Determination of Tensile Properties.

ISO R 1183 (1987) Plastics: Methods for Determining the Density and Specific Gravity of Plastics, excluding Cellular Plastics.

ISO R 1431 (1972) Vulcanized Rubber Determination of the Resistance to Ozone Cracking under Static Conditions.

Koerner, R. (1990) Geomembrane Properties and Test Methods, Designing with Geosynthetics (Second Edition), Prentice and Hall, Englewood Cliffs, N.J., 363-402.

Koerner, R., Halse, Y. and Lord, A: (1990) Stress Cracking Testing (Chapter 11), Geomembranes: Identification and Performance Testing, RILEM TC-103, Chapman and Hall.

NF T 54102 (1971) Feuilles. - Détermination des Caractéristiques en Traction.

NF T 54110 (1971) Feuilles Minces - Détermination de la Témperature Limite Conventionnelle de Non-Fragilité.

Peggs, I.D. and Tisinger, L.G. (1990) Chemical Compatibility Testing of Geomembranes (Chapter 13), Geomembranes: Identification and Performance Testing, RILEM TC-103, Chapman and Hall.

Rigo, J.M., Degeimbre, R. and Wiertz, J. (1990) Biological Resistance Testing (Chapter 14), Geomembranes: Identification and Performance Testing, RILEM TC-103, Chapman and Hall.

Rollin, A.L., Lafleur, J., Marcotte, M., Dascal, O. and Akber, Z. (1984) Selection Criteria for the Use of Geomembranes in Dams and Dykes in Northern Climate, Proc. Int. Conference on Geomembranes, Denver, June.

Rollin, A. (1984) Geomembranes, Proceedings of the Canadian Symposium on Geotextiles, Geogrids and Geomembranes, Toronto.

UEAtc (1982) Union Européenne pour l'Agrément Technique dans la Construction. Directives Génerales UEAtc pour l'Agrément des revetments d'Etancheité de Toitures, Fasc.CSTB 1812.

Van Zanten, R.V. - Editor (1986) Geomembranes, Geotextiles and Geomembranes in Civil Engineering, Balkema, 575-640.

Venesia, S. (1983) Membrane Sintetiche per Impermeabilizzazione, Internal Report ENEL-CRIS N° 3093, February.

3 GEOMEMBRANE SEAMING TECHNIQUES

A. L. ROLLIN
Ecole Polytechnique, Montreal, Quebec, Canada
D. FAYOUX
Plavina, Solvey Group, Brussels, Belgium

3-1 Geomembrane seams

The main property of a seam is that it must ensure a continuous seal between two geomembrane sheets to prevent liquid to escape through the impervious layer installed in the field. Consequently discontinuity, unbonded areas and lack of adhesion between the sheets must not be found within a seam.

Also geomembrane seams must resist stresses imposed on the impervious layer during installation procedures and the service of the work. The seam quality bond strengths (peel and tear) must be stronger than acceptance criteria set by designers.

Upon the material used and the expected performance required, one of the many seaming techniques is usually selected. Since many commercial seaming machines are available, the types, geometry and consequently strength of seams are quite different.

For polyethylene (PE) geomembranes, a heat fusion welding technique must be used since polyethylene sheets can be united by fusion of the polymer. The available techniques are hot air, hot wedge, overlap extrusion, fillet extrusion and ultrasonic.

For poly vinyl chloride (PVC) geomembranes, the hot air and hot wedge welding techniques can be used. But PVC sheets can also be seamed using solvent such as tetra hydrofurane (THF) and rarely contact adhesives.

Elastomers can be seamed using contact adhesives and tapes while thermo plactic elastomers can be united using solvent adhesives. Finally, bituninous sheets can be seamed using direct flame that melt the bituminous mixture.

In this paper, only heat fusion welding techniques will be analysed and data presented on the influence of important welding parameters (geomembrane properties and conditions, climatic conditions and welding machine parameters). These techniques are used extensively with both polyethylene and polyvinyl chloride sheets which represent the majority of installed geomembranes.

Since the geometry of a welded seam depend on the technique used, a classification of heat fusion welds should be made as shown schematically in Figure 3.1. Single and dual welds are obtained with hot air, hot wedge and ultrasonic techniques while overlap and fillet extrusion welds are obtained respectively by each extrusion technique.

The recommended dimensions of single and dual seams are presented in Figure 3.2. Basically a heat fusion weld should have a width greater than 30 mm with a free overlap of 50 mm at the bottom sheet and an overlap of 5 times the sheet thickness at the upper sheet.

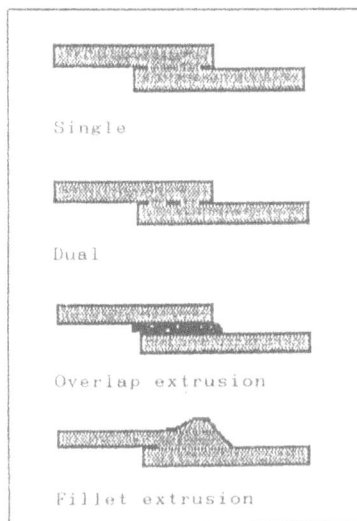

Single

Dual

Overlap extrusion

Fillet extrusion

Fig. 3.1 Types of heat fusion welds

Single

Dual

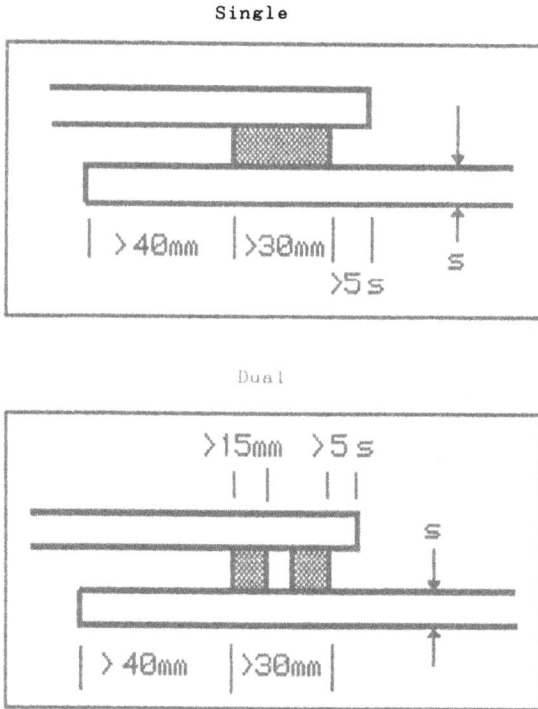

Fig. 3.2   Schematic of seam geometry

3-2   USUAL FIELD WELDING TECHNIQUES

High density polyethylene (HDPE) and polyvinyl chloride (PVC) geomembranes can be united by fusion using one of the many available techniques such as the hot air, the hot wedge, the overlap extrusion (between the sheets), the fillet extrusion (at the sheets interface) and the ultrasonic techniques.

Basically, a quantity of energy or heat is necessary to melt the polymer at the sheet's interface. This quantity of heat can be transferred (by conduction, convection and/or radiation) to the sheets to be welded from a heating element, by air, by a hot extrudate or even generated by friction. The techniques then differ from one another by the method used to transfer the necessary energy.

61

These techniques can be used to weld geomembranes in the field. Most of the commercial apparatus have been designed to seam adequately under moderate field constraints and sheets' parameters. The quality of the welds depends largely on the expertise of the trained operators, the reliability of calibration and the quality of the apparatus instrumentations.

3-2.1 Hot air technique

Heated air from an electrical element is blown between the two sheets to be bonded together in order to melt an interface strip (width > 28 mm for automatic apparatus) from each sheet using an air blower. Heat is then transferred from the flowing air to each sheet in such a rate that polymer melting at the interface occurs. This quantity of heat must be controlled carefully to avoid under or overheating of the sheets. In the field, this heat transport depends on the sheet numerous parameters and the atmospheric conditions. The average temperature of the heated air is 375° C.

The weld is secured by applying a pressure on the heated area of the sheets. This is usually performed using motorized rollers on automatic apparatus or manually on other machines. The seams that can be obtained are single and dual welds of thickness equal to or less than the total thickness of the two sheets.

Such automatic commercial welders are portable units welding at an estimated average rate of 1.0 metre per minute. The actual welding rate depends on sheet parameters and climatic conditions.

3-2.2 Hot wedge technique

The energy necessary to melt the sheets interface is generated by electrical elements placed directly between two sheets. Rollers are used to both drive the machine and to applied pressure on the heated strip of the sheets. The heating elements must be at all time placed in between the sheets in such a way as to avoid any contact between the heating element and the sheets or to avoid excessive melting of the polymer. Average temperature of the heating element is approximatively 375° C but differs upon the sheet formulation.

Because heat is directed locally, preheating of the sheets is an alternate method used to avoid any thermal shock along the seam that causes rearrangement of the molecules resulting in a fragile structure. The heat necessary to condition the sheets is usually supplied by hot air.

Commercial welders are available with one or two heating wedges to produce a single or double weld type as shown on Figure 3.1. The estimated speed to weld HDPE sheets of 1 to 3 mm thickness ranges between 1 to 2 metres per minute.

3-2.3 Overlap extrusion

A hot polyethylene extruded strip of thickness equal to 60% of the sheet thickness and of temperature greater than 230° C is placed between the two sheets. A portable extruder with its die placed between the sheets is then used to produce a strip of polyethylene of identical formulation to the sheets. To avoid thermal shock that will weaken the polymeric structure along the edge of the weld, preheating of the sheets is usually performed using an electrical heating element placed between the sheets. The energy necessary to melt the polymer at the sheets interface is then generated from the preheating and from the extruded strip itself.

Pressure must be applied uniformly to insure good bond between the extruded strip and both sheets. It must be applied for a length of time necessary to cool down both sheets and the extruded polymer. Finally, a surface cleaning treatment (grinding and hot air) is often used to prepare the welding area in order to increase the machine performance.

Figure 3.1 shows a weld thicker than the ones produced by hot air and hot wedge techniques. It is equal to or less than 2.6 times a single sheet tickness. The welding speed depends strongly on the sheets' thickness and temperature and the on site atmospheric conditions. The estimated maximum welding speed that can be reached with thinner sheets (1.0 mm) is 2 metres per minute and reliability of weld quality can be achieved with thick geomembranes such as 2.5 to 3.0 mm. The large and heavy machines are excellent to perform longitudinal seams but are not usually used for repairs works, patches, at corners and at connecting areas.

3-2.4 Fillet extrusion

A hot polyethylene fillet as produced by a portable
extruder may also be installed at the interface of
two geomembranes. The hot PE fillet (of formulation
identical to the sheets) is placed at the edge of the
upper sheet to be welded as shown schematically on
Figure 3.1. Basically the energy necessary to melt the
polymer of the upper sheet and the polymer at
the lower sheet is generated from the hot fillet and
from preheating of the sheets with hot air. To
avoid thermal shock that will weaken the polymeric
structure along the edge of the weld, preheating of the
sheets must be performed with this type of welding
technique. The bond is secured by applying a weak
pressure on the weld.

The fillet temperature is approximately 250° C and the
maximum welding speed is approximately 0.6 metre
per minute bonding sheets of thickness ranging from 1.0
to 3.0 mm. This technique is the only known for
repair works and at locations difficult to weld.

3-2.5 Ultrasonic technique

The ultrasonic technique is a welding method
introduced recently on the market. Polyethylene and
polyvinyl chloride can be melted using ultrasonic
waves. Under resonnance, a metallic horn in contact
with a geomembrane can produced enough energy by
friction to melt the polymer. Upon the vibration
frequency of the horn, usually between 20 to 40
KHz, the interface of a sheet can be melted. The
technique consists in placing an ultrasonic horn
between the two sheets in such a way as to maintain
a contact between the horn and the sheets. Because
the melted polymer is displaced in a saw tooth
pattern (crown and valley pattern) , a pressure must
be applied by motorized rollers to secure an
uniform bond between the sheets.

Recently a prototype has been introduced in field
applications and preliminary tests performed at
Ecole Polytechnique have indicated that HDPE sheets
of thickness ranging from 1.0 to 1.5 mm can be
succesfully welded at a maximum speed of 1.0 metre per
minute. It can be used to performed longitudinal
seams but the available prototype has not been
designed to perform repairs works or in areas
difficult to reach.

64

At this stage of development of the apparatus, more data are needed to consider welding succesfully thinner and thicker HDPE sheets. Finally, the performance life of the horns is limited because of polymer sticking on it and consequently must be changed periodically.

3-3 WELDING PARAMETERS

The quality of a geomembrane field seam depends on many factors related to the membrane itself, the atmospheric conditions, the project design and the welding technique used.

The geomembrane parameters influencing the welding conditions are its compounding (specific gravity, carbon black content and melt index), its thickness, its surface neatness (presence of greases, water or soil particles) and the seam width. Also atmospheric conditions such as temperature (cold and hot), wind and air humidity will influence the welding quality of sheets. But the more important factors are related to the welding machine and its calibration set by the trained operators such as the welding speed, the quantity of heat transferred to the sheets, the applied pressure and the alignment of the sheets relative to the welding machine.

3-3.1 Membrane parameters

3-3.1.1 Geomembrane quality

The quantity of heat needed to melt the polymer on both sheets' surfaces depends greatly on the molecular structure of the polymer and the composition of the compound. Sophisticated quality control programs must be implemented to ensure homogeneity of the product and to detect any poor dispersity. An optimum value for additives content must be used in the making of the sheet to prevent risk of producing fragile seams and to ease the welding process.

3-3.1.2 Thickness

Generally thick geomembranes are easier to weld than thinner ones more vulnerable to overheating responsible for poor quality bonds and undulation in the material.

65

On the other hand, the quantity of heat to supply to thicker sheets become so large that the resulting welding speed is greatly affected. Most of the welding techniques used are then limited to a range of sheet thicknesses in order to weld at a reasonable speed. For example, the maximum attainable welding speed to seam HDPE sheets with an ultrasonic machine is plotted versus the sheet thickness in Figure 3.3. These data have been gathered in a research program performed at Ecole Polytechnique of Montreal. It can be observed that acceptable peel strength resistance of the welds were obtained at a maximum welding speed of 1.3 m/min for sheets of 1.0 and 1.4 mm thick. This maximum welding speed value is decreased as the sheets' thickness increases or decreases. This behaviour has been also observed with others welding techniques.

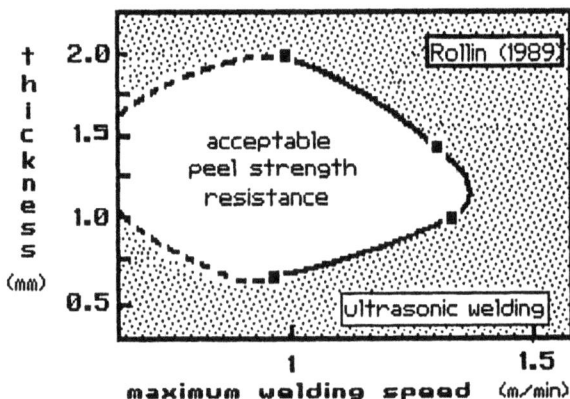

Fig. 3.3 Sheet thickness influence on welding speed

3-3.1.3 Surface neatness

The surface of sheets to be welded must be cleared of greases or oily products, soil particles and water. As shown by Peggs (1987), Fayoux (1988) and Rollin (1989) and as presented on Figures 3.4 to 3.7, the presence of these pollutants on the sheets' surface is lowering the bonding strength of welds. Preheating and cleaning of the sheets' surface to be welded using a hot air flow and/or solvents is often necessary to dry the surfaces and to remove greases and dirt particles.

66

3-3.1.4 Seam width

Geomembranes are welded together to form a continuous seam wide enough to resist shear and peel forces from thermal expansion and from soil movements. Also each seam must be sealed bonded to limit liquid infiltration into the soil underneath the geomembrane layer. For narrow width seams, the contact area might not be enough to offer the necessary resistances while it is more difficult to uniformely transferred¯ heat and applied the necessary pressure to secure the bond of a larger width seam. An optimum seam width estimated at 30 mm for automatic hot air technique (Figure 3.2) should be selected and of 50 mm for manual produced welds.

As an example, experimental data collected by Peggs (1987) are presented on Figure 3.8 for fillet extrusion welding of 2.5, 2.0 and 1.0 mm thick HDPE sheets. In all cases as the width of the extruded fillet decreases from the reference width (probably 30 mm width), seams of lower peel strength values were produced. Similar behaviour was observed with seams larger than the reference width.

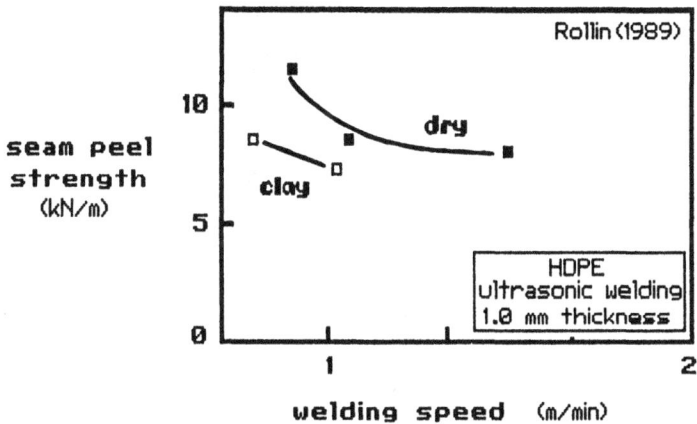

Fig. 3.4   The influence of clay particles
on seam quality

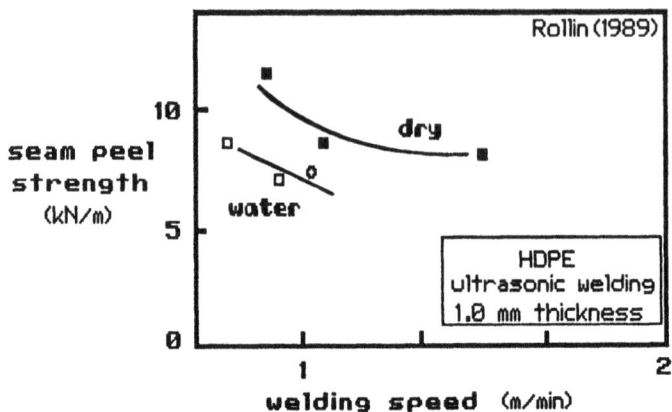

Fig. 3.5   The influence of water on HDPE
seam quality

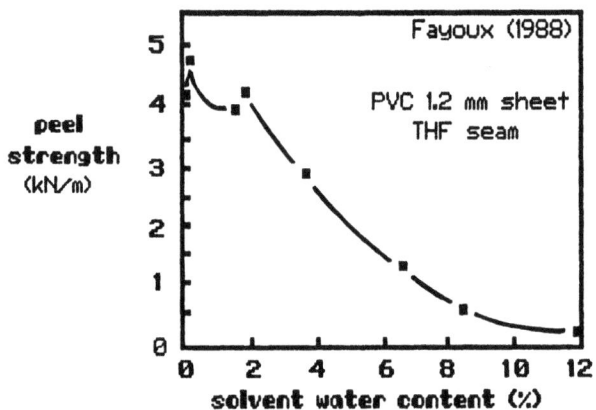

Fig. 3.6   The influence of water on PVC
seam quality

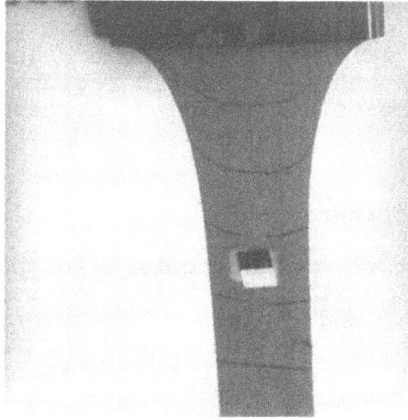

Fig. 3.7   The influence of surface pollutants
on seam quality

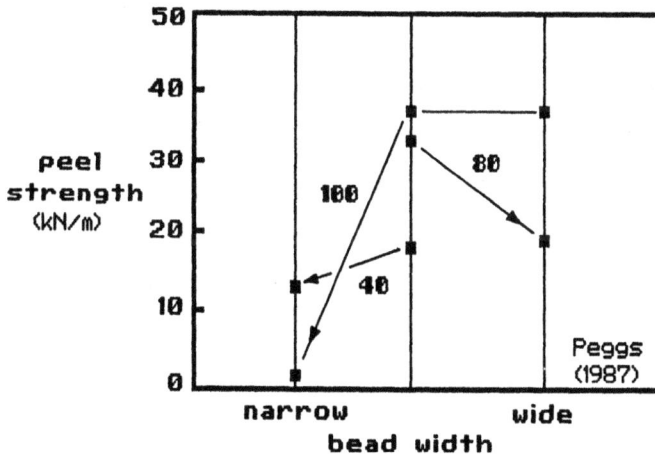

Fig. 3.8   Seam width influence on fillet
extrusion welding

3-3.2 Atmospheric conditions

Atmospheric conditions such as temperature, humidity, rain, and wind can drastically influence the geomembranes' welding. Since geomembranes are often installed in extreme climatic regions, care must be taken that the field welding process is performed adequately.

3-3.2.1 Temperature

The temperature of sheets fluctuates constantly during their installation with lower temperatures in the morning and higher temperatures in the afternoon. Their maximum values for HDPE sheets can be as high as 50 °C higher than the atmospheric temperature because of their black colour that increases their sun radiation absorption. In nordic climates, geomembranes can reach freezing temperatures because they are usually installed on the ground for long period of time prior to the welding.

As an example of the influence of the ambient temperature on the quality of seams, experimental results obtained at 20 and 8° C during a research program performed at Ecole Polytechnique of Montreal using an ultrasonic machine are presented on Figure 3.9. Field welds made at a temperature of 20° C had greater resistance to peel than the ones performed on sheets at 8° C. In a similar way, the influence of the sheets' temperature on the quality of seams can be observed on Figure 3.10. The peel strength of seams performed on sheets at temperatures lower than 20° C is decreased.

It is then very important that the temperature of the geomembranes be monitored during the welding of the sheets and that adjustments of the welding parameters be performed adequately.

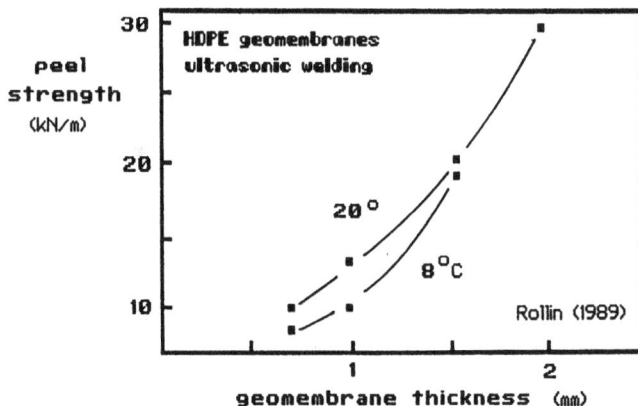

Fig. 3.9   The influence of ambient temperature
           on seam quality

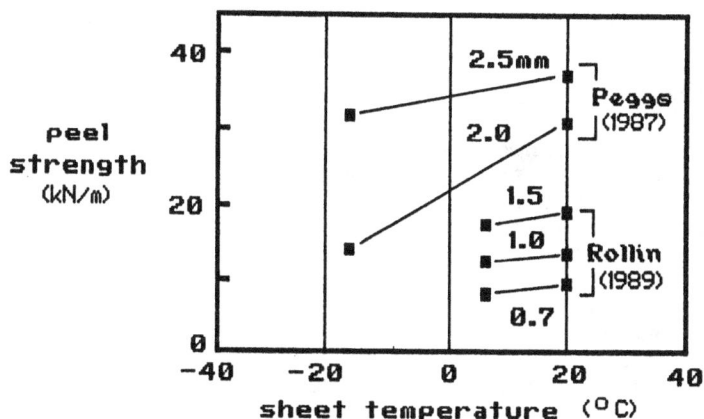

Fig. 3.10   The influence of sheet's temperature
            on seam quality

3-3.2.2   Air humidity

It is believed that the air moisture also affects
the welding parameters since in very humid
regions, sheets' surface will always be wet. The air
used in the preheating step or used in the hot air
technique will need to carry a greater amount of energy
to reach the desired temperature to dry the sheets. No
scientific investigation of this parameter has
been done until now to clearly identify and quantify
its influence on seam quality.

3-3.3 The welding parameters

Some of the welding parameters influencing the quality of geomembrane's seams are the welding speed, the quantity of heat transferred to the sheets, the applied bonding pressure, the extrudate thickness and temperature, the position of the pressure rolls, the time pressure is applied and the sheets alignment.

The peeling resistance of seams differs from one seaming techniques to another. In order of resistance for PVC materials (from weaker to greater resistance) according to Fayoux (1988) : 1- solvent; 2- hot air manual welding; 3- hot air machine welding; and 4- hot wedge machine. Fayoux (1988) and Rollin (1989) generated data respectively for PVC geomembranes and for HDPE geomembranes to support that maximum peel resistance can be found for a welding temperature, a welding speed and an applied pressure.

3-3.3.1 The welding speed

As already mentioned, a maximum welding speed can be reached using a specific machine calibrated to perform under a set of field conditions. This speed can be defined as the highest welding speed the machine can reach to produce a seam of acceptable quality. Unfortunatly each technique and apparatus are limited to relatively low welding speeds. As an example, experimental results are presented on Figures 3.11 and 3.12. They were conducted in a plant using HDPE 1.0 mm (40 mils) thickness geomembranes with an overlap extrusion technique and in field for PVC 1.2 mm sheet using a hot wedge machine. The maximum welding speed was determined to be at approximately 2.5 m/min (7.5 ft/min). At welding rate greater than this value, the seam peel strength were identified as non acceptable using the acceptance requirement criterion.

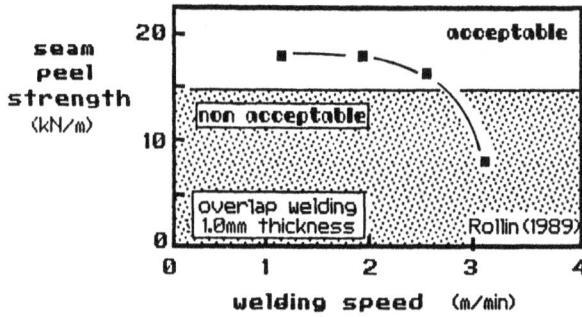

Fig. 3.11 The influence of welding speed
           on HDPE seam quality

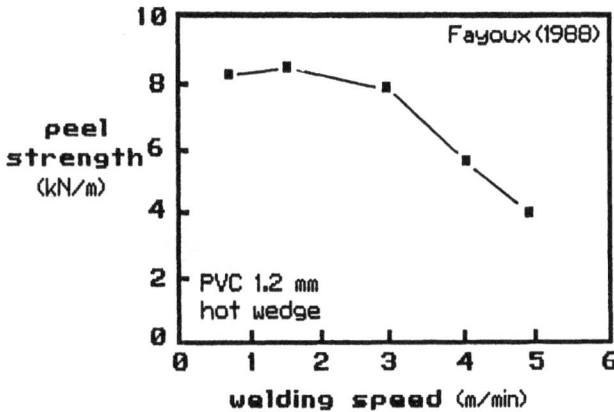

Fig. 3.12 The influence of welding speed
           on PVC seam quality

3-3.3.2 The preheating temperature

Preheating is usually needed to perform a good quality
seam with the extrusion techniques. This step is
necessary to increase the sheets temperature to
avoid any thermal shock when the hot extruded
polymer comes into contact with the sheets. Preheating
is also used to transfer a certain quantity of
heat to the polymer blowing out the moisture in
the overlap region and removing pollutants.

As an example, a typical welding curve for an overlap
extrusion welding of a 1.0 mm thick HDPE
geomembrane is presented in Figure 3.13. Acceptable
peel strength seams were performed with a preheating
temperature ranging from 80 to 125 °C while weaker
seams were produced at lower and higher temperatures.

73

Many authors have indicated the changes of sheet properties at seam edges resulting from insufficient preheating prior to seaming (Charron(1989)). Welding machines are usually locally melting the polymer of the sheets creating an edge of hot and cold frontiers resulting in rearrangement of the molecular structure and usually concentrating stress cracks for HDPE membranes or fragilation of the material for both PVC and HDPE membranes. The weakest point in a seam is then at the edge of the lower sheet. For PVC, weaker tearing resistance were found by Fayoux (1988) from concentration of stresses generated from geometry and thermal shock.

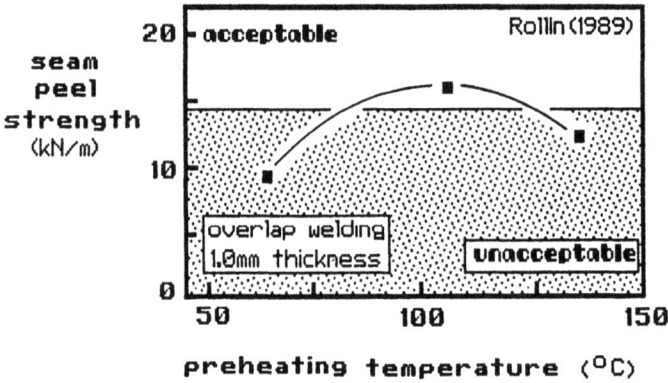

Fig. 3.13 The influence of the preheating
temperature on HDPE seam quality

3-3.3.3 The die temperature

In a similar fashion, the temperature of the extruded polymer must be closely monitored to transfert the quantity of heat necessary to melt the polymer. This temperature depends primarily on the melting point of the compound used and to a less extent upon the geomembrane temperature and the preheating conditions used.

As an example, a typical welding curve is presented on Figure 3.14 for an overlap extrusion of a 1.0 mm HDPE geomembrane. At very low temperatures, not enough heat is transferred to the sheets to produce acceptable welds.

74

Identical results are obtained at too high temperatures because a too large quantity of polymer is melted whitin the sheets and the extruded strip of fillet cannot be placed adequatly between the sheets since it has loss any consistancy. Die temperature must then be controlled in a range of temperatures.

For PVC geomembranes, the optimum machine temperature is reached just before the burning point. Care must then be exercised to observe burning conditions and homogeneity of the welds since test results can be acceptable even when PVC sheets are burned.

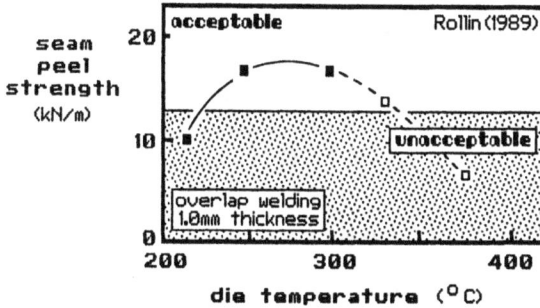

Fig. 3.14 The die temperature influence on HDPE seam quality

3-3.3.4 The air temperature

The air temperature in hot air welding technique must be closely monitored to transfer the quantity of energy necessary to melt the polymer. This temperature depends on the melting point of the polymeric compound of the sheets and climatic conditions prevailing in the field.

An example is given in Figure 3.15 for a hot air welding of a 1.2 mm thick PVC geomembrane. Eventhough the tensile resistance of the seam is not affected by the welding temperature in the observed range, the peel resistance is highly sensitive. Acceptable seam peel strength were obtained with air temperature ranging from 360 to 420 °C. It is interesting to observe that both peel and tensile resistances of the seams were lower in value than the tensile resistance of the sheet itself with respectively relative value of 20% and 70%. These values are quite different from results obtained with HDPE geomembranes.

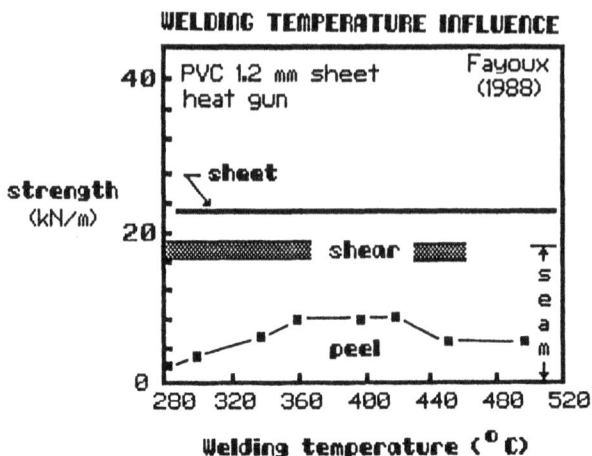

WELDING TEMPERATURE INFLUENCE

Fig. 3.15   The influence of air temperature
on PVC seam quality

3-3.3.5 The welding pressure

Since all the welding techniques are basically
melting the polymer at sheets' surface, most require
that pressure be applied for a certain period of
time during the cooling of the seam to bond the
sheets together. This is usually performed using
motorized rollers.

The value of the applied pressure depends on the welds'
thickness and prevalent field conditions. As an
example, experimental results are presented on
Figure 3.16 indicating the extent of peel strength
variations of welds produced using the overlap
extrusion technique to bond 1.0 mm thickness HDPE
sheets. Weak bonds were achieved whenever too low
or very large pressures (machine gage pressure) were
applied. Microanalysis of performed seams suggested
that under high pressure, the cooling process is
affected to such an extent that the weld structure is
weakened.

Not enough attention has been paid to the
importance of controlled cooling process and the
pressure influence on the seam quality. It is often
the parameter responsible for the weakening of seam.
Not only the value of the pressure is important but the
duration of its application as well as when it is
applied.

The pressure is usually applied using rollers that must be parallel to one another to obtain good quality welds. The gap between the rolls must be selected to ensure the appropriate squeezing of the sheets forcing the molten polymer to spread between the sheets. This step can be badly performed if the rollers are not parallel resulting in uneven thickness of the weld and creation of unbonded areas as shown schematically on Figure 3.17.

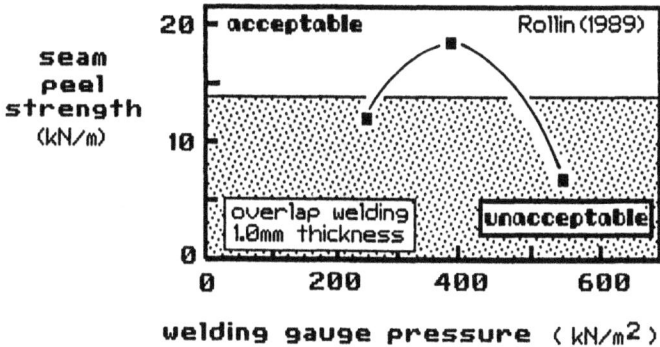

Fig. 3.16   The welding pressure influence on
             HDPE seam quality

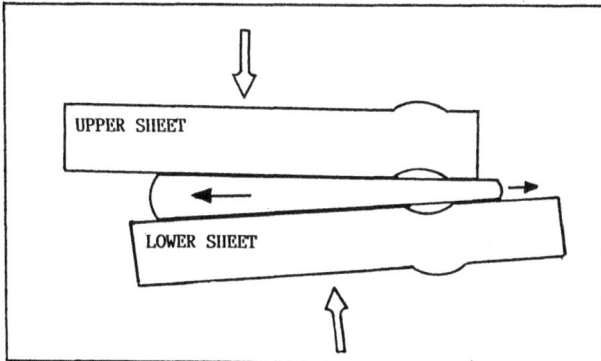

Fig. 3.17   The influence of unaligned sheets
             on seam quality

3.4 Calibration of the welding apparatus

Since many factors can influence a seam quality, calibration of the welding apparatus is very important. Usually a field welding apparatus must be calibrated every morning prior to the starting of a seam. This is accomplished by trial runs on pieces of geomembrane samples and field testing is mandatory to make sure that all the apparatus parameters are set satisfactory.

This calibration must be performed with the assigned operator of the day and should be kept steady for the recorded climatic conditions. Unless disturbing climatic conditions suddently appear, or a change of operator is done, or electrical problems are encountered, or faulty welds are identified, usually the seaming of geomembranes is performed without the need of calibrating again the apparatus.

3-5 Conclusion

The described welding techniques can be used adequately to weld PVC and PE geomembranes in the field. Most of the commercial equipments have been designed to seam succesfully under moderate field constraints and sheet parameters. The reliability of the quality of the welds in the past depended largely on the expertise of the trained operators because of lack of reliability of calibration and lack of instrumentation. Significant improvements are still needed from an execution and quality assurance point of view since the manipulation portion of virtually all machines is viewed as the weakest link.

3-6 References

Fayoux, D., (1988) "Technical Report on Welding Techniques", Plavina, Solvey Group, June 1988.

Peggs, I.D. and Rose, S., (1987) "Practical Aspects of Polyethylene Geomembrane Seam Welding", Geotechnical Fabrics Report, January 1987, pp 12-16.

Rollin, A.L., Vidovic, A. and Denis,R., (1989) "Evaluation of HDPE Geomembrane Field Welding Techniques: Need to Improve Reliability of Quality Seams", Proceedings Geosynthetics'89, San Diego, February 1989, pp 443-455.

Charron, R.M., (1989) "Polymer for Synthetic Lining Systems: Some Molecular Structure Property - Application Relationship", Proceedings Geosynthecs'89, San Diego, February 1989, pp 408-420.

4 NON-DESTRUCTIVE AND DESTRUCTIVE SEAM TESTING

A. L . ROLLIN
Ecole Polytechnic, Montreal, Quebec, Canada
D. FAYOUX
Plavina, Solvey Group, Brussels, Belgium
J. P. BENNETON
CETE, Ministère de l'équipment, Lyon, France

4-1   GENERAL

Acceptance criteria of geomembrane seams differ greatly from one work site to another and it is presently impossible to reach unanimity of opinion on value of bond strength both for PVC (poly vinyl chloride) and PE (polyethylene) seams. The major functions that a seam must performed should then be considered and analysed.

The main property of a seam is that it must insure a continuous seal that prevent liquid to escape through the impervious layer. Consequently discontinuity, unbonded areas and lack of adhesion between sheets must not be found within a seam.

Geomembrane seams must resist stresses and elongation imposed on the impervious layer. These stresses are imposed during the installation and the service of the work:

a) during installation:
- unrolling of the sheets on the site
- installation under water (tensile and tear resistance whenever high water flow rate)
- wind action
- rip rap protection over the membrane
- weak support
- others

b) during service
- imposed elongation
  - water pressure on membrane weakly supported
  - concrete pressure in tunnel
  - weight of the material
- imposed stresses
  - by wind action
  - water speed and turbulence
- induced stresses
  - thermal contraction
  - polymer ageing

Procedures must then be taken to ensure that faulty seams are identified during the non-destructive quality control programs (plant and field), at the implementation stage and during the initial period of used.

Seam quality control must be performed to obtain information on both the continuity of the seam and its mechanical resistances. Since non-destructive tests (visual observation, set point, air lance, ultrasonic impedance and pulse echo, vacuum box, high voltage spark, pressurized dual seams and microscopic analysis) do check the continuity of a seam but do not quantitatively measured mechanical resistances, peel and tensile tests must be performed regularly.

Calibration of the non-destructive method apparatus must be performed prior to the field program using prepared samples with faulty seams to implement the non-destructive testing program.

4-2 NONDESTRUCTIVE TECHNIQUES FOR ASSESSING SEAMS

Nine nondestructive test methods can be used to assess geomembrane seams in the plant and the field. These methods are detecting faulty seams or measuring the continuity of a seam without measuring the relative strength of the bond.

A nondestructive seam test is defined as a test performed in a plant or in a field on a continuous seam without destroying nor altering the seam itself. There is no need to collect samples since those tests are performed along the seam laid down on a floor or on a soil.

Three of the methods (mechanical point stress, air lance and ultrasonic impedance plane) are not

extensively used since they assess qualitatively the
seams while the pressurized dual seam, vacuum box,
ultrasonic pulse echo and high voltage spark
techniques are used in most of the geotechnical
applications to identify seam discontinuities.
These techniques are presented in Figure-4.1.

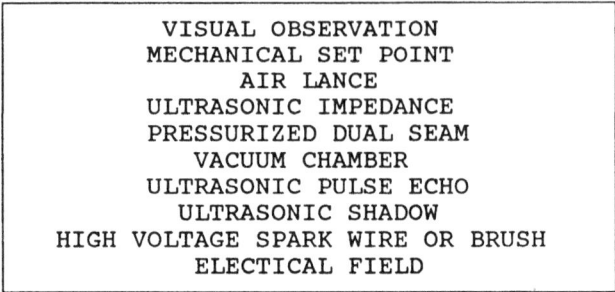

```
VISUAL OBSERVATION
MECHANICAL SET POINT
AIR LANCE
ULTRASONIC IMPEDANCE
PRESSURIZED DUAL SEAM
VACUUM CHAMBER
ULTRASONIC PULSE ECHO
ULTRASONIC SHADOW
HIGH VOLTAGE SPARK WIRE OR BRUSH
ELECTICAL FIELD
```

Fig. 4.1   Nondestructive test methods

4-2.1 Visual observation

Unbonded areas can be detected by visual observation
along the seam. Also overheated seams can easily be
detected by observing ondulations in the sheet seams.

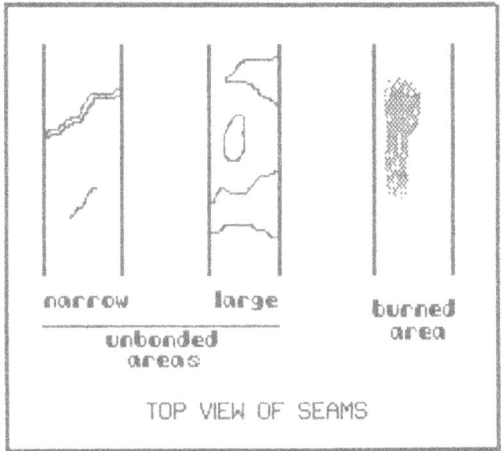

Fig. 4.2 Faulty transparent geomembrane seams

The detection can drastically be increased if transparent polymeric sheets, such as PVC geomembrane, are used. As shown schematically in Figure 4.2, cracks, unbonded areas and burned areas can easily be observed through transparent welded seams.

4-2.2 Mechanical point stress

Unbonded area can be detected by sliding a dull tool along the edge of a weld. The tool can easily lift unbonded areas of thin geomembranes but cannot detect unbond area within the width of a weld.

4-2.3 Air lance

The air lance technique detects unbonded area using air. Under 350 kPa of pressure, air is blown through a nozzle and is applied along the edge of the weld in an effort to lift unbonded areas. The impinging air stream inflates the unbonded areas. This technique cannot detect an unbond area within the width of a weld and is not used extensively with stiff HDPE geomembrane. The maximum seam inspection rate is approximately of 4 m/min.

4-2.4 Ultrasonic Impedance Plane

The Ultrasonic Impedance Plane (UIP) technique uses the accoustic impedance of the polymer. A well-bonded weld possesses a certain accoustic impedance different than that of a weld with an unbonded area. A continuous wave of 160 to 185 kHz is transmitted through the seam by mean of a transducer in contact with the geomembrane and a characteristic dot pattern is registered on a monitor. The location of the dot pattern on a screen indicates the quality of the bond. Under field conditions, this technique has not been favored.

4-2.5 Pressurized dual seam technique

As shown on Figure 4.3, this method establishes an
air pressure in the channel of a double weld seam.
The seam is sealed at two locations and a needle,
connected to a pressure gauge and an air supply, is
inserted into the channel between the welded areas. A
200 kPa pressure is applied and the pressure is
monitored for a certain time. Benneton (4.1)
recommended that the pressure be maintained at 0,1
MPa for 3 minutes for PVC geomembranes while Fayoux
(4.2) is recommending a maximum pressure of 2.5 bar.

A leak is detected along the length of a seam if
pressure cannot be established or held at the
original value. A leak can be detected from the
noise originating from the escaping air otherwise
a leak is very difficult to locate because air can
escaped from both sides and all along the length
tested. A long lasting procedure must then be performed
to detect and locate the leak usually using one of
the other methods. On the other hand, this technique
allows rapid testing of good quality seams.

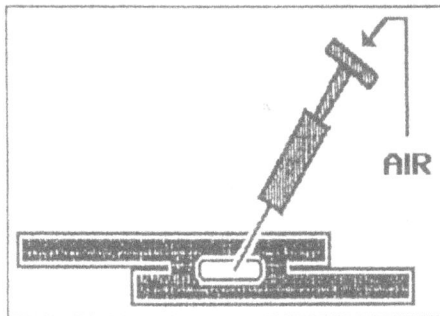

PRESSURIZED DUAL WELD

Fig. 4.3   Schematic of the pressurized
dual seam technique

4-2.6 Vacuum chamber technique

This test is performed by applying a vacuum to a
soaped section of a seam. The vacuum is applied
through a chamber equipped with a vacuum gauge, a
clear glass view panel in the top and a soft rubber

gasket on the periphery of the open bottom. The inspection is performed by moving along the soaped seam the sealed vacuum chamber. A vacuum, set between 122 and 244 mm of mercury (Fayoux (4.2) recommends between 0.2 to 0.4 bar), is applied inside the chamber by use of a vacuum pump. Any unbonded area across the width of a weld can be detected by observing the formation of bubbles in the chamber as schematically shown on Figure 4.4. This technique is used extensively and the working speed is approximated at 1.5 to 3.0 metres per minute.

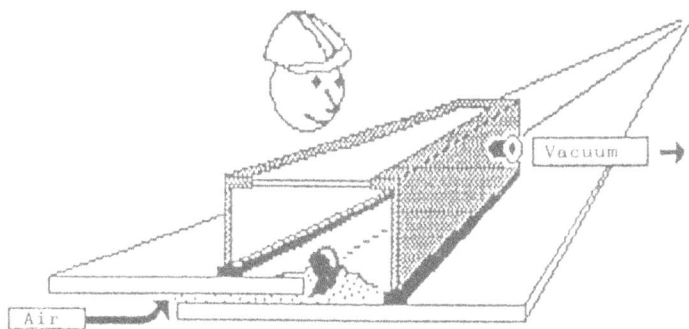

Fig. 4.4   Schematic of the vacuum chamber technique

4-2.7 Ultrasonic pulse echo technique

This test is performed by forcing a high frequency sound wave through a seam overlap to detect discontinuities. A commercially available frequency generator capable of producing frequencies in the range of 5 to 15 MHz is used. The contact send/receive transducer head is is capable of being moved at a rate of 1.5 to 2.0 m/min along the surface length of the seam area. The transducer head is so designed as to give continuous surface to surface thickness measurements once calibrated. The transducer head must assured good contact with the lining surface by providing a continuous contact medium (water) at the interface between the transducer and the seam. The ultrasonic signal is capable of being viewed on a monitor, as shown schematically on Figure 4.5, as well as activating an audible alarm when a discontinuity is detected. The method is limited to detect large unbonded areas, larger than half the transducer diameter (usually of 5 mm) (Benneton (1990)).

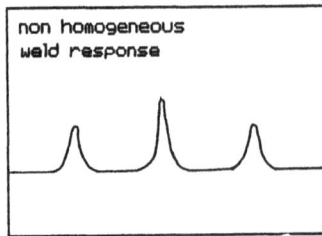

```
non homogeneous
weld response
```

Fig. 4.5   Ultrasonic pulse echo signal

4-2.2.8 High voltage spark technique

A continuous metallic tape or wire is inserted between the two sheets to be bonded and then connected to a metallic point. A high voltage (15 to 30 kVolts) electrical current is applied at the end of the metallic conductor and the set point is slowly moved along the seam as shown schematically on Figure 4.6. Any leakage to ground can be detected by sparking and an audible alarm can be activated from the unit.

The spark intensity depends on the voltage applied, de size of the unbonded area (width and length), the testing speed (set point) and to a less extent the air moisture. Benneton (1990) has carried tests using a voltage equal to 30.000 volts at a testing speed of 3 m/min to find the followings limitations:

length of the spark between 30 to 40 mm
detection of unbonded areas of width greater than 1,5 mm
unbonded areas of width smaller than 0,7 mm not detected
difficulty to detect unbonded areas of width ranging from 0,7 to 1,5 mm

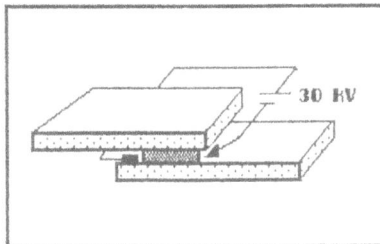

Fig. 4.6   Schematic of the high voltage spark technique

4-2.9 High voltage electrical brush

This method is using the same principle as the high
voltage spark technique. No metallic strip or wire is
inserted bertween the two sheets since the seams are
laid down on a ground. Any leakage through the seam
or hole in the sheets can be detected by sparking
from the metallic brush as shown on Figure 4.7. An
audible alarm can be activated from the portable unit.

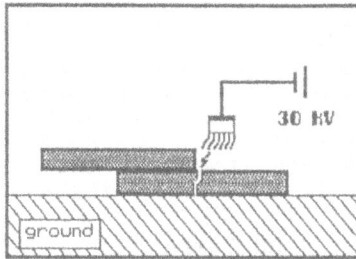

Fig. 4.7   Schematic of high voltage electrical brush

4-3 DESTRUCTIVE TEST METHODS

Destructive tests must be performed in fields and in
laboratories to support proper welding machines'
calibration and to quantify the bond strength of
performed seams. Usually they are done whenever a
welding machine is started or new set of conditions
are meet on a site. Also samples collected, as control
specimens, are submitted to laboratory testing to
determine the quality of the work performed.

Normally tests should have been developped to
evaluate seam properties of geomembranes used in
geotechnical applications but, because of the urgent
need to assess the bond strength of work performed in
situ, seam strength of geomembranes are determined
using two customary destructive tests (the shear
and the peel tests) as shown on Figure 4.8 For HDPE
membranes, these methods are modified version of ASTM
standards developped to measure the tensile and peel
properties of very thin plastic sheetings' seams.

As we shall demonstrate, the data obtained from these tests should be used with care and in many cases, these tests must be supplemented by microscopic analysis to evaluate proper quality of welds and detect changes in the polymeric structure (detection of unbonded areas, determination of brittle failures and observation of slow crack growth phenomenon). In the future, a better quality control test will have to be developped to produce proper immediate failure.

FOR HDPE	TENSILE RESISTANCE	ASTM D-638
		ASTM D-3083
	PEEL RESISTANCE	ASTM D-413
		ISO 6133
FOR PVC	TENSILE RESISTANCE	ASTM D-3083
		ASTM D-882
		DIN 1910/3
	PEEL RESISTANCE	ISO 6133
		ASTM D-413

Fig. 4.8 Customary destructive test methods

4-3.1 Tensile test

4-3.1.1 HDPE seams

Destructive seam testing becomes more technical when high density and very low density polyethylene (HDPE and VLDPE) seams are evaluated. Many seam tests acceptance criteria are simply oriented to determine whether the seam is of adequate strength at the time of installation. No thought is given to those parameters that may provide information on the probable durability of the seam and the adjacent geomembrane. Guidelines for test procedures and appropriate test parameters for HDPE were presented by Peggs (1985, 1987).

The dynamic test method recommended by the National Sanitation Foundation as part of its standard test 54 on Flexible Membrane Liners (revised 1985) is a modified version of the ASTM D-3083 test method (Standard Specification for Flexible Poly Vinyl Chloride Plastic Sheeting for Pond, Canal and Reservoir Lining of thicknesses between 0.20 to 0.76 mm). The shear test simulates the stresses seen in service induced by mechanical and thermal contraction on a seam and the results are used to evaluate if it fails before the liner itself. The NSF 54 standard identifies minimum requirements for Bonded Seam Strength by comparing the tensile strength at yield of a welded specimen to the tensile strength at yield of the base material (the acceptance requirement coefficient F_T). The required coefficient value usually recommended is such that the yield stress of a HDPE geomembrane seam exceeds 80% of the tensile yield strength of the base material.

The standard test used to determine the tensile yield strength of the base material (sheet without a seam) is the ASTM D-638 (Standard Test Method for Tensile Properties of Plastic). For HDPE material, the test uses a standard dumbbell-shaped test specimen stresses at a rate of 50 mm/min. The specimen overall length is 115 mm, with overall width of 19 mm, length at the narrow section of 33 mm and width at narrow section of 6 mm. A typical shear-stress curve of a HDPE sheet showing the relative location of the yield and failure points is presented in Figure 4.9. The result can be recorded in Newtons per square metres of cross section of the specimen or in Newtons per metre of width of the specimen at both the yield and failure points.

The recommended NSF 54 test for non-reinforced materials uses a strip specimen 1 inch (25.4 mm) wide and of length permitting a grip separation of 4 inches (100 mm) plus the width of the seam. The weld is to be centered between the clamps (see Figure 4.10) and the rate of separation of the clamps is 2 inches/min (50 mm/min). The result are recorded in Newtons per metre of seam width and the mode of failure (failure at seam edge, at grip edge, in seam interface) must be recorded as well as all test parameters such as grip separation, crosshead speed, base material thickness, seam width, temperature and humidity.

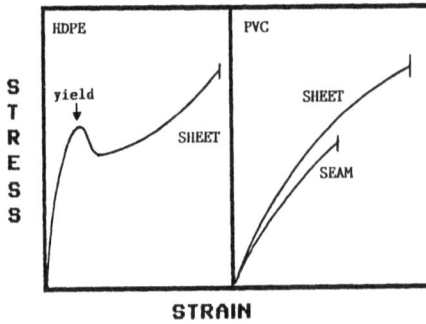

Fig. 4.9   Typical tensile curves of HDPE
            and PVC geomembranes

The weld submitted to this shearing action is usually
more rigid than the sheet such that the elongation of
the specimen is within the base material itself. It
is then very daring to compare tensile   strength   of
specimens with different shapes and dimensions   to
assess the weld acceptability. It becomes evident that
the acceptance requirement coefficient obtained from
this   procedure   cannot   be used as an adequate
acceptance criterion for seam quality. Peggs   (4.3),
Peggs   (4.4), Rollin (4.5), Rollin (4.6) and Rollin
(4.7) clearly demonstrated that point of view.

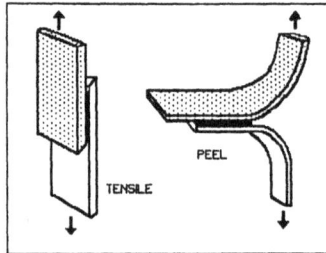

Fig. 4.10   Schematic of specimens used in
            shear and peel tests

90

In the shear test, it is not adequate to require failure in the geomembrane at a tension or stress exceeding a minimum value. When failure occurs in the geomembrane, it must do so in a ductile manner and not in a brittle manner. It is therefore necessary to measure elongation during the shear test. Elongation is measured as the crosshead displacement expressed as a percentage of the initial distance between the edge of the seam (in the center of the gage length) and the nearer grip. This distance should be the same on each side of the seam. Experience indicates that elongation at failure should exceed 50%.

If failure does occur in a brittle manner, it is most probable that if the liner became overstressed in service, even at a stress less than 45% of the yield stress, it could fail by stress cracking (see Chapter 12). The features that cause brittle failure and, potentially, stress cracking are excessively deep grinding groves oriented parallel to the seam and excessive thermal energy input during seaming. The latter may oxidize the HDPE and/or cause a critical reorientation of the microstructure at the edge of the melted and resolidified material. It is also possible that secondary crystallization may occur in the heat affected zone of the adjacent geomembrane (see Chapter 13).

4-3.1.2 PVC seams

A typical shear-stress curve of a PVC sheet showing the relative location of the failure point and the tensile curve of a seam is shown on Figure-4.9. The result should be recorded in in Newtons per millimeter of width of the specimen. It should be noted that the elongation at rupture is approximatively at 300% of elongation and that the tensile resistance of the seam is lower than the sheet itself (approximatively 70 to 80 % of the sheet). These results were obtained using DIN 1910/3 test standard.

NSF-54 recommends to measure the tensile resistance of PVC sheets using ASTM D-882, method A procedure. The specimen width is 25.4 mm and the elongation speed must be set at 500 mm/min. Similarly to the testing of HDPE geomembranes, the NSF-54 is recommending to use modified ASTM D-3083 standard to measure the bonded seam strength. The factory seams requirements are presented in Table-4.1.

The acceptance criteria for a 1,1 mm thick PVC membrane is that the factory seam breaking strength (average value of five measurements) is greater than 80% of the sheet itself (NSF-54 report (1983)).

Table 4.1   NSF-54 PVC geomembrane and seam requirements

SHEET THICKNESS

gauge nominal	10	20	30	45
mils	9,30	19,0	28,5	42,75
mm	0,25	0,50	0,75	1,10

MINIMUM TENSILE   ASTM D-882

breaking factor (pounds/in width)	23	46	69	104
elongation at break (%)	250	300	300	300

FACTORY SEAM STRENGTH

Tensile (ppi width) ASTM D-3083 mod.	18,4	36,8	55,2	83,2
Peel adhesion (min. lb/in) ASTM D-413	FTB or 10	FTB or 10	FTB or 10	FTB or 10

Table 4.2   Recommended PVC 1,0 mm geomembrane
            and seam requirements

SHEET
           thickness (mm)   1
           breaking load (average 5 values)(lb/in)...... > 90
           elongation at break (%) .... > 300

HOT WEDGE SEAM
           seam width (mm) .... > 30

           tensile resistance at 20 in/min (508 mm/min)
           - rupture outside the seam
           - breaking load (average of 5 values) > 60 %
               of the sheet breaking load with minimum
               difference of 80% between the lowest
               value and the average value.
           - brittleness: elongation at rupture (%)... > 180

           peel resistance at 2 in/min (51 mm/min)
           - peel strength (average of 5 values) > 20%
               of the sheet breaking load with minimum
               difference of 80% between the lowest
               value and the average value.

Results of tests performed on 189 seam specimens collected in situ are presented in Figures 4.11 and 4.12 (Benneton (1988)). It is interesting to note that both automatic and manual seams give identical results.

number of
observations

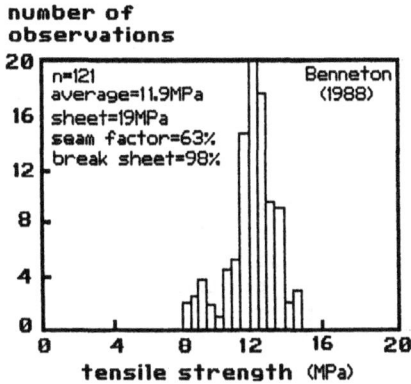

Fig. 4.11   Measured tensile strength of PVC
            automatic seams

number of
observations

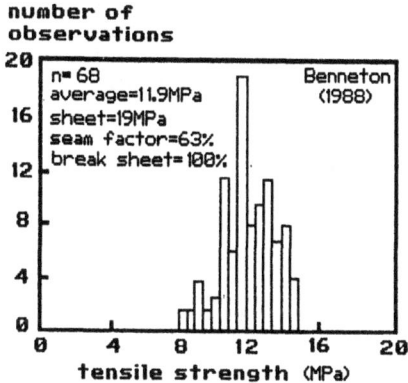

Fig. 4.12   Measured tensile strength of manual
            PVC seams

The described approach is used by installers to ensure that the seams are not weaker than the geomembranes. Unfortunatly the tensile test performed on seams do not measure correctly the tensile resistance because of the difference in thickness between the seam and the sheet. The seam and the sheet deformations are different such that the recorded strain/stress curve (as shown in Figure 4.9) represents an average behaviour.

93

At the edge of the seam, a very high transverse strain is imposed creating triaxial stress causing the rupture of the specimen by shear along the seam and not necessarily by traction alone. This phenomenon is more important for material of large elongation a lot such as PVC geomembranes. Fayoux (1988) has measured transverse strains for PVC seams of the order of magnitude equal to their peel strength.

The seam factor as measured with the described tensile test do not correspond to the stress transmitted by the seam since the geometry of the specimen and the testing parameters inluence the measure. For example a poor quality seam has a lower transverse strain resulting in a higher seam factor.

Because of this phenomenon, the German criteria is now using a seam factor of 0,9 for HDPE membranes (semi-cristalline polymers) and 0,6 for PVC membranes (high amorphous polymers). In Switzerland, the standard SIA-280 only stipulates that the rupture must not be in the seam and that no slipping occured. In France, the French committee on geotextiles and geomembranes is studying the influence of the geometry of the specimen on the seam resistance while in Belgium, R. Degeimbre from University of Liege is developping a bursting test on large slit to evaluate seam resistance. More research is needed to develop the needed test.

4-3.1.3 VLDPE seams

The testing of very low density polyethylene (VLDPE) geomembranes is more difficult because the stress/strain curve does not show a dinstinct yield point and because it has such high break elongation values that on most tensile testing machines the specimens can not be pulled to failure. Another complication is that when seam peel specimens are tested they behave much like a regular tensile test specimen, but one containing an extra "lump" of material in the center of the gage length. Material for elongation, which occurs on both sides of the seam, is provided by the seam (the lump) and this gives the appearance that the seam is peeling, however, the seam is not peeling. As a general guideline, if the material yields up to the seam and continues to yield, then the seam will not peel. If the load is high enough, the seam should be considered acceptable.

There is a major difficulty with defining an acceptable load since the maximum load at a distinct yield point is not available for measuring. If VLDPE geomembrane is treated in a similar manner than the HDPE geomembrane, then a yield strength can be defined using an offset strain of approximately 12% or a tangent to the strain/stress curve may be drawn at the lowest slope after the yield region and extrapolated back to zero strain to produce an equivalent yield stress. An offset strain of 12%, when applied to an HDPE stress/strain curve, would correspond to the yield point.

4-3.2 Peel test

4-3.2.1 HDPE seams

To complement the tensile resistance of a seam, a peel test is used to evaluate the adhesion strength between two welded geomembranes or between the extrude polymer and the sheets. The test is accomplished by applying a dynamic load such that the interfaces are subjected to a peeling force that attempts to separate the adhered surfaces of the weld (as shown schematically on Figure 4.10).

In North America, it is accomplished using the recommended NSF-54 test which is a modified version of ASTM D-413 standard (Standard Test Methods for Rubber Property - Adhesion to Flexible Substrate). The specimen strip Type A, 1 inch (25.4 mm) wide, is subjected to 180° peel at a constant speed of 2 inches/min (50 mm/min).

Test result should include peel strength in Newtons per metre of seam width (lb/inch) as well as the mode of failure (in base material, at seam interface). The peel specimen should be inserted in the machine so the grips are not closer than 1 inch (25.4 mm) to the edge of the seam. If the material is insufficient to allow the grip to be at this position, the actual location of the grips must be recorded.

An acceptance criterion has been proposed by NSF and is refered to as Film Tearing Bond (FTB) criterion: the specimen must fail before the welded interfaces separate. But as pointed out by Peggs (1985), the use of a peel strength acceptance requirement coefficient (F_p) for HDPE geomembrane seams, defined as the ratio of the peel strength at break to the tensile strength at yield of the base material, is an improvement over the FTB criterion.

The require coefficient value usually recommended is such that the peel strength at break of a HDPE geomembrane seam exceeds 60% to 80% (upon seam types) of the tensile strength of the base material. But it does define the necessity for a modified peel test performed as a sharp wedge penetration test.

Recent studies on the determination of seam resistance to shear and peel forces (Peggs (1985), Rollin (1989b)) have indicated that the seam peel strength test do evaluate seam quality more adequatly than the shear test method. Still microscopic analysis of cross sections of welds that have passed both the shear and peel acceptance criteria often has identified unbonded areas as well as changes in molecular structure that represent potential hazards (Charron (1989), Peggs (1989b), Peggs (1989a), Halse (1989a), Halse (1989b) and Peggs (1989b)).

In the peel test, it is not sufficient to require that the seam fail by a film tearing bond mechanism because the definition of FTB is open to interpretation. There is no definition of how much, if any, peel separation is permissible before the failure finally occurs through the geomembrane. Since it has been observed (see Chapter 12) that seam peel separation can introduce crazes, the precursors of stress cracks, into the separated surfaces it is necessary that seam peel separation be minimized if not eliminated. The amount of separation should be quantified as a function of the originally bonded area subjected to the peel test. While separations of zero percent are achieved and desireable, in practice values of between 10 and 20% are used.

During a peel test, the load at failure should be monitored. In practice the load at failure is required to exceed aproximately 60% of the measured yield load of the membrane, or 60% of the seam shear load, but in actuality peel strengths in excess of 80% of the shear strength are achieved.

If a peel specimen fails in the deposit of an extruded fillet weld, but at a load exceeding the minimum, it can be considered acceptable. The features that cause it to fail in the weld should also produce a weld failure in the shear specimen at very low, therefore unacceptable, elongation value.

4-3.2.2 PVC seams

In North America, the peel resistance of a PVC seam is accomplished using the recommended NSF-54 test which is a modified version of ASTM D-413 standard (Standard Test Methods for Rubber Property - Adhesion to Flexible Substrate). The specimen strip Type A, 1 inch (25.4 mm) wide, is subjected to 180° peel at a constant speed of 20 inches/min (500 mm/min). Test result should include peel strength in Newtons per millimetre of seam width (lb/inch) as well as the mode of failure (in base material, at seam interface). The peel specimen should be inserted in the machine so the grips are not closer than 1 inch (25.4 mm) to the edge of the seam. If the material is insufficient to allow the grip to be at this position, the actual location of the grips must be recorded. Both edges of the seam should be subjected to the peel.

An acceptance criterion has been proposed by NSF (see Table-4.1) and is refered to as Film Tearing Bond (FTB) criterion: the specimen must fail before the welded interfaces separate. But as pointed out by Peggs (1985), the use of a peel strength acceptance requirement coefficient (F_p) for geomembrane seams, defined as the ratio of the peel strength at break to the tensile strength at yield of the base material, is an improvement over the FTB criterion. The required coefficient value usually recommended is such that the peel strength at break of a factory PVC geomembrane seam exceeds 10 lb/in (2 daN/cm) corresponding to 10% of the tensile strength of the base material for a 45 mils thick geomembrane. But it does define the necessity for a modified peel test performed as a sharp wedge penetration test.

In Europe others peel tests are used: the DIN-1910/3 and ISO-6133 on analysis of multi-peak traces in determinations of tear strength and adhesion strength. Acceptance criteria, as presented in Table 4.2 for a 1,1 mm sheet hot wedge seam, are used by many installers. For example, the peel resistance at rupture should be greater than 20% (4.6 daN/cm) of the rupture strength of the base material for a 1.2mm sheet (Fayoux (1988)) and greater than 4 daN/cm for automatic seams and 3 daN/cm for manual seams for 1.5mm thick sheets (Benneton(1988)). Similar criteria was used by Hemond (1983) on quality control of more than 3200 seam specimens (manual welds) in tunnel works. To be valid, the specimen must fail outside the seam. Finally it is recommended that 5 specimens be tested and that the lowest recorded value should not be smaller than 80% of the average value.

Results are presented in Figures 4.13 to 4.15 (Benneton (1988)) using 189 samples. Six specimens of 2.5cm width were prepared from each sample of 40cm in length. Only 14 seams had peel strength lower than the acceptance criteria (30% automatic hot wedge seam and 70% manual hot air seam).

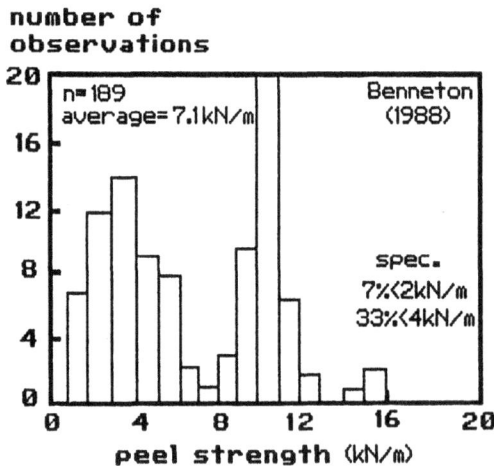

Fig. 4.13 Measured peel strength of PVC seams

number of
observations

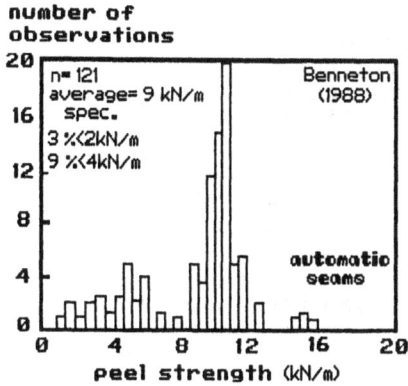

Fig. 4.14   Measured peel strength of PVC automatic
hot wedge seams

number of
observations

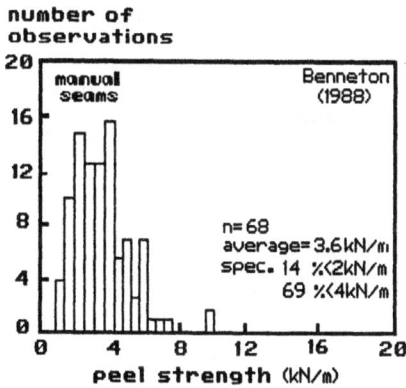

Fig. 4.15   Measured peel strength of PVC manual
hot air seams

As shown in Figure-4.16, the dotted peel strength curve represents a good quality seams without partial unsticking of the sheets before reaching the maximum peel strength. In practice, peel separation can occur before the failure finally occurs through the membrane with resulting curve as shown in Figure-4.16 (full line). The interpretation of the multi peaks curves to determine the peel strength can be achieved using the ISO-6133 standard (determination of a median value) or determining an average value using integration of the curve. Results obtained by Benneton (1990) and similarly by Rollin (1989d) indicated that the average peel strength can differ from the median peel strength by as much as 40%. Since geomembranes are thermal welds, it is believed that the ISO-6133 standard developped to determine the peel strength between a plastic sheet and a metallic or textile support cannot be used. Instead the peel strength of PVC geomembranes should be determined using an arithmetic average value obtained from the integration of the curve strength versus time.

The integration should be done between the first peak (or first deflexion) and the last peak as shown in Figure-4.17. If the rupture in the membrane occurs at the initial step of the test, the maximum strength should be used while if the rupture occurs after reaching the maximum value, the average peel strength should be retained. Finally if separation occurs during the initial step of the test, the average strength must be calculated integrating from the first peak to the last peak.

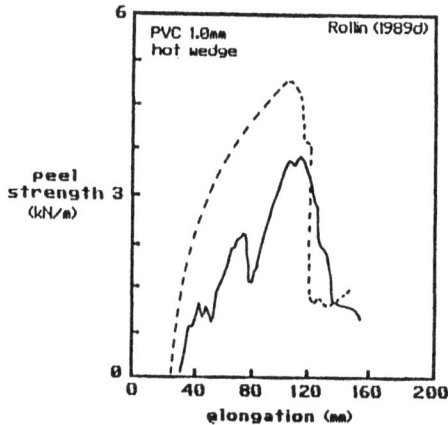

Fig. 4.16   Peel strength curves of a PVC
geomembrane seam

100

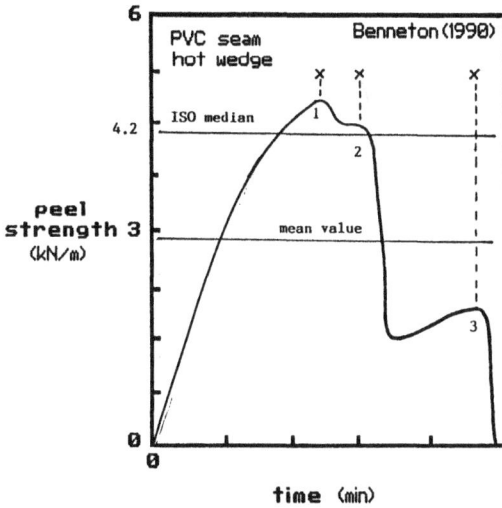

Fig. 4.17   Peel strength curve of a PVC
            geomembrane seam

4-3.3 MICROSCOPIC ANALYSIS

Microscopic analysis of cross section of seams
is used to characterize molecular structures,
identify micro stress cracks within the bonded sheets,
detect unbonded areas and observe slow crack growth
phenomena. It can be performed using an optical
microscope and/or a scanning electrons microscope
(see Chapter 12).

4-3.3.1 Optical microscopic analysis

The examination of cross sections of welded
geomembranes using an optical microscope can be
performed on very thin slices used extensively by
researchers (Peggs (1989c), Charron (1989)). To
obviate the difficulties encountered with this
specimens' preparation, optical fibres have been used
to light up surfaces of black polyethylene samples
obtained from sharp knife cutting (Rollin (1989a)).
The specimens, of approximately 2.0 cm wide with
length equivalent or greater than the weld itself,
are directly installed under the microscope.

Using minimum resolution of 100-200 nm and magnification up to 2000 X, cross sections of seams can be observed and photographed to detect important defects within the seam that could not be suspected following shear and peel tests. A cross section of a weld obtained using the ultrasonic welding technique and including flaws is presented in Figure-4.18. A carefull examination of photographs permits the location of unbonded areas, flow pattern of the melted polymer, micro craks, pattern of the extruded polymer on both edges of the weld and presence of channels inside the weld. Since the polymeric bond was satisfactory at the edge of the seam, its shear and peel resistances were large enough to obtain acceptable values using customary tests.

Macro and micrographic examinations can easily be achieved with an optical microscope. It has less applicability for microfractography which is a micrography analysis of fracture surfaces because of its limited resolution and depth of field. The analysis of rough surfaces can be achieved only by a scanning electron microscope.

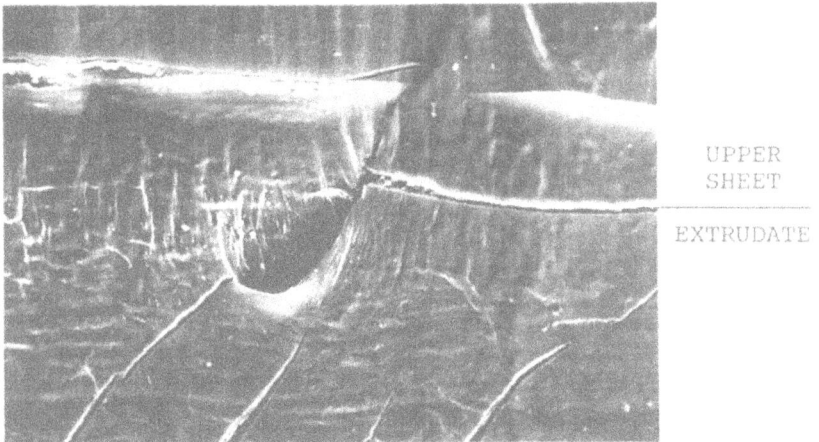

Fig. 4.18 Microphotograph of an HDPE weld

4-3.3.2   Scanning Electron Microscope analysis

The scanning electron microscope is the most versatile instrument for investigation of microstructural changes and identification of micro stress cracks within welds. Its minimum resolution range is 4-5 nm with a maximum useful magnification up to 60000 X. To use adequatly this technique, any cross section of a seam must be coated with a paladium-gold mixture to improve its electrical conductivity and reflectivity.

To demonstrate the performance of such a technique and its need to supplement destructive testing, a microphotograph taken at 1500 X of a cross section of a seam is presented in Figure-4.19. This specimen has passed both acceptance criteria for shear and peel resistance but had failed on impact during field installation. A microanalysis revealed cracks throughout the sheet thickness. Crack characteristics and particularly their mode of propagation, provided information on the factors responsible for their initiation and development.

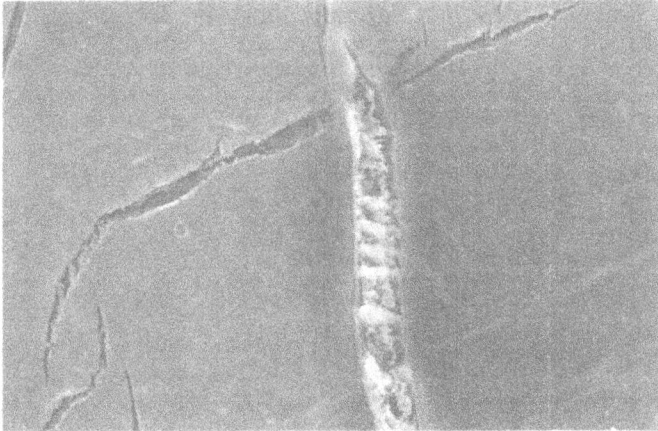

Fig. 4.19   Microphotograph of cracks in HDPE seam

4-4   Conclusion

Geomembrane seams must be checked for continuity and mechanical strength. The continuity of a seam is performed by using one of the many non-destructive techniques presented in this paper. More work is needed to improve these methods or to develop other techniques more reliable and rapid. It is presently a time consuming procedure that increase considerably the cost of a project.

The mechanical bond resistance is measured for HDPE and PVC seams by performing tensile and peel tests. The rupture strengths obtained for the seams are compared to the the yield or rupture strength of the base material and a value is set as acceptance criterium. Unfortunatly no common consensus has been reached on the criteria to accept because of lack of knowledge and also because of the many different standard tests used.

Recently many cases have been reported of long term failure on seams by GRI Halse (1989b), Geosyntec Peggs (1988) and GEOSEP Rollin (1989b). These findings can be alarming if not well documented on existing conditions prevailing during the installation procedures and during the service life of the works. As shown in this paper, microscopic analysis is certainly a complementary test that can be used to assess seam quality and to detect faulty seams not identified by the tensile nor peel tests. Unfortunatly this technique is costly and apparatus are not easily available. Tremendous energy must then be focused on developping a more appropriate test to measure seam fragility than measuring the elongation of specimens.

4-5   References

Benneton, J.P., (1988), "Geomembranes caracteristiques et controles des soudures: projet tunnel St-Germain de Joux", Technical Report CETE de Lyon, June

Benneton, J.P., (1990), "Evaluation de la qualite des soudures par essais non-destructifs", CETE report, Ministere de l'Equipement, Lyon, France

Charron, R.M., (1989), "Polymers for Synthetic Lining Systems: Some Molecular Structure-Property-Application Relationships", proceedings of Geosynthetics'89 Conference, San Diego, February, pp 408-420

Fayoux, D., (1988), "Technical Report on Welding Techniques", Plavina, Solvey group, June.

Halse, Y.Lord, A. and Koerner, R., (1989a), "Stress Cracking Morphology of HDPE Geomembrane Seams", proceedings ASTM symposium on Microstructure and the Performance of Geosynthetics, in press, Orlando, January

Halse, Y., Koerner, R. and Lord, A.E., (1989b), "Laboratory Evaluation of Stress Cracking in HDPE Geomembrane Seams", proceedings GRI-2, Durability and Aging of Geosynthetics, Journal Geotextiles and Geomembranes, Elsivier, May

Hemond, G., (1983), "Application de feuilles synthetiques en etancheite amont-extrados: controle d'etancheite en tunnel", SOLEM report, Centre d'etude des tunnels, UTC/CEB, Geneva, November 11

National Sanitation Foundation Standard 54, (1983), "Flexible Membranes Liners", Ann Arbor, November

Peggs, I.D. and Little,D., (1985), "The Effectivemess of Pell and Shear Tests in Evaluating HDPE Geomembranes Seams", proceedings Second Canadian Symposium on Geotextiles and Geomembranes, Edmonton, pp. 141-146

Peggs, I.D., (1987), "Evaluating Polyethylene Geomembrane Seams", proceedings of Geosynthetics'87 conference, New Orleans, February, pp 505-518

Peggs, I.D., (1988), "Failure and Repair of Geomembrane Lining Systems", Geotechnical Fabrics Report, November, pp. 13-16

Peggs, I.D. and Carlson, D., (1989a), "Brittle Fractures in Polyethylene Geomembranes", proceedingd ASTM symposium on Microstructure and the Performance of Geosynthetics, in press, Orlando, January

Peggs, I.D. and Charron, R.M., (1989b), "Microtome Sections for Examing Polyethylene Geosynthetic Microstructures and Carbon Black Dispersion", proceedings of Geosynthetics'89 Conference, San Diego, February, pp 421-432

Peggs, I.D. and Carlson, D.S., (1989c), "Stress Cracking of Polyethylene Geomembrane: Field Experience", proceedings GRI-2, Durability and Aging of GEosynthetics, Journal Geotextiles and Geomembranes, Elsevier, May

Rollin, A.L., Vidovic, A., Denis, R. and Marcotte, M., (1989a), "Microscopic Evaluation of HDPE Geomembrane Field Welding Techniques", proceedings ASTM synposium on Microstructure and the Performance of Geosynthetics, in press, Orlando, January

Rollin, A.L., Vidovic, A., Denis, R. and Marcotte, M., (1989b), "Evaluation of HDPE Geomembrane Field Techniques: Need to Improve Reliability of Quality Seams", proceedings of Geosynthetics'89 Conference, San Diego, February, pp 443-455

Rollin, A.L., Vidovic, A. and Ciubotariu, V., (1989c), "Assesment of HDPE Geomembrane Seams", proceedings of Second International Landfill Symposium, Porto Conte, Sardinia, October

Rollin, A.L., (1989d), "Controle de conformite pour les soudures de geomembranes en chlorure de polyvinyle", RILEM meeting, Porto Conte, Sardinia, October

GEOMEMBRANES – IDENTIFICATION AND PERFORMANCE TESTING

5 TENSILE TESTS FOR GEOMEMBRANES

J. P. GOURC
IRIGM University of Grenoble 1, France
H. PERRIER
Cer, Rouen, France

5.1 Introduction

Knowledge of the mechanical behaviour of geomembranes, particularly their behaviour under tensile testing, is of extreme importance for civil engineers in charge of designing waterproofing systems. The choice of geomembrane confronts a wide variety of geomembranes available on the market, but also the tensile testing standards. This article refers essentially to the vast coordinated study conducted by IRIGM - University Joseph Fourier - Grenoble, for the French Committee of Geotextiles and Geomembranes (GOURC J.P. et al., 1986, 1987, 1989) : 17 different geomembranes were tested and 12 types of specimens were considered as part of an inter-laboratory (Ponts et Chaussees and Institut Textile de France principally). The experiment extended with tensile test on welded specimens. Finally, one type of specimen was recommended for all the geomembrane materials.

5.2 Tensile test program

5.2.1 Geomembrane materials
Generally, we consider 3 major classes of geomembranes, bituminous products, thermoplastics and elastomer rubbers. We give the results for products belonging to each of these classes (cf table 5.1).

Table 5.1   Geomembranes

Classe	N°	Thickness (mm)	Mass per unit area (g/m²)
Bituminous	A1	4,3	5146
	A1'	4,0	4740
Thermoplastic			
PVC	B2	1,4	1336
Reinforced PVC	B3	1,2	1437
PVC	B2'	1,0	1250
Reinforced PVC	B3'	1,5	1778
HDPE	B12	2,5	2330
Rubber Elastomer			
EPDM	C3	1,45	1666

5.2.2   Standard specimens (figure 5.1)
We considered three classes of specimens, "strip" specimens
(table 5.2.1), "dumb-bell" specimens (table 5.2.2) and
"three-dimensional" specimens (table 5.2.3).

5.2.2.1   "Strip" specimens
These specimens are rectangular measuring a height of h_o
between the clamping jaws and having a constant width b_o.
The strip "e2" is normalized for bituminous products
(NF G 07-001), strip "e3" for thermoplastics
(NF T 54-102). While Standard DIN proposes, for
thermoplastics, a strip of the same $h_o = 120$ mm, but which
is narrower (DIN 53455 : $b_o = 15$ mm). The ASTM standard
requires $h_o = 250$ mm and   5 mm < b_o < 25,4 mm for thin
plastics sheets (thickness   < 1 mm).
    We have added two wide strips to the test program,   "e1"
and   "e1**", by analogy with the specimen recommended for
geotextiles.

5.2.2.2   "Dumb-bell" specimen
For these specimens, b_o corresponds to the minimum central
width, h_o to the height between the jaws and h_c to the
height corresponding to the minimum width of b_o
(Figure 5 - 1).
    Specimen "e4" is a specimen type 2 of Standard ISO R527,
used for elastomer rubber, whereas specimen "e5" is a
specimen type 1 to the same standard ISO R527, used for
thermoplastics, in parallel to strip specimen "e3".
    Initially, specimen "e6" was recommended by the Labora-
toire Central de Ponts et Chaussees - France - for
bituminous products.

Table 5.2.1

legend	Refér.	Sample	b_0	h_0	h_c	$\frac{\Delta h}{\Delta t}$	$\frac{1}{\Delta t}\cdot\frac{\Delta h}{h_0}$
			mm			mm/min	% /min
e1	Type		200	70			
e1**	C.F.G.G		500	70	–	35	50
e 2	NF G07001		50	200	–	100	50
e 3	NF T54.102		25	120	–	60	50

Table 5.2.2

legend	Refér.	Sample	b_0	h_0	h_c	$\frac{\Delta h}{\Delta t}$	$\frac{1}{\Delta t}\cdot\frac{\Delta h}{h_0}$
			mm			mm/min	% /min
e 4	ISO R527(1966) Type 2		6	70	33	35	50
e 5	ISO R527(1966) Type 1		10	115	60	57,5	50
e 6	Type L.C.P.C standard		25	85	60	42,5	50
e 7	Type E.P Montréal		100	100	50	50	50

111

Specimen "e7" is used by the Montreal Polytechnical School (Rollin et al., 1985) to all geomembrane materials.

For thick plastics sheets, the ASTM standard D 638 proposes 5 types of specimens similar to the "e4" and "e5" type.

5.2.2.3 Specimens with welded seams

The problem of welding quality is a fundamental problem.

Several types of tests (cf Chapter 3 - Rollin et al.) will permit welds to be studied. Tensile testing is one of these test which complements the peeling test.

In this light, the specimens are modified : each sample is provided with a central weld seam of constant length h_s = 40 mm. This increases by an equivalent amount the initial distance between the clamping jaws which become $(h_o + h_s)$ (Figure 5.2).

5.2.2.4 "Three-dimensional" specimens

To simulate the sollicitation conditions encountered in situ, several laboratories propose a "performance test" permitting multi-directional pulling on the geomembrane. A test on sleeve "e_c" is particular (Laboratoire Regional des Ponts et Chaussees de St. Brieuc - France) in that the membrane takes the form of a cylinder with a circular directing line and bursting test "e_s" (Centre du Machinisme Agricole du Genie Rural des Eaux et Forets Antony - France ; AKZO - the Netherlands ; Geosynthetic Research Institute Philadelphia - USA) in which the membrane takes the form of a quasi-spherical cap. (cf. Chapter 6 - Hoekstra).

Table 5.2.3

leg	THREE DIMENSIONAL SPECIMEN	(mm)
e_c		ϕ_o = 100 b_o = 360
e_s		d_o = 50 d_o = 500

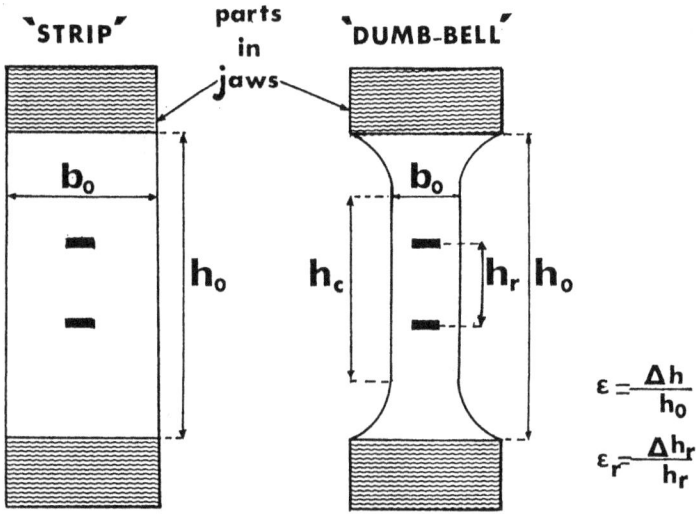

Figure 5.1   Typical shape of tested samples

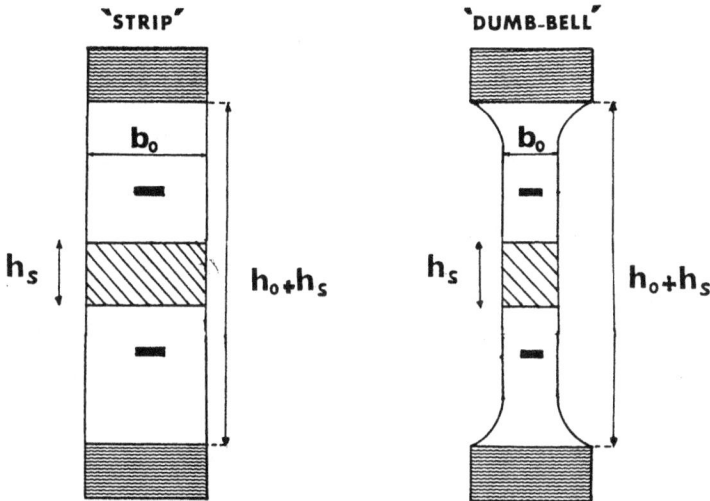

Figure 5.2   Typical shape of tested samples with seams

5.2.3   General testing conditions
Because of the mechanical sensitivity of the geomembranes
to the ambient atmosphere, all the result were obtained at
a temperature of 20°C and for a 65 % relative humidity.

The tensile testing machine used for the "strip" and "dumb-bell" tests will allow the jaws to move at constant speed. Obtaining maximum pulling force needed to break the specimen is generally not a problem. Conversely, some geomembranes are highly extensible (beyond 500 % relative elongation) requiring a considerable movement of the jaws not always compatible with the stroke of the tensile test press ram. As far as this rate of elongation is concerned, or the jaw spreading speed, the recommended value varies greatly depending on the reference standard. In order not to multiply the variable parameters of the study, the speed has been set down for all the tests at ($h_o/2$) per minute. The same speed has been preserved for specimens with weld seams in spite of their different heights (h_o + h_s), because the weld seam is generally barely extensible.

Conversely, this cannot be complied with for "three-dimensional" specimens for which elongation is obtained by the staircase increase of the hydraulic pressure.

The clamping jaws of the specimen form a large section of the equipment. It is impossible to guarantee a total blocking of the specimen for all the geomembrane types, in particular since excessively tight clamping would lead to a premature risk of the specimen being pulled breaking near the point of jaws.

Figure 5.3.1 shows the principal of self-blocking jaws used for this study. Observe that in the case of geomembranes which are not particularly flexible, such as HDPE, two steel bars take the place of the geomembrane winding (figure 5.3.2.).

5.2.4 Measurements

The pulling force T on the strips or dumb-bells is measured during the elongation at a constant speed. To obtain a comparable speed for different specimen geometries, we use the force per the unit of width :

$$\alpha = T/b_o \ (kN/m)$$

The spread of the clamping jaws Δh is measured and gives use an "average deformation" :

$$\epsilon = \Delta h/h_o$$

However, the parameter ϵ may not be particularly representative if a slippage of the geomembrane occurs between the jaws. In addition, for "dumb-bell" specimens the width of which is not constant, the deformation may be heterogeneous between the narrow central section and the nearly parts of the jaws.

For these reasons, we advise central measurement of the geomembrane's elongation. The system used here is an optical system that measures the spread (figure 5.1).

This gives us the "central deformation" :

$$\epsilon_x = \Delta h_x / h_x$$

For specimens with a weld seam, the distance between the test area is considered as equal to $(h_x + h_s)$. If the seam is assumed to be non-extending, it can be considered that :

$$\epsilon = \Delta h / h_o \qquad \text{average deformation}$$

$$\epsilon_x = \Delta (h_x + h_s) / h_x \qquad \text{central deformation}$$

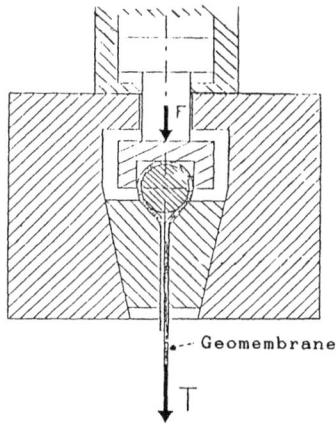

Figure 5.3.1 Clamp jaws used for current geomembranes

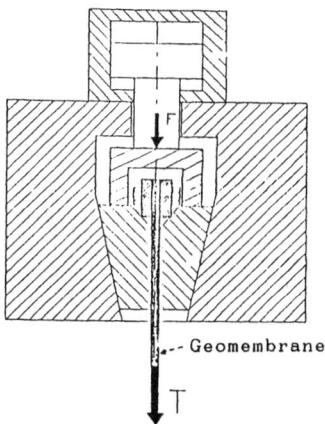

Figure 5.3.2 Clamp jaws used for HDPE geomembranes

In addition, to allow for the inevitable dispersion of the test results, particularly concerning breakage (α_f, ϵ_f), we advise a minimum of five specimens per test type. The results show below are all the average results obtained from 5 to 7 specimens.

5.3   Influence of test conditions

5.3.1   Elongation speed $\dot{\epsilon}$ and height of specimen h_o

It should be observed that for our tests, we have maintained a constant speed of $\dot{\epsilon} = (\Delta h/h_o)/\Delta t = 50$ % per minute. This means considering a speed $\Delta h/\Delta t$ of clamping jaw spread which increases with h_o.

However, in one particular case we tested this parameter. The geomembrane material chosen is a PVC (B2 of table 5.1).

We test the specimens - strip width b_o = 25 mm but height h_o varies between 20 mm and 140 mm (h_o = 120 mm corresponding to the specimen "e3" - cf table 5.2.1).

We carry out two types of tests on these specimens :

- test at $\dot{\epsilon}$ = 50% per minute
  (corresponding to our general test condition)

- test at $\Delta h/\Delta t$ = 50 mm per minute

The diagrams below show the evolution of the tensile force α_f and the mean deformation α_ϵ on breakage as a function of height h_o of the specimen (figures 5.4.1 and 5.4.2). Case h_o = 100 mm verifies the two conditions for $\dot{\epsilon}$ and $\Delta h/\Delta t$ at the same time.

It will be observed that :

- the test at constant $\dot{\epsilon}$ reduces less sensitivity of α_ϵ to the variation of h_o, than does the test at $\Delta h/\Delta t$ constant.

- the influence of h_o is greater than the influence of $\dot{\epsilon}$ on α_ϵ, decreasing systematically when h_o increases (up to h_o = 80 mm).

- it is instructive to replace the point obtained on broad strips "e1" and "e1**" (cf table 5.2.1), therefore for very different widths (b_o = 200 mm and b_o = 500 mm respectively) and $\dot{\epsilon}$ = 50 % per minute.

The study presented here, carried out on PVC alone, obviously does not permit a choice of test speed but demonstrates that it is more significant to compare test results at a constant $\dot{\epsilon}$ than at a constant $\Delta h/\Delta t$ for different specimen geometries.

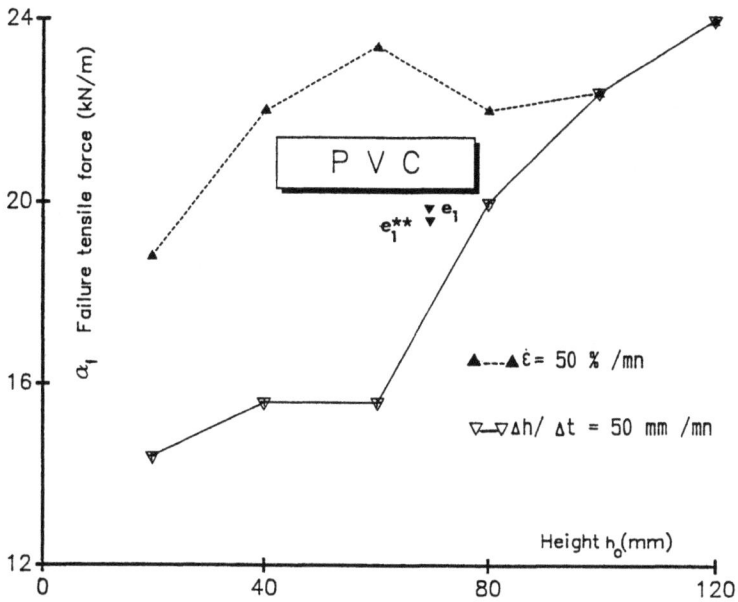

Figure 5.4.1   Influence   of h_o and displacement   rate   upon
                breaking tensile strength (PVC - (B2) -)

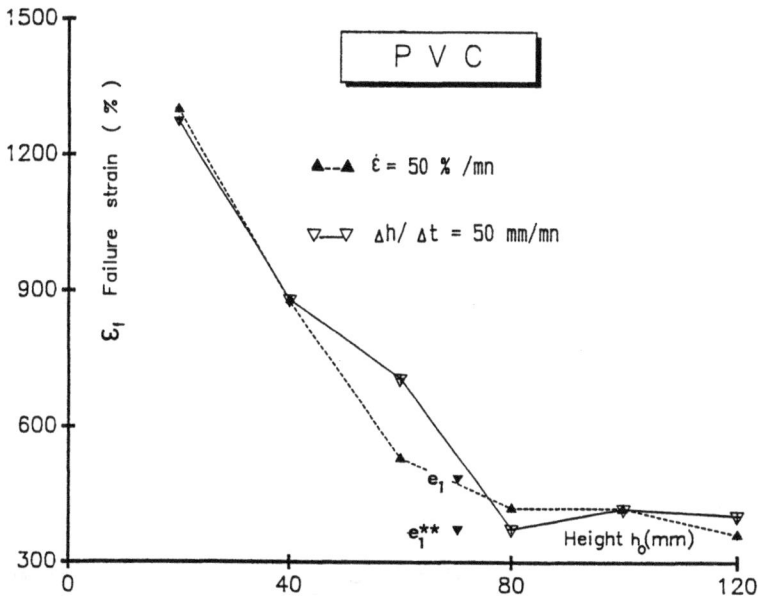

Figure 5.4.2   Influence   of h_o and displacement   rate   upon
                maximal strain at breaking point

117

5.3.2 Plane specimen geometry

Because of its shape, a "dumb-bell" specimen will not provide elongation which is spread evenly over the entire height h_o of the specimen ; this form concentrates the elongation into the middle areas. The use of test patterns as shown in paragraph 5.2.4 gives a measurement of deformation between 2 points located at 20 mm either side of the center.

The example of PVC - B2 - examined previously is demonstrative (figure 5.5) : for the same tensile strength α dispersion between ϵ for a broad strip ("el**" or "el"), or between a narrow strip "e3" or a dumb-bell "e6", but the diagrams obtained as a function of ϵ_x are very similar, giving the same behaviour in the middle zone.

A dumb-bell is therefore as valid a way of obtaining the rheological behaviour of a material as a strip, as long as the "central deformation" of the specimen is used.

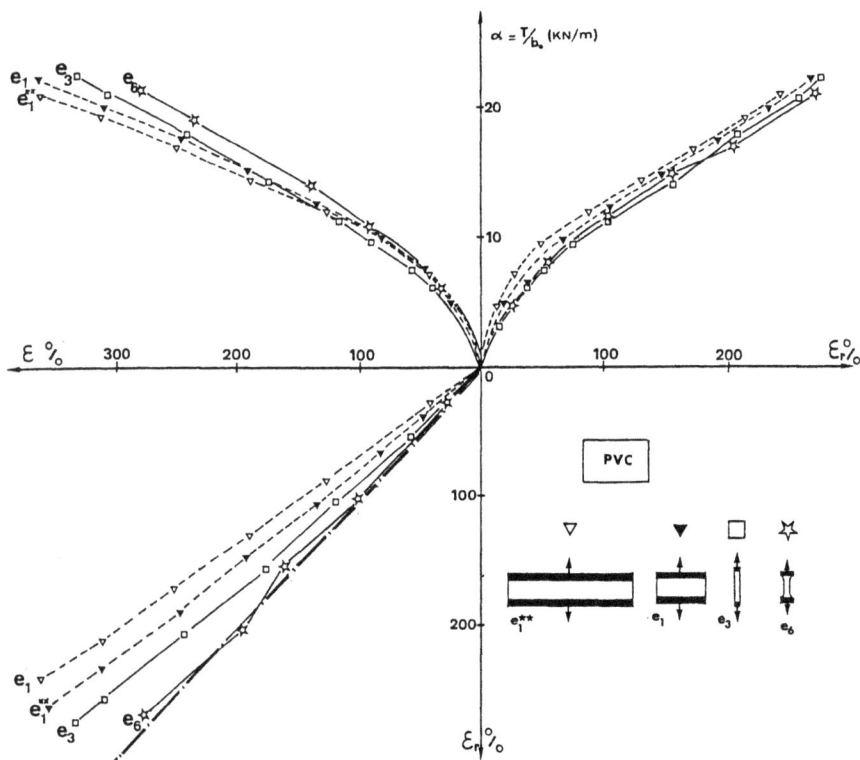

Figure 5.5 Influence of geometry of the specimen on the tensile behaviour (PVC - (B2) -)

The dumb-bell offers certain advantages :

- easier clamping of the specimen in the jaws: the same for T obtained for a strip and a dumb-bell of the same width b_o spread over a greater jaw width for the dumb-bell (greater than b_o), thus decreasing the risk of sliding in the jaws. Figure 5.6 shows this arrangement for a bituminous membrane (A1' - cf table 5.1) : for a broad strip "e1" and dumb-bell "e6" the behaviour in the central zone is identical ($\alpha = f(\epsilon_x)$). Conversely, the great deviation between ϵ and ϵ_x are obtained for the strip translates the sliding of membrane "e1" in the jaws, (photo 5.1).

- the dumb-bell allows the breakage to be located in the central zone, thus decreasing the unfavorable in-fluence of the clamping jaws.

Conversely, the use of "dumb-bell" type specimens calls for some precautions :
The cutting of the specimen requires a cutting die. Cutting with a cutter causes micro-notching leading to the premature localization of the break (photo 5.2 : case of specimen "e7" of HDPE). This also appears in diagrams of figure 5.7 : indeed, we have a die-cutter for specimens "e6" but not for dumb-bells "e7" which are cut with the cutters.
Thus, for PVC (-B2' cf table 5.1), the break (α_r, ϵ_r) is obtained early for "e7" with respect to dumb-bell "e6" or strips "e1" or "e2".

5.3.3 "Three-dimensional" specimens
We compare the results obtained with these specimens to those obtained on "plane" specimens for several types of geomembranes (cf paragraph 5.2.2.4) :

PVC - (B2) - figure 5.8
Reinforced PVC - (B3) - figure 5.9
Bituminous products - (A1) - figure 5.10

These three-dimensional tests will be considered as a simulation of real conditions, justifying their interest, all the more so in that these results are difficult to correlate with the results obtained on plane specimens.

5.4 Choice of "dumb-bell" specimen

As demonstrated in paragraph 5.3, the "dumb-bell" specimen gives, in the central zone, behaviour equivalent to that of a broad strip.
It offers the advantage of being able to limit the in-fluence of the jaws (better clamping, localized breaking far from the jaws).

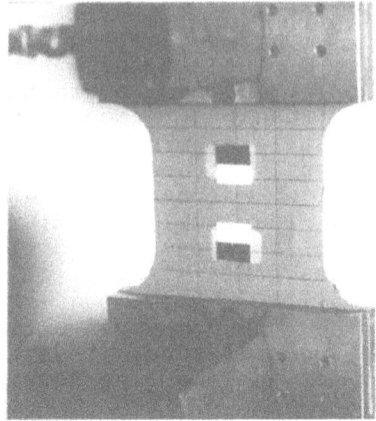

Photo 5.1

Sliding of the membrane
between clamp jaws - (PVC) -

- specimen "e7" -

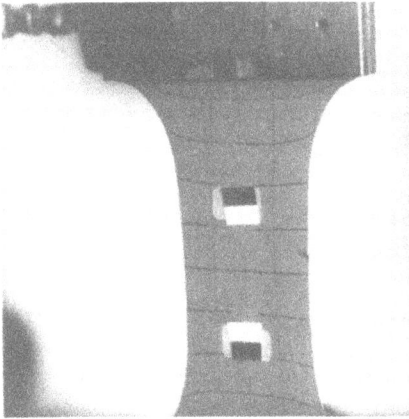

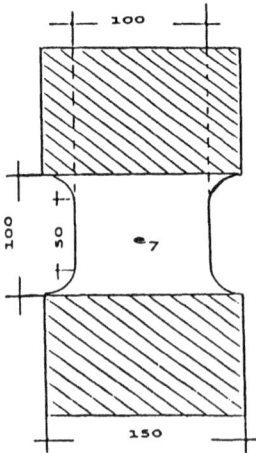

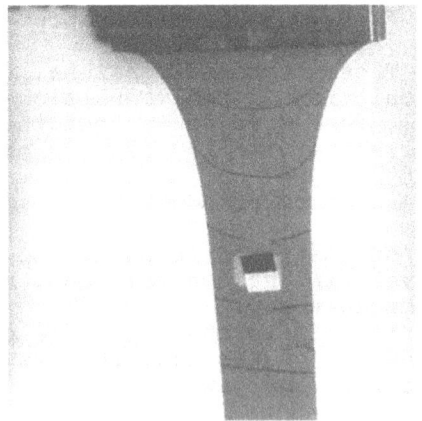

Photo 5.2

Premature rupture
of a "e7" specimen of HDPE

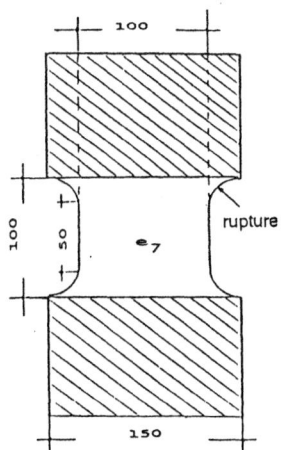

121

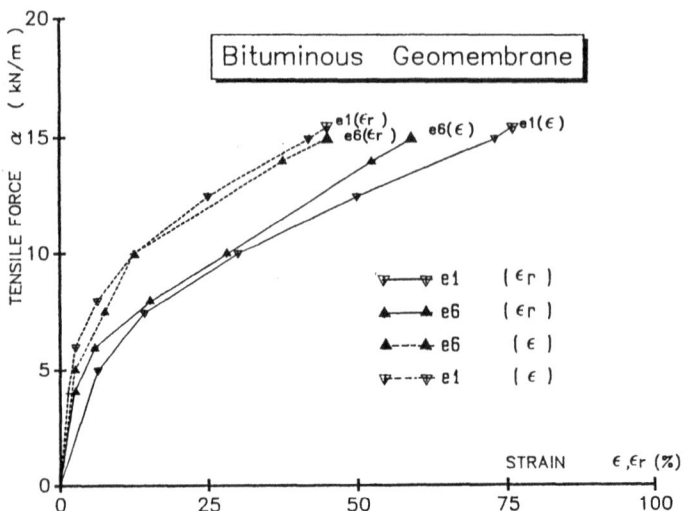

Figure 5.6   Tensile force upon average and central strain
              for two geometries of specimen
              (bituminous - (A'1) -)

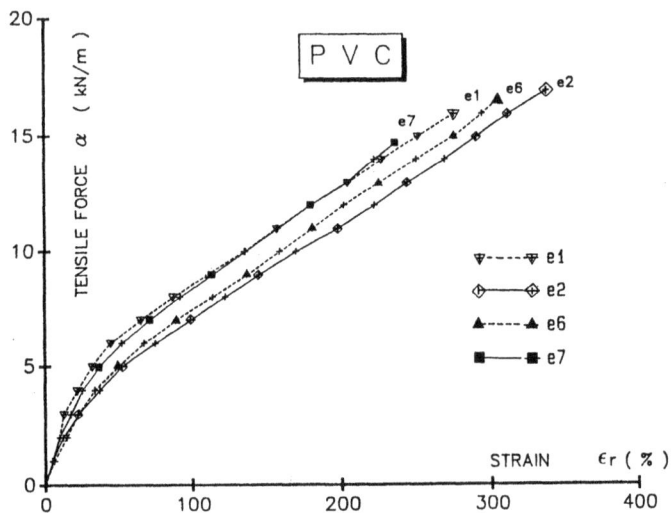

Figure 5.7   Influence of the geometry of the   specimen   on
              the tensile behaviour - (PVC - (B2) -)

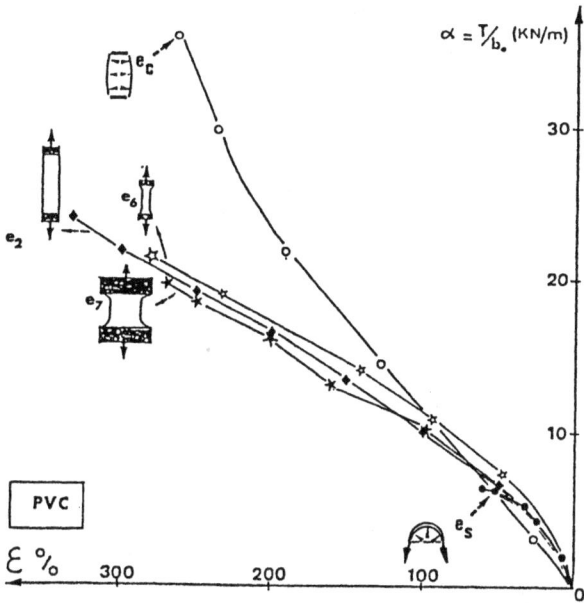

Figure 5.8   Tensile   behaviour   for   monodirectional   and
              multidirectional tests - (PVC - (B2) -)

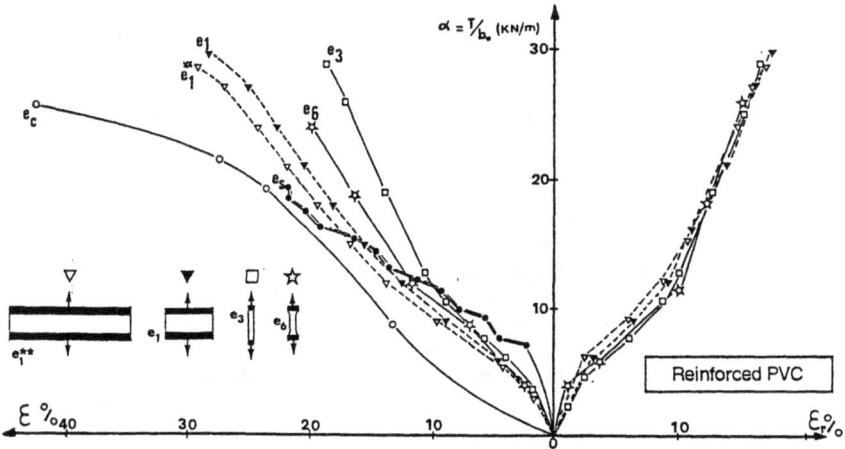

Figure 5.9   Tensile behaviour for monodirectional and   mul-
             tidirectional tests - (Reinforced PVC-(B3) -)

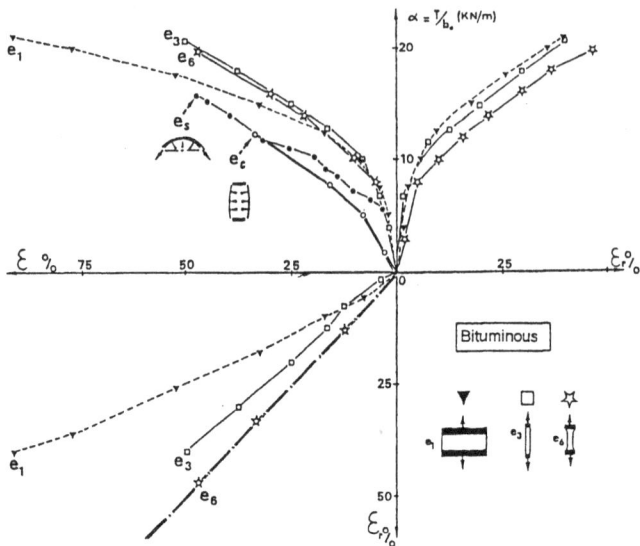

Figure 5.10

In addition the dumb-bell specimen is compact and could be placed easily in a closed enclosure for testing under different atmospheric conditions.

Conversely, this type of test absolutely requires the utilization of a die cutter. That is why it is impossible to determine the optimum dumb-bell at the present time and to choose between "e6" and "e7" since there is no "e7" die cutter.

Figures 5.11 and 5.12 make it possible to compare the results obtained for a thermoplastic and an elastomer on the basis of the "e6" dumb-bell, compared with the specimen habitually recommended for this type of material.

figure 5.11 Reinforced PVC - (B3) - "e2" and "e6"
figure 5.12 EPDM - (C3) - "e4" and "e6"

Note that, to obtain a significant result with a geomembrane reinforced with integrated synthetic threads, the standard specimen must have a minimum width b_o (far greater than the spacing between threads).

A conclusive result is obtained with HDPE, (figure 5.13) which has extremely particular rheological behaviours. In the field of minor deformations, this material has a pic corresponding to yield point threshold. Then, the yield initializes at a given point of the specimen and is propogated through neighbouring zones (photo 5.3). Elongation at the breaking point can exceed 500 %. Thus, the initial height h_o of the specimen must be limited at the risk of exceeding the maximum displacement of the traction machine in extension (specimen "e2" often cannot be used for this reason). From this standpoint, specimen "e6" with its relatively low height is advantageous.

124

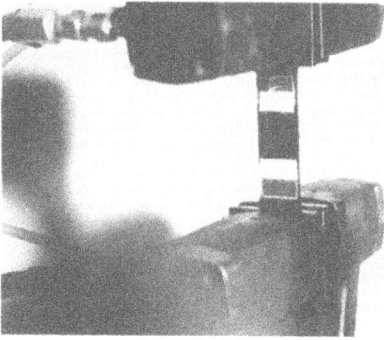

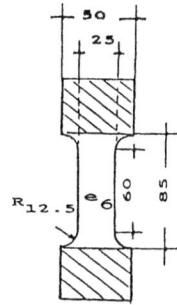

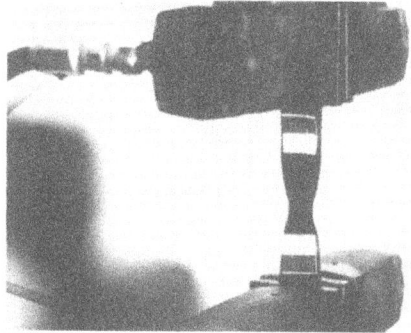

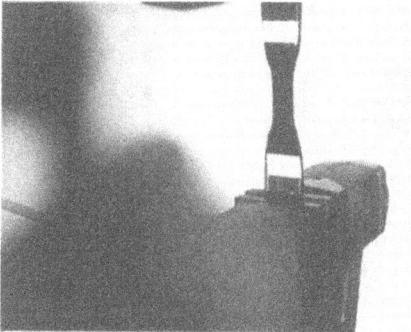

Photo 5.3

Tensile test on HDPE "e6" specimen : flow stage by stage

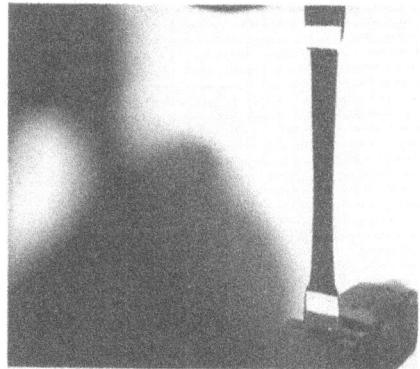

125

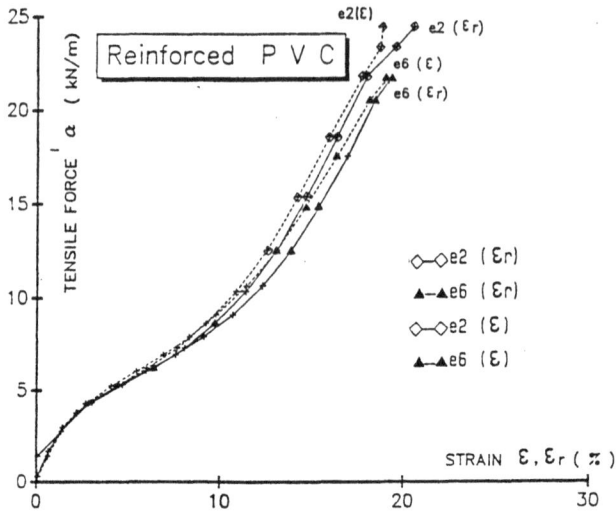

Figure 5.11 Tensile behaviour for the proposed geometry specimen "e6" and the standard specimen "e2"- (Reinforced PVC - (B'3) -)

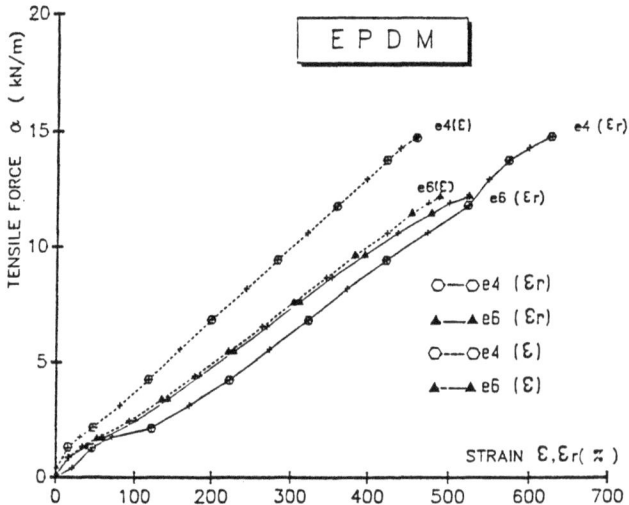

Figure 5.12 Tensile behaviour for the proposed geometry specimen "e6" and the standard specimen "e4" - (EPDM - (B'3) -)

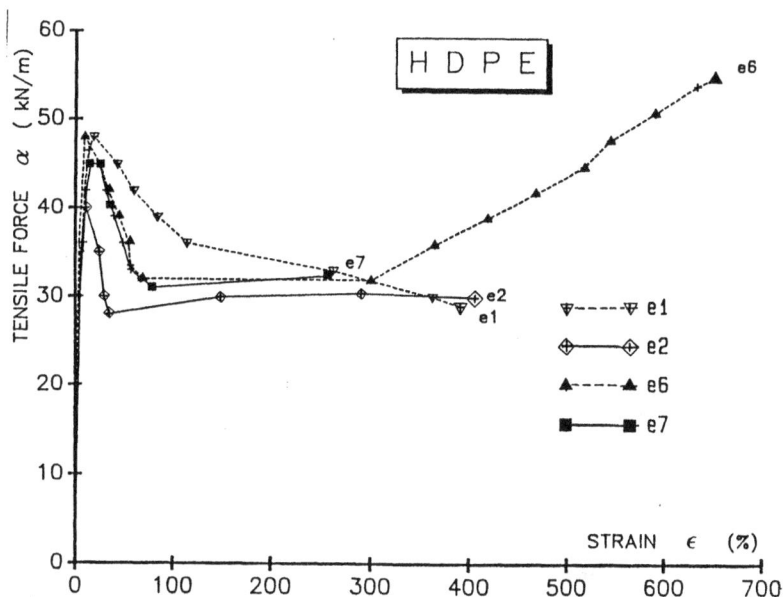

Figure 5.13 Influence of the geometry of the specimen on the tensile behaviour - (HDPE - (B12) -)

In addition, the present study shows that only dumb-bells "e6" produces a hardening at the end of the test. This behaviour characterizes the end of the yield in the zone at a constant width (height h_c - figure 5.1) and the propagation of this yield to around the areas near the jaws.

The premature breakage of "e7" can be explained, in the same way previously, by the fact that there is · no die cutter. As far as the specimens "e1" and "e2" are concerned, the break is obtained near the clamping jaws for a deformation that is substantially less than that of "e6".

Observe that the use of deformation measuring patterns at the center is not of interest in this case except up to the yield point.

5.5 Adaptation of the tensile test to the weld study

The tensile test alone is insufficient to define the quality of a weld. Conversely, it definitely contributes to it. Observation of the behaviour of the welded part, location of the initial break point, the resistance drop under the traction obtained are all important elements.

As an example, we demonstrate (photo 5.4, figure 5.14)
the compared diagrams obtained with or without a weld for
PVC - (B'2) - according to the conditions specified in
paragraph 5.2.2.3. : diagrams with or without welds will
be very close to each other. However, the break in the
welded specimen will begin around the weld and deformation
at the break will therefore be less than for an unwelded
specimen (welding factor concept).

The case of a bituminous geomembrane - (A1') - is
particular : the length h_s of the weld recommended here is
far less that the length used in situ (figure 5.15). The
break in the welded specimen is then obtained by sliding at
the weld.

This enables us to consider an evolution of the tensile
test whereby samples taken in situ may be tested i.e.,
samples with welds having a variable height h_s (up to
150 mm). This would lead us to consider specimens having a
variable total height ($h_o + h_s$).

But, although the cutting of strips with a variable
height ($h_o + h_s$) poses few problems, the cutting of variable dumb-bell ($h_o + h_s$) is costly if we impose the
method of cutting with a die cutter (several die cutters
available).

However, we have considered purposing a new fixed height
dumb-bell geometry ($h_o + h_s$) big enough to integrate wide
welds : $h_o + h_s = 285$ mm.

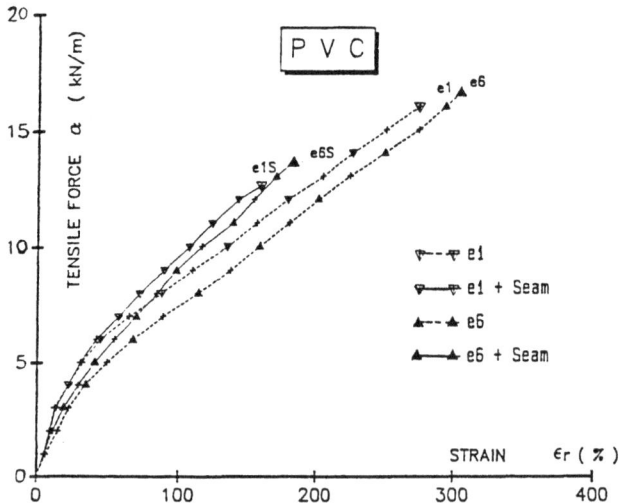

Figure 5.14 Tensile behaviour for samples with and without
welding seam - (PVC - (B'2) -)

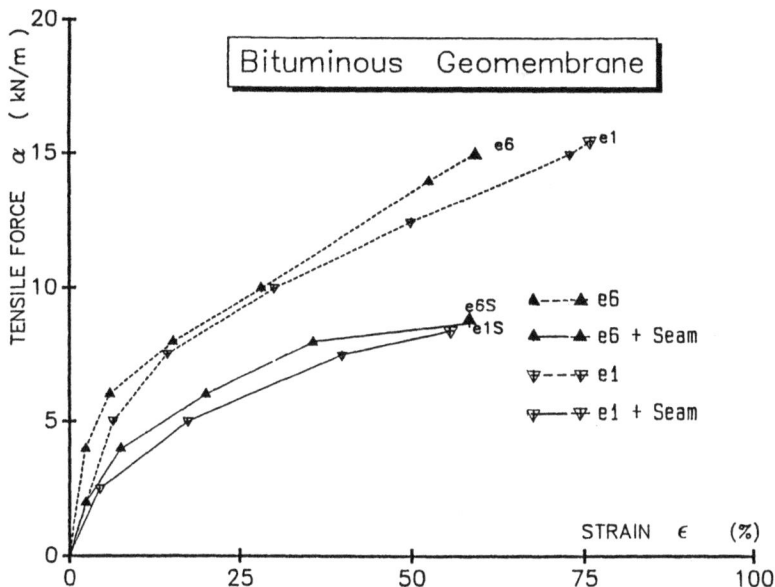

Figure 5.15 Tensile behaviour for samples of bituminous -
(A'1) - with and without welding seam

In reality, this is a welded "e6" specimen (figure 5.2)
whose height h_s at constant width is increase by 160 mm
(figure 5.16) : it will be referred to "e6S + 160".
For a weld with a constand height h_s, what will the
behaviour modification be as cause by the greater height of
the specimen ?.   We demonstrate the results in the
following diagrams for two materials :
            figure 5.17   PVC - (B2')-
            figure 5.18   HDPE - (B12)- .
In both cases, the increased height leads to increased
rigidity of the specimen :   for PVC, α_r increases and ϵ_r
decreases with increased height.
This result is totally comparable with that obtained in
5.3.1   (to be correlated with the modification of the test
rate $\Delta h/\Delta t$, because $\dot{\epsilon}$ is maintained constant).
The influence observed on the non-welded free height (in
this case 85 mm and 245 mm) is obtained when the two welds
of different height h_s are compared with the same specimen
"e6S +160". Thus, this type of specimen will not allow the
objective comparison of two geomembrane materials having
different weld heights. Complementary studies are underway
on this point.

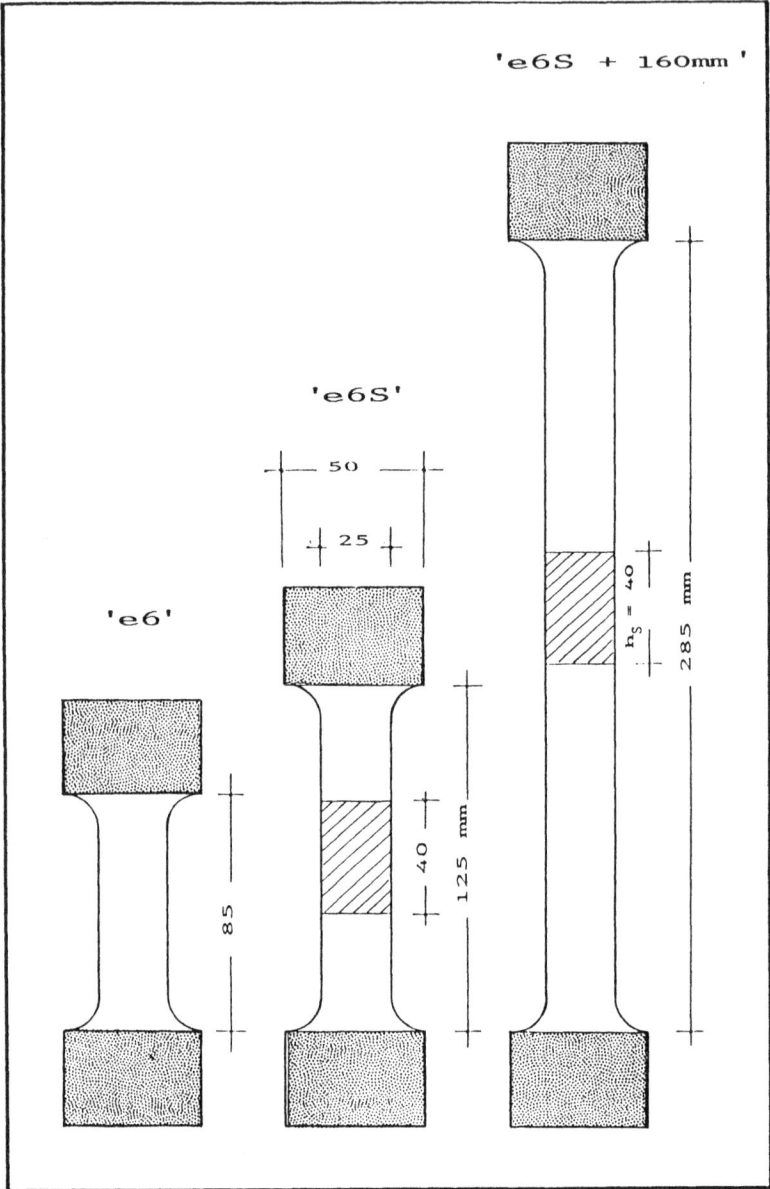

Figure 5.16 Proposed geometry for specimens with welding
seam, "e6s" and "e6s + 160 mm"

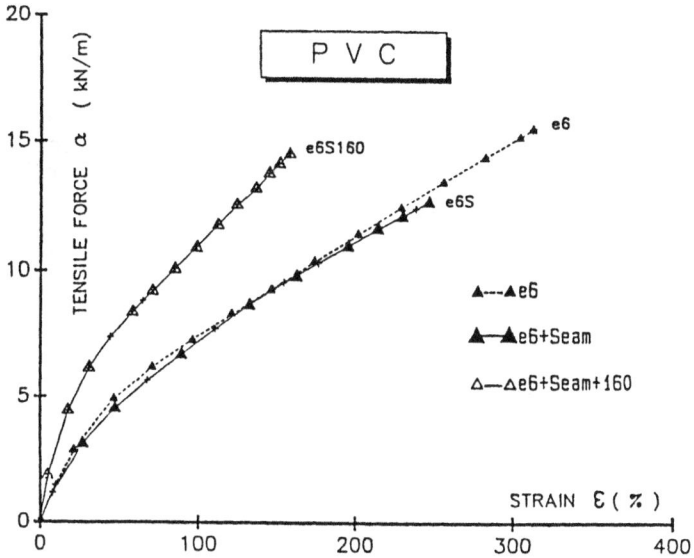

Figure 5.17 Tensile behaviour for samples with and without welding seam, - (PVC - (B'2) -)

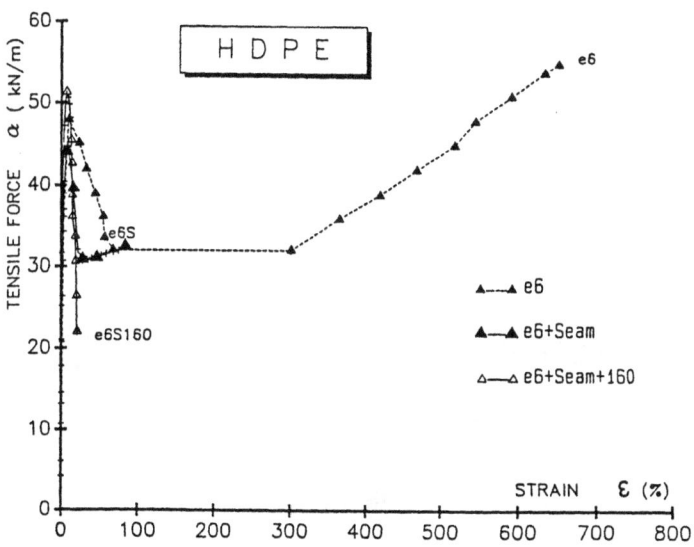

Figure 5.18 Tensile behaviour for samples with and without welding seam - (HDPE - (B12) -)

131

5.6   Conclusion

The vast tensile test program on geomembranes undertaken by the Comité Français des Géotextiles et Géomembranes has led to the recommending of a single specimen for all the geomembrane classes :   the "dumb-bell" shape is preferred to the "strip" shape and the choice has gone to the "e6" specimen already used in France for certain materials.
The study has been extended to the case of welds on geomembranes.

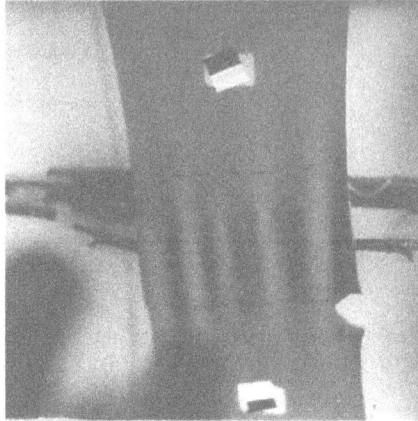

Photo 5.4   Initiation of sample tearing near the seam,   for a PVC "e7" specimen

Acknowledgements

J. PUIG
Laboratoire Régional des Ponts et Chaussées de Toulouse -

J.P. BENNETON
Laboratoire Régional des Ponts et Chaussées de Lyon -

B. LECLERCQ
Institut Textile de France de Lyon -

M. DRUON
Laboratoire Central des Ponts et Chaussées de Paris -

Comité Français des géotextiles et géomembranes

References

Gourc J.P., Leclercq B., Benneton J.P., Druon M., Puig J. and Delorme F. (1986),
"Tensile Strenght Behaviour of Geomembranes", 3rd International Conference on Geotextiles, Vienna, Austria.

Gourc J.P., Leclercq B., Benneton J.P., Druon M., Puig J., (1987),
"Contribution à la détermination d'un essai de traction standardisé pour géomembranes," 1st International Congress RILEM, From Materials Science to Material Engineering, Paris, France.

Gourc J.P., Perrier H., Leclercq B., Benneton J.P. (1989),
"Standard Tensile Test for Geomembranes," 2nd International Landfill Symposium, Porto Conte, Italy

Rollin A.L. (1985)
"Testing of Geomembranes," 2nd Canadian Symposium on Geotextiles and Geomembranes," Edmonton, Canada.

Rollin A., Fayoux D., Hemond G.
"Seaming techniques, sampling and destructive testing geomembranes : Identification and performance Testing"
Rilem -Chapter 3 - (1990)

Hoekstra
"Burst testing"
"Geomembranes : Identification and Performance Testing
Rilem - Chapter 6 - (1990)

ASTM - D 638-89 Standard test method for Tensile Properties of Plastics

ASTM - D 882-83 Standard test methods for Tensile Properties of thin plastic sheeting

ASTM - D 3083-76 Standard specification for Flexible polyvinyl chloride plastic sheeting for pond, canal, and reservoir lining

ASTM D 412-80 Tests for rubber properties in tension

ISO R527-66 Plastics : determination of tensile properties

DIN 53455 Plastics sheets : determination of tensile properties

NF G 07-001 - Textile : determination of tensile properties

NF T 54-102 Plastics sheets : determination of tensile properties

Symbols

α : tensile force per unit width

α_f : tensile force per unit width at failure

b_o : initial width of the strip

T : tensile force on the sample

h_s : height of the seam

h_o : height of the sample between the clamps

h_x : distance between the optical makes

ϵ : average strain in the tensile direction

ϵ_x : central strain in the tensile direction

ϵ_f : average strain in the tensile direction at failure

$\dot{\epsilon}$: strain velocity

t : time

6 BURST TESTING

S. E. HOEKSTRA
Akzo Industrial Systems bv, Arnhem, The Netherlands

6.1 Introduction

In several countries the need has been felt for a three-dimensional stress strain "performance" test simulating in the laboratory such stresses on geomembranes as occur in the field due to an uneven settlement of the supporting soil.

This type of out-of-plane loading conditions occurs at the bottom of reservoirs and landfills, and for liners used for covers over subsiding landfills and other compressible subgrade materials.
Depending on its friction to the adjacent layers and its resistance to deformation a smaller or greater part of the geomembrane outside the settlement will participate in the deformation. Also the speed of settlement and the ambient temperature will influence the behaviour of the geomembrane.

During the past 5 years in many countries test equipment has become available to simulate these conditions.
Fayoux and Loudière (1984) mention the CEMAGREF experiments with an apparatus in which geomembrane samples (diameter 290 mm) are fixed over an orifice with a diameter of 12.5, 25 or 50 mm.
Comparitive tests with this apparatus, were made by Bernard (1987).
In the Fed.Rep. of Germany the working group AK14 of the DGEG (Deutsche Gesellschaft für Erd- und Grundbau) published recommendations for the use and testing of geosynthetics in which a bursting test for samples with a diameter of 500 mm is suggested. DVWK (1986).
Comparative tests on the same materials as were tested by Bernard (1987) were made according to the AK14 recommendations at the Akzo laboratories in Arnhem by Hoekstra (1987).
At the Int. Conf. on geomembranes in Denver, Rollin et al. (1984) reported bursting tests on samples of 150 mm diam. at different temperatures.
At the same conference Steffen (1984) presented the results of experiments in a diam. 1000 mm vessel in which by photographic means the deformations are evaluated.
Later a similar bursting test series on HDPE and modified HDPE (CHD) geomembranes was completed by STUVA research

institute (1988). The same STUVA-institute made an
elaborate investigation in which a newly designed apparatus
with a diameter of 1000 mm was used to study the behaviour
of HDPE and CHD geomembrane samples in a simulated soil
settlement under water pressures of 250 kPa.
At the Denver Conference Steffen (1984) presented similar
studies employing a more simple apparatus.
Recently Koerner, R.M., Koerner G.R. and Lin Hwu B. (1990)
of the Geosynthetic Research Institute (GRI) at Drexel
University (USA) reported the test results from a 610 mm
(24 inch) diameter vessel on a variety of different
geomembranes. Also the effect of load rate and
reproducibility were mentioned.
Degeimbre (1983) studied the performance of large 600 x
1130 mm samples with rounded edges seamed in the middle of
the longest side.
The deformation of elastomeric, thermoplastic and reinforced
bituminous geomembranes were investigated.

6.2 **Review of equipment and testprocedures**

6.2.1 **CEMAGREF-Paris**

Equipment

The apparatus is schematically shown in Figure 6.1.

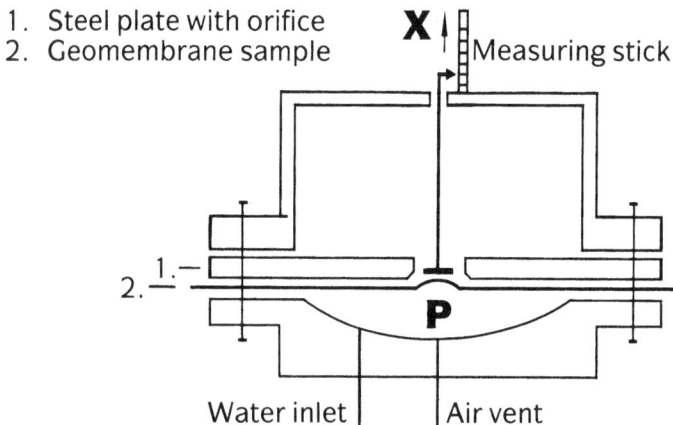

1. Steel plate with orifice
2. Geomembrane sample

Figure 6.1 Apparatus for bursting tests
at CEMAGREF (F.)

The centre of the geomembrane sample with a diameter of 350 mm is pressed into the circular orifice with rounded edges of the supporting steel plate.
The deformation is measured at the top. Plates with an orifice diameter of 12.5, 25, 50 and 200 mm are available.

Test procedure

Water is admitted. After the air is replaced the pressure is raised in steps of 100 kPa per minute.
During the experiment the height of deflection at each pressure level is measured. Also the temperature and the number of seconds the sample "survives" the bursting pressure is noted.

6.2.2 Akzo Industrial Systems bv, Arnhem

Equipment

The apparatus as recommended by the German AK14 working committee for routine bursting tests is shown in Figure 6.2

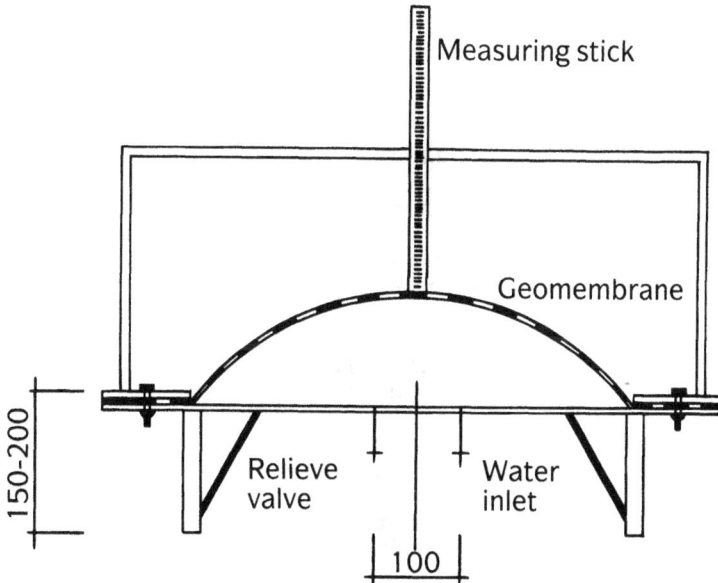

Figure 6.2   Apparatus for bursting tests
             at Akzo Industrial Systems bv (NL)

A round sample with a diameter of 700 mm is placed upon a 5 mm thick steel base plate with an inwards rounded water inlet and air outlet.
The geomembrane is fixed watertight by means of sixteen 10 mm diam. steel bolts between the base plate and a 100 mm wide, 10 mm thick steel ring with a rounded edge at the inside.
If a bituminous liner is tested around the edge of the sample an aluminium foil is folded to prevent the bitumen sticking to the apparatus.
By feeding water into the space between sample and base plate the sample is subjected to an increasing three-dimensional load. The water pressure and the height of the dome are recorded.

Test procedure

First the sample is mounted. Before the beginning of the actual test any residual air between the base plate and the geomembrane should be pressed out through the relieve valve. By means of a graduated ruler extending down through the frame that encloses the pressure unit, the distance between the top of the frame and the centre of the sample is recorded.
At the beginning of the test the water pressure is set to 10 kPa (0.1 bar) and after 1 minute the distance between the centre of the sample and the top of the frame is measured. After 1 more minute the pressure is increased to 20 kPa (0.2 bar), and so on.

With some elastic, low modulus geomembranes the water supply can be too slow to fill the continuously increasing volume at the required pressure level within the time mentioned. In this case the time intervals should be decreased sufficiently, but maintained equal for a certain type of material. Eventually the pressure steps can be made smaller.

The water pressure is increased in steps of 10 kPa every two minutes until bursting occurs. Besides a visual observation of the sample during the test, the pressure and height of deformation are recorded at each interval. Each material should be tested in triplicate.
With bituminous membranes it may be necessary to retighten the bolts in the steel ring during the experiment to maintain sufficient waterproofness at the edge.

Calculation

Assuming a spherical deformation and no slippage of the
sample in the steel ring the strain ϵ (%) and stress
σ(kN/m') in the sample can be calculated from the data
obtained by using equation 6.1.

$$1 = \frac{2\pi R\alpha}{360} = 0.01745 \ R\alpha \qquad (6.1)$$

From the available data

Distance centre of sample to base plate	H	(mm)
Sample radius	r	(= 250 mm)
Inner diameter steel ring	lo	(= 500 mm)
Water pressure	p	(MPa)
Thickness of the sample	d	(mm)

all necessary parameters:

Radius of the sphere	R	(mm)
Length of the spherical segment	l	(mm)
Strain	ϵ	(%)
Stress	σ	(kN/m' or MPa)

can be calculated as shown below (see Fig. 6.3):

$$\tan \gamma = H/r \qquad (6.2)$$

$$\gamma = \tfrac{1}{4} \alpha \qquad (6.3)$$

this gives:

$$R = \frac{r}{\sin \tfrac{1}{2} \alpha} = \frac{r}{\sin 2 \gamma} \qquad (6.4)$$

and with eq. 6.1, 6.3:

$$1 = 0.0698 \gamma . R \qquad \text{and} \qquad (6.5)$$

$$\epsilon = (1/lo - 1).100 \qquad (6.6)$$

139

Further is:

$$\sigma' = \sigma.\sin \tfrac{1}{2} \alpha \qquad \text{and} \qquad (6.7)$$

$$\sigma'.d.2\pi r = \pi.r^2.p \qquad \text{or}$$

$$2d.\sigma' = p.r \qquad (6.8)$$

with eq. 6.7:

$$2d.\sigma.\sin \tfrac{1}{2} \alpha = p.r$$

and with eq. 6.4:

$$2d.\sigma = p.R \qquad \text{or}$$

$$\sigma = \frac{p.R}{2d} \text{ (MPa) or} \quad \frac{p.R}{2d} \text{ x d (kN/m')} \qquad (6.9)$$

Remark

For reinforced geomembranes the thickness of the material is not relevant for its stress-strain behaviour. It is therefore recommended to express σ in kN per meter width rather than in N per unit of cross-section or MPa.

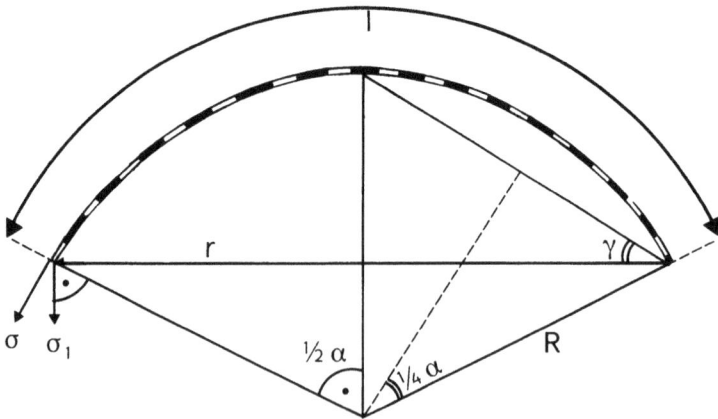

Figure 6.3 Data to be calculated

6.2.3 Geosynthetic Research Institute, Philadelphia

Equipment

The test vessel in this laboratory with a diameter of 610 mm (24 inch) is shown schematically in Figure 6.4. The hydrostatic pressure is applied from above the geo- membrane forcing deformation to occur in the empty lower portion of the vessel. The GRI system is capable of sustaining 2 MPa internal pressure.
The equipment does not allow for visual observation of the sample during the test.

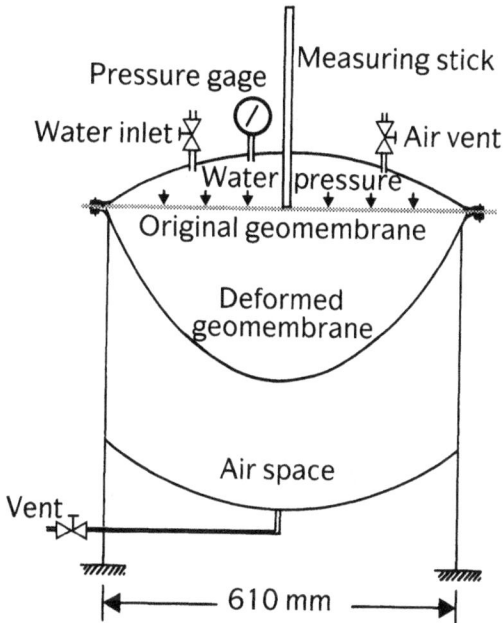

Figure 6.4 Apparatus for bursting tests at GRI (USA)

Test procedure

The geomembrane specimen is draped over the base, the lid is sealed above it, and water is introduced until the lid is full. The centerpoint measuring stick is initialized, the air vent in the upper lid is closed, the vent in the lower base is opened, and the test is ready to commence.

Water is introduced above the geomembrane at a constant
flow rate thereby mobilizing gradually increasing
hydrostatic pressure of 70 kPa/min. The geomembrane deforms
at a constant rate downward into the empty base of the
pressure vessel while hydrostatic gage pressure and
centerpoint deflection are regularly measured. From these
data corresponding stress and strain values are calculated.
After failure occurs, which is signaled by an abrupt
decrease in pressure accompanied by a dramatic noise, the
test is completed. The pressure vessel lid is then removed
and the mode of failure of the geomembrane is observed.

Calculation

Stress-strain curves are calculated under the assumption
that the geomembrane deforms with gradually decreasing
radius in the form of a spheroid or in the form of an
ellipsoid when its center-line deflection exceeds the
radius of the sample. Typically, HDPE and all reinforced
geomembranes are characterized by a spherical deformation,
whereas PVC, CPE and the elastomeric liners can be
characterized by assuming an ellipsoidal deformation.

6.2.4 Steffen Laboratory - Essen-Kettwig

Equipment

The test vessel with a diameter of 1000 mm is shown in
Figure 6.5. Samples with and without a seam can be tested
at pressures up to 1.5 MPa.
Since the pressure is obtained by compressed air the
apparatus should be safely constructed.
The deformation can be followed through the sightglasses in
the top.
The behaviour of the membrane during the test can be
evaluated with the aid of two photocameras and a network of
square chalk markings upon the sample (see Photograph 6.1).

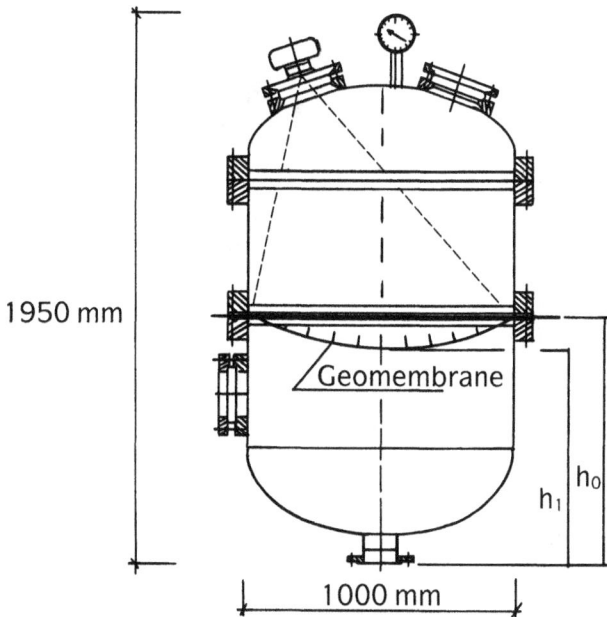

Figure 6.5 Apparatus for bursting tests
at Steffen-Laboratory (FRG)

Test procedure

The geomembrane is fixed between the lower and middle
section of the vessel. After installation of sticks and
tapes for the observation of vertical and in plane
deformation the lid is placed and air pressure is applied
at a rate of approx. 10 kPa/min.
Local differences in deformation are evaluated by means of
the pictures from the cameras, the measuring auxilaries and
the markings on the sample.
Each set of data is correlated with the corresponding air
pressure and used for the calculation of stress-strain
curves.

Photograph 6.1   HDPE geomembrane deformation test just
                  before failure

6.2.5 **STUVA, Cologne**

Equipment

Based upon the experience obtained by Steffen (1977,
1984) the STUVA research institute developed a rather
elaborate series of testing equipment to study the
behaviour of a liner under a combination of three-
dimensional stresses and the friction of its surfaces with
the surrounding soil layers.
STUVA's objective was to determine whether the deformation
of the liner in the settlement area is homogeneous and -
like Steffen - to establish how much extra material from
the peripheral zone participates.
Depending on geomembrane properties such as stiffness and
surface roughness as well as on the soil structure, the
availability of such extra material will result in a lower
level of strain on the membrane (see Figure 6.6).

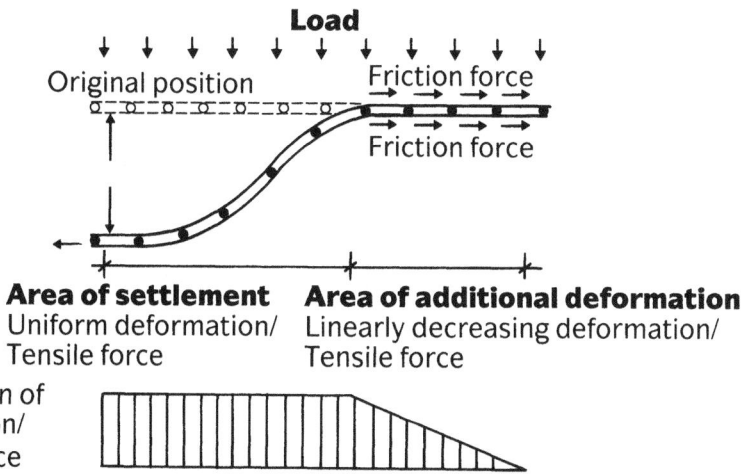

Figure 6.6   Deformation of geomembranes in an area of settlement

Since the liner will be more reliable at a lower stress, the assumption was that a geomembrane, as a horizontally positioned base liner, should have as low a friction with the surrounding mineral layers as possible.
In order to study these phenomena STUVA constructed an apparatus capable of measuring the sliding of a 500 x 1000 mm geomembrane sample between two mineral layers in which the lateral contraction can be restricted mechanically (see Figure 6.7). The movement of the sample between the supporting and the covering soil is measured by means of threads laying in length direction upon the sample and at their end cemented to the membrane (see Figure 6.8).

Cross section

① Geomembrane 500×1000 mm) Friction free sliding frame

Top view

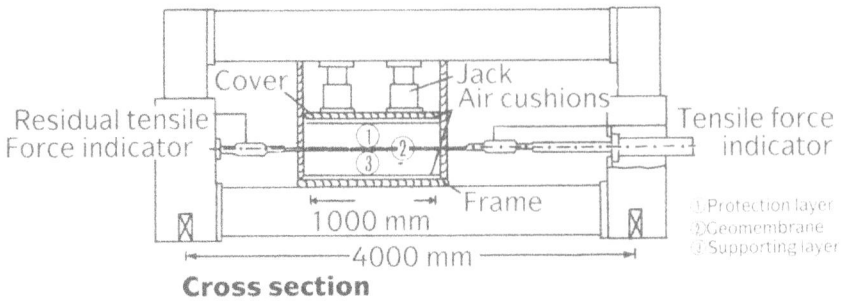

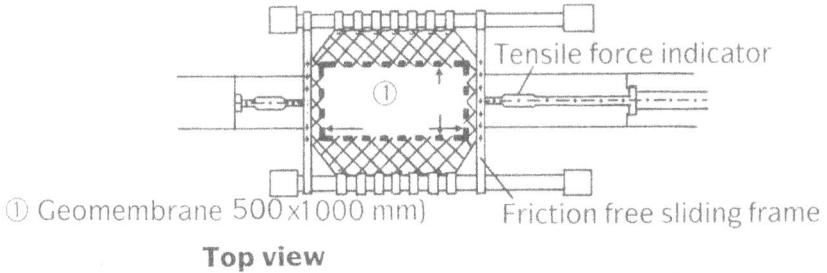

Figure 6.7 Apparatus for tensile tests with friction and restricted lateral deformation at STUVA (FRG)

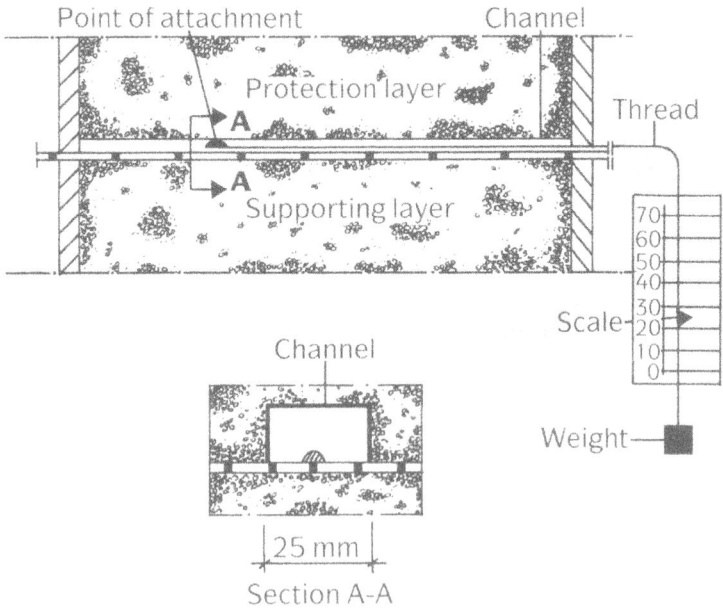

Figure 6.8 Indication of deformation pattern by means of threads

146

Furthermore a 1000 mm-diameter vessel was built in which the sample can be studied during a simulated settlement of the supporting soil through 4 sight glasses and three vertical measuring sticks (see Figure 6.9).

The local strain variations in the sample can be observed by means of two perpendicular measuring tapes. One, weighted by a lead strip, follows the membrane into the settlement while the other, fixed between the flanges of the vessel, is used as a reference to the grid made upon the membrane and as an aid in determining the diameter of the settlement.

The same vessel is also used for "normal" bursting tests under water pressure. Because of the cushion of compressed air the apparatus should be safely constructed.

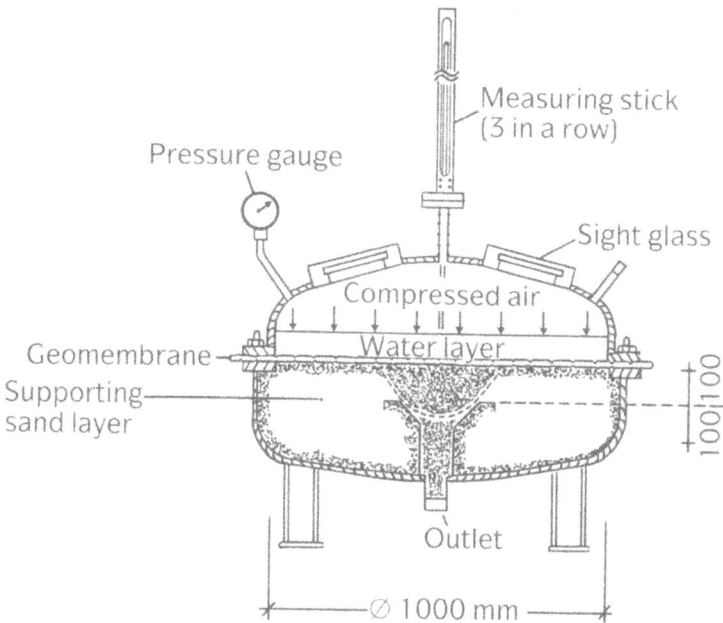

Figure 6.9   Apparatus for settlement tests at STUVA (FRG)

6.3 Test Results

The results reported by the various laboratories form a dispers mixture of data that only allow a rough comparison. A selection of the available information is compiled in table 6.1., 6.2. and 6.3.

Table 6.1   Bursting characteristics of various geomembranes without yield

Geomembrane	Test conditions		Test results	
Type (thickness)	Sample diam. (mm)	Laboratory	Stress at burst (MPa)	Strain at burst (%)
PVC				
(1.2 mm)	50	CEMAGREF	8.3	135
(1.2 mm)	500	AIS	7.3	115
(2.0 mm)	1000	Steffen	6.0	55
(0.5 mm)	610	GRI	6.0	75
CPE				
(1.2 mm)	50	CEMAGREF	7.3	231
(1.2 mm)	500	AIS	>6.4	>173
(2.4 mm)	1000	Steffen	3.3	68
Modified HDPE				
(2.1 mm)	250	Steffen	4.8	55
(2.2 mm)	500	Steffen	6.4	56
(2.2 mm)	1000	Steffen	5.9	58
(2.0 mm)	1000	STUVA	6.2	>50
(2.0 mm)	frame	STUVA	6.4	>80

Remarks

- The expansion of the 500 mm Ø CPE sample was too large to realise burst at the maximum rate of water flow.

- The yield point of the modified HDPE is much less pronounced than that of the pure HDPE material (see table 6.2). For this reason not the data at yield but at burst are given.

148

Table 6.2   Bursting characteristics of reinforced
             bituminous geomembranes

	Geomembrane	Test conditions		Test results	
	Type (thickness)	Sample diam. (mm)	Laboratory	Stress at burst (MPa)	Strain at burst (%)
	Bituminous			(kN/m')	
fabric reinforced	(5.0 mm) (5.0 mm) (5.0 mm)	200 500 1000	CEMAGREF AIS Steffen	41.7 27.2 28.0	34 30 13
non-woven reinforced	(4.0 mm) (4.0 mm)	50 500	CEMAGREF AIS	15.1 11.7	49 24

Remark

- For reinforced (bitumininous) geomembranes the reinforcement is responsible for the mechanical behaviour, the
  thickness of the geomembrane is not relevant and the
  stress has to be expressed as a force per width (kN/m').

Table 6.3   Bursting characteristics at yield of various
             HDPE geomembranes

	Geomembrane	Test conditions		Test results	
	Type (thickness)	Sample diam. (mm)	Laboratory	Stress at yield (MPa)	Strain at yield (%)
	HDPE				
	(1.0 mm)	610	GRI	20	10
	(2.0 mm)	1000	Steffen	17	10
	(2.5 mm)	250	Steffen	14.0	15.4
	(2.7 mm)	500	Steffen	15.4	15.0
	(2.7 mm)	1000	Steffen	16.3	9.9
	(2.0 mm)	1000	STUVA	16	12
	(2.0 mm)	frame	STUVA	18	12
	(2.0 mm)	500	Steffen	18.0	18.7

149

6.4   Discussion of test results

Sample diameter

By comparing the Akzo/CEMAGREF data from table 6.1 it
becomes clear, that for all products the stress and strain
values from the diam. 50 mm samples (CEMAGREF) are higher
than those from the diam. 500 mm tests (Akzo).
Another study about the influence of sample diameter was
made by Steffen. He compared the behaviour of various HDPE
liners in pressure vessels with a diameter of 250, 500 and
1000 mm respectively (see table 6.2.).
Independent of sample diameter all membranes show a yield
area of about 20-30 mm in the final stage of the test. As a
consequence because of the larger percentage of the sample
being overstretched in the yield zone the overall strain
level in the diam. 250 mm samples is higher than that from
the diam. 500 mm and 1000 mm tests.
At the edge of the diam. 250 mm vessel the strain decreases
to 0 % . In the 500 mm vessel however, the chalk quadrants
close to the edge still show a strain level of about 10 % .

From the test on the 1000 mm samples it became clear, that
only the central part with a radius between 200 and 300 mm
participates in the bursting process.

In this large apparatus also tests on samples with a seam
have been made. Special care should be taken to obtain a
good seal at the edge. With a good seam bursting data
similar to the normal product are obtained, but if the
sample is mechanically or thermally damaged burst will
occur at a much lower pressure.

All observations indicate clearly that at this stage with
the limited available knowledge a minimum sample diameter
of 500 mm should be recommended to properly simulate the
performance of the geomembrane material in practice.
The larger the diameter the better an inhomogeneous
material will be detected (!).
Important in this respect is the necessity to use the
strain values at yield for those products that posses a
yield point.
If tests on samples with a seam are made a minimum diameter
of 500 mm is demandatory.

The HDPE-data mentioned in table 6.3 show the stress/strain
values at yield from experiments with samples of various
sizes, thicknesses and producers. These data confirm that
no significant reduction in strain
values at yield occurs by increasing the sample diameter
over 500 mm. Also the results obtained by Steffen (1984)
and those found with the frame at the STUVA institute lead
to this conclusion (see Figure 6.10).

With this frame in special cases it could be useful to
determine more realistically than in a circular bursting
test apparatus the stress-strain characteristics of
geomembranes with restricted lateral contraction.
It should be noted that also with this frame the HDPE
sample yields at a strain of about 12 % .

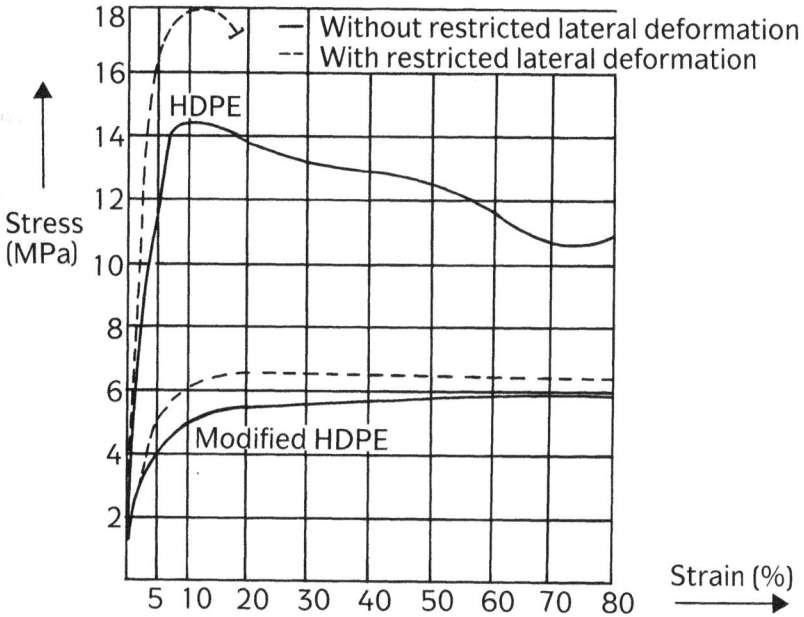

Figure 6.10   Stress-strain curves with and without restricted lateral
deformation   (Rate of deformation: 10% per minute)

Material performance

From recent observations by Steffen and STUVA (1988) it
has become clear that at least for materials with a yield
point the earlier published theory by Steffen (1977, 1984)
about the behaviour of geomembranes in uneven soil
settlements has to be corrected.
The original thinking was, that the area activated by a
liner to spread the stress forces initiated by the
settlement is directly related to the E-modulus of the
geomembrane material (see Figure 6.6)
Observations from Steffen indicate however, that for a high
modulus HDPE liner only the central 500 mm circle of a
1000 mm sample participates in the bursting. This

observation is confirmed by the STUVA studies. In a settlement test as illustrated in Figure 6.9 it was found that the HDPE material at a distance of about 150 mm from the edge of the settlement does not participate anymore (see Figure 6.11).

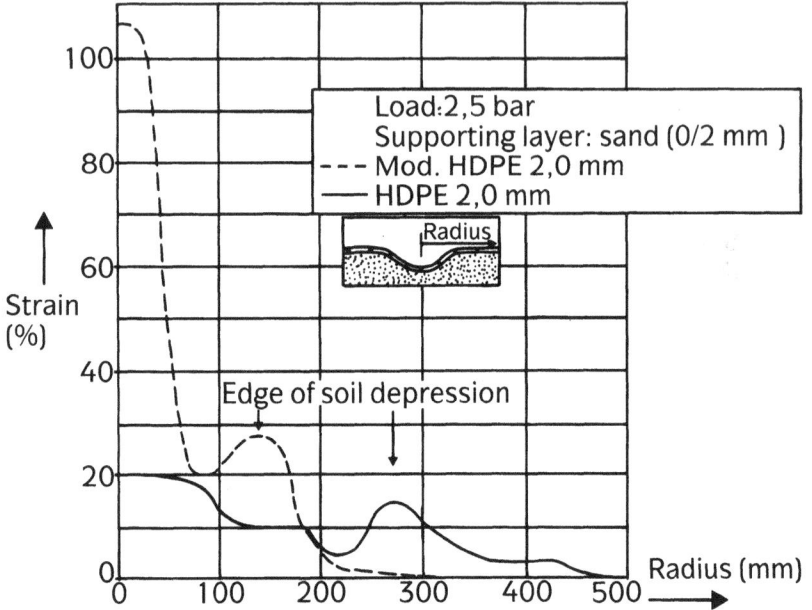

Figure 6.11   Distribution of the geomembrane deformation in settlement tests under a fixed load

Both Steffen and STUVA observe an irregular decrease in strain even in the normal burst tests without the additional influence of friction at the surface of the geomembrane (see Figure 6.12).
However, since all these experiments were made on geomembranes with a yield point, it is quite possible that for example reinforced liners and other materials without a yield point perform differently.
This means that conform the concept illustrated in Figure 6.6 a homogeneous material with a low E-modulus will deform strongly over a relative small area whereas for example a bituminous or thermoplastic membrane with a

high-modulus reinforcing fabric will be capable to lower
its overall strain level linearly by inclusion of a large
area around the settlement in its deformation.

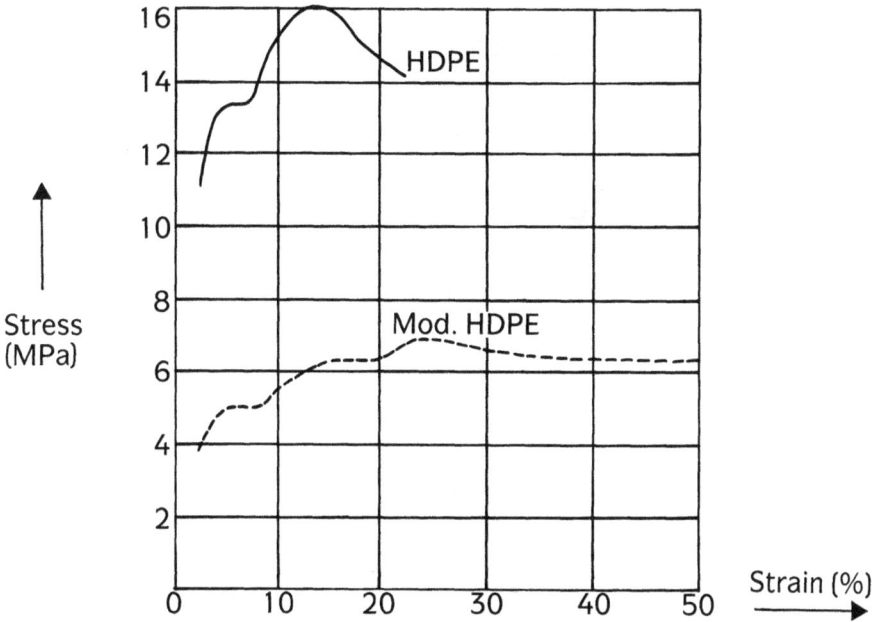

Figure 6.12   Calculated stress-strain curves from bursting tests

6.5   **Recommendation for a future routine burst test apparatus**

Equipment

Based upon the available information given in the
previous sections of this chapter it should be assumed that
results from samples with a minimum diameter of 500 mm will
be comparable. This leads to the conclusion that with the
present experience, especially for samples with seams, a
burst testing apparatus with a minimum diameter of 500 mm
as described in paragraph 6.2.2. should be recommended. A
visual impression of the equipment is given in the pictures
6.2, 6.3 and 6.4.

Because of the risk of explosion the use of air to
inflate the sample is not recommended unless the equipment
is designed according to the local safety regulations.

Photo report of a bursting test on a Ø 500 mm PVC sample

6.2

6.3

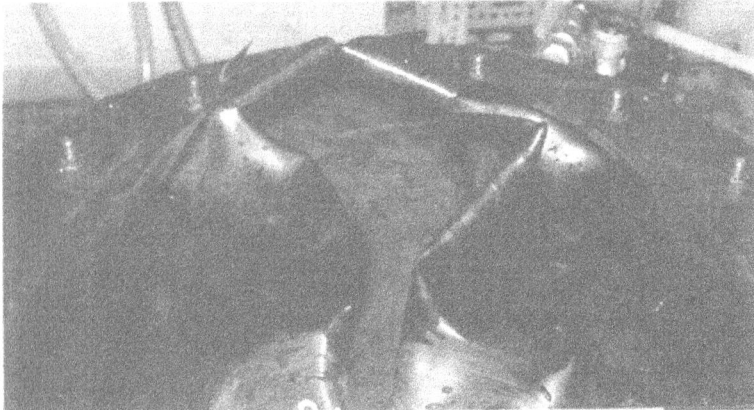

6.4

Test procedure

Special attention should be paid to the description of
the performance of the sample as observed during the test.
In this respect the deformation downwards into a pressure
vessel is less attractive.

It also is recommended to put together in one graph a
plot of the stress-strain curves at strain levels below
50 % for each of the individual tests. The calculation
should be made under the assumption of a spherical
deformation or an ellipsoidal if the centre-line deflection
exeeds the radius of the sample.

Furthermore it seems desirable to raise the pressure in
fixed steps at fixed time intervals or continuously
increasing with time.
The temperature and the rate of expansion during the
experiment should be recorded.
Together with the stress-strain values at burst resp. at
yield, these parameters should be included in the test
report.

At present only limited experience with burst testing at
various temperatures is available (Rollin, 1984). Valuable
additional information could be derived from a more
detailed study about the behaviour of geomembranes at low
and elevated temperatures.

6.6   References

AK-14, DVWK Schriften Heft 76, "Anwendung und Prüfung von Kunststoffen im Erdbau und Wasserbau" ("Application and testing of geosynthetics in civil engineering"), Paul Parey, (Berlin, 1986) 245, 246.

BERNARD, C., Rilem TC 103 MHG, "Note sur les essais d'éclatométrie avec mesure de déformation", CEMAGREF, (Antony, 1986).

DEGEIMBRE, R., "Les membranes d'étanchéité pour les constructions hydrauliques", Annales des Travaux Publics de Belgique, (avril 1985).

FAYOUX, D., and **LOUDIèRE, D.,** "Interaction between geotextiles and geomembranes-laboratory tests", 2nd Int. Symp. Plastic and rubber waterproofing in civil engineering, (Liège, 1984), 3A Add. 1.1 - 1.9.

HOEKSTRA, S.E., Rilem TC 103 MHG, "Results of a comparative bursting test on various lining materials", Akzo Industrial Systems bv, (Arnhem, 1987).

HOEKSTRA, S.E., Rilem TC 103 MHG, "Review of bursting experiments on geomembranes", Akzo Industrial Systems bv, (Arnhem, 1988)

KOERNER, R.M., KOERNER, G.R., and **LIN HWU, B.,** "Three dimensional, axi-symmetric geomembrane tension test", ASTM D-35 Symposium, (Las Vegas, 1990). Rollin, A.L.,

LAFLEUR, J., and **MARCOTTE, M.,** "Selection criteria for the use of geomembranes in dams and dikes in northern climate", Int. Conf. on geomembranes, (Denver, 1984), 415-419.

STEFFEN, H., "Anforderungen an flexible Dichtungsbahnen bei Erdbauwerken im Bereich des Umweltschutzes", R.I.L.E.M. - I.C.P. - C.E.P. Int. Symposium in Plastic und Rubber Waterproofing in Civil Engineering, (Liege, 1977), III. 18.1 - III. 18.11 .

STEFFEN, H., "Report on two dimensional strain stress behaviour of geomembranes with and without friction", Int. Conf. on geomembranes, (Denver, 1984), 181-185.

STUVA (Studiengesellschaft für unterirdische Verkehrsanlagen e.V. "Untersuchung der anwendungsbezogenen Möglichkeiten eines Verfahrens zur Prüfung des mechanischen Verhaltens von Deponiebasisabdichtungen aus Dichtungsbahnen" ("Study into the possibilities to develop realistic tests for the mechanical behaviour of geomembranes as base liner for landfill sites"), (Köln, March 1988).

7 PUNCTURE TESTING

R. FROBEL
R. K. Frobel and Associates, USA
J-M. RIGO
Geotextiles and Geomembranes Research Centre, University of
Liege, Belgium

7.1 Introduction

One of the most underrated properties of geomembranes is
puncture resistance and one of the least specified test
methods is puncture ! This is totally illogical when
considering the integrity of the geomembrane for the life
of the containment, especially the containment of wastes
and the protection of our environment.
During installation, the geomembrane can be punctured by
tools, machinery, aggregate or protrusions in the
subgrade. During placement of the soil cover, falling
rocks or equipment pressure may puncture the geomembrane
and once covered, holes are not visible. During the
service life of the facility, the geomembrane could be
punctured by gradual subsidence and piercing eventually
forming holes. Uncovered geomembranes are always subject
to animal traffic, ice action, floating debris, or
aggregate falling down slope. Consequently, a high
resistance to puncture is an important geomembrane or
geocomposite property. Figures 7.1 and 7.2 illustrate
field stress and puncture.
    The current methods fall under either index properties
or performance properties test methods.
Index tests are generally small scale (i.e. small specimen
size) tests that are used by manufacturers to gauge the
quality of goods produced during the manufacturing
process. Also, the design engineer or material specifier
may rank various products on a given test property or
method. They are typically methods that are fast,
inexpensive and easy to perform. Index tests, however, are
not generally useful in predicting the ability of a
geosynthetic to withstand installation stresses and
in-service conditions.
Performance tests, on the other hand, attempt to simulate
in the laboratory the conditions and stresses that a
geosynthetic will experience in actual field applications.
These tests are generally slow, expensive and sometimes
difficult to perform. They are, however, necessary to
accurately predict field performance and it is recognized
that performance test results can and are being used in
the design process.

It must also be said that many puncture test methods were initially developed specifically for geotextiles and related products. Most of these might also be used to characterize geomembranes.

Fig. 7.1   Geomembrane stress caused by point loading
            of aggregate against the geomembrane

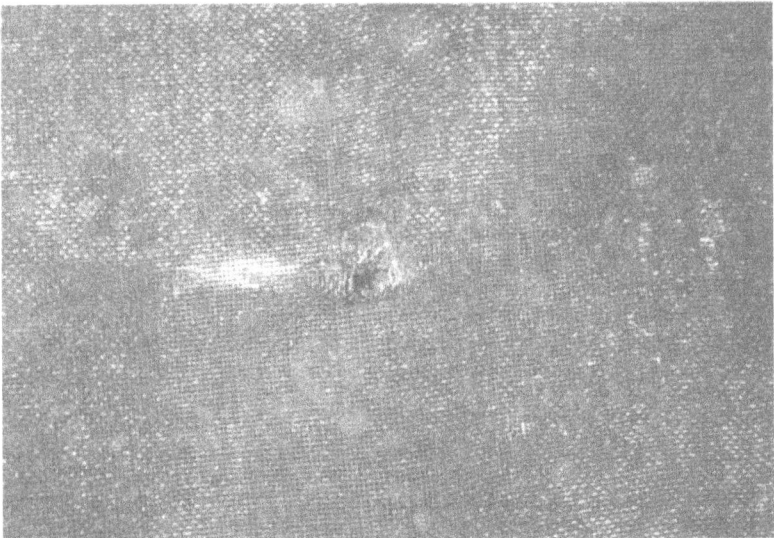

Fig. 7.2   Example of a site induced puncture
            to a geotextile

7.2 The state of the art and classification of puncture testing methods

The state of the art in puncture testing currently consists of both the index and performance test methods, however, the commonly referenced tests are the index test methods. All of these vary with size of probe, test speed and clamping methods.

In reviewing the puncture test methods, it is particularly important to note the rate of deformation and its effect on puncture resistance of geomembranes. Since the geomembrane can and will be punctured by the action of impact (i.e. dropping aggregate, tools, equipment) or by slow propagation (i.e. loading the geomembrane over time when subjected to a point stress),the speed of test becames important. Also some geomembranes are viscoelastic and some are semicrystalline, thus the rate of deformation will have significant effect on the stress or point load required to puncture them. Matrecon (1988) studied the rate of deformation on puncture resistance using Puncture Test FTMS 101C, Method 2065 which is described later in this chapter. Figure 7.3 illustrates results of testing two materials, one a 75 mm thick PVC and the other a 1.15 mm thick EPDM at speeds ranging from 5 to 50 mm/minute. The results show the difference in puncture resistance values versus rate of deformation. When semicrystalline HDPE was tested in two thicknesses (1 mm and 2 mm) the range of results showed a much greater susceptibility to speed of testing (rate of deformation) than did the thermoplastic and crosslinked materials. In studying the effect of thickness, Matrecon (1988) found that the thickest semicrystalline materials were more susceptible to loading rate, however for the same loading rate, there was a linear relationship between maximum stress and thickness. Thus, it is very important to note the rate of deformation used in puncture testing as it relates to type of material and actual or design conditions.

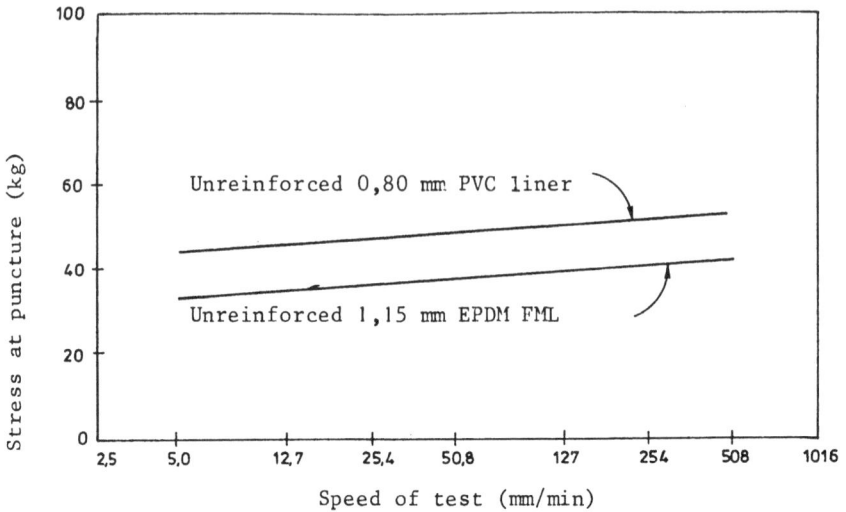

Fig. 7.3   Example of rate of deformation effect on
puncture resistance (Matrecom, 1988)

7.2.1   Index tests

7.2.1.1   Static puncture index tests
A) ASTM D 4833-88
   This standard is entitled : "Standard test method for
   Index Puncture Resistance of geotextiles, geomembranes
   and related products".
   During this test, the specimen is clamped without
   tension between circular plates of a ring   clamp
   attachment   secured   in   a   tensile/compression   test
   machine. A force is exerted against the center of the
   specimens unsupported portion (45 mm diameter) by a
   steel rod attached to a load indicator until rupture of
   the specimen occurs. The   rod tip is 8 mm in diameter
   with a flat surface and 45 degree chamfered edges.   The
   maximum force   recorded is   the value   of the   puncture
   resistance of the specimen. Test speed is set at 300 ±
   10 mm per minute. Fig.   7.4   gives some details on   the
   test device.
   A similar testing procedure has been developed in
   Finland by Rathmayer (1986) for the characterization of
   geotextiles.

45 mm ± 0,025 mm DIA

100 mm ± 0,025 mm DIA

all holes 8 mm DIA
equally spaced

37 mm ± 0,025 mm RAD

TOP VIEW

facing surfaces
grooved with O-ring or
coarse sanopaper fixed
to plates

geotextile
and/or geomembrane

Test Fixture Detail (not to scale)

130 mm

18 mm   bevel

50 ± 1 mm   8 mm ± 0,1 mm

16 machine   0,8 mm x 45° chamfer
finish

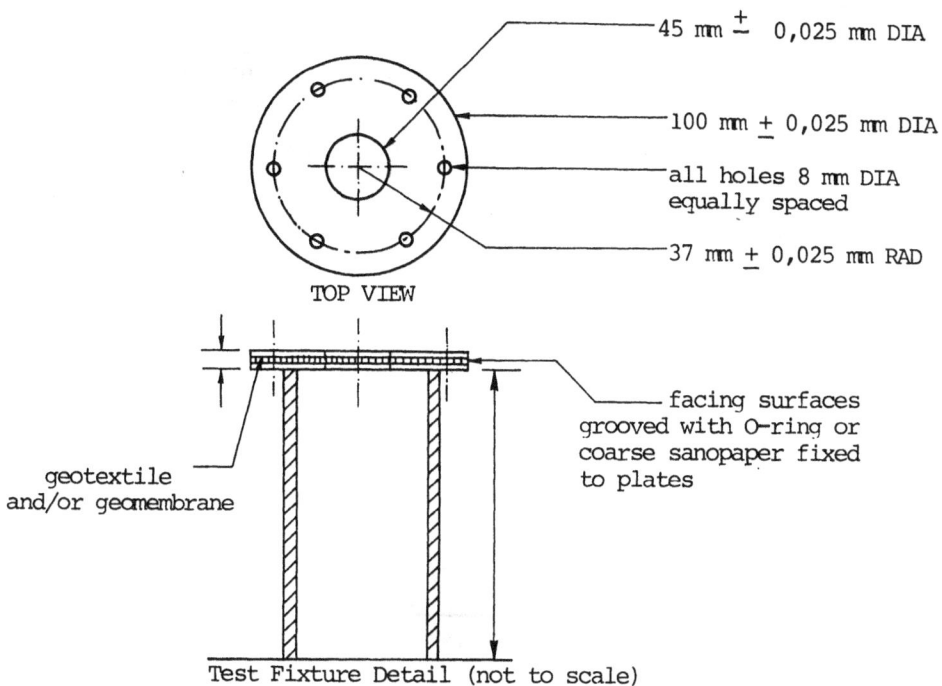

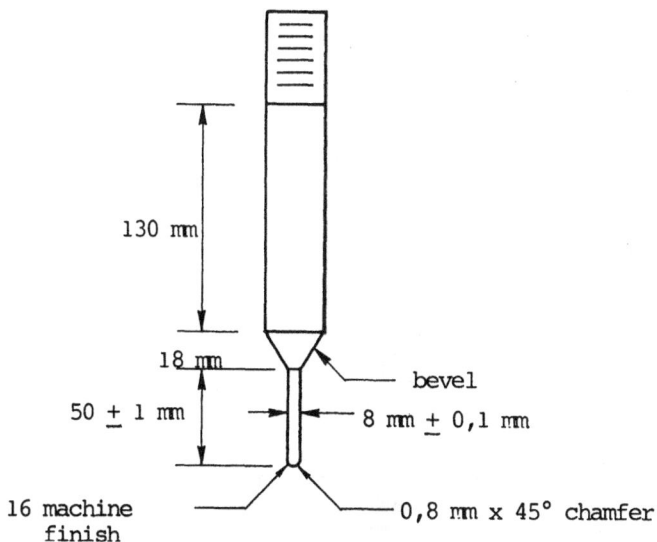

Fig. 7.4   ASTM D 4833 Test Probe Detail (not to scale)

B) FTMS 101 C - Method 2031-1980

This standard is also called "Tetrahedron point test".
In this method, a 100 mm by 250 mm test specimen is
looped around a tetrahedron point with the ends of the
looped specimen fixed in an upper clamp of a CRT
tensile test machine. The lower clamp or base of the
machine fixes the point centered on the specimen at the
bottom of the loop (see fig. 7.5). The upper moving
clamp is operated at 300 ± 25 mm per minute until the
puncture head ruptures the specimen. The load required
to puncture is recorded.

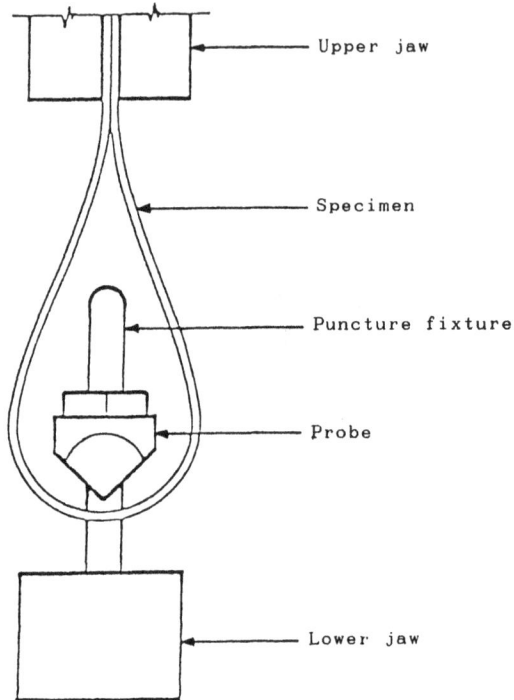

Fig. 7.5   FTMS 101 C Method 2031 Tetrahedron point
           (not to scale)

C) FTMS 101 C - Method 2065 - 1980

This is a puncture resistance and elongation test on a
3.2 mm radius probe. For this test procedure, a tapered
rod probe is pushed through a specimen clamped between
two plates with a 25 mm diameter hole. The 12.8 mm
diameter, 127 mm long tapered rod is pushed through the
specimen at a speed of 500 mm per minute. The taper is
50 mm long with a 3.2 mm radius at the end. The maximum
puncture resistance and elongation is recorded.
Fig. 7.6   illustrates this test device.

162

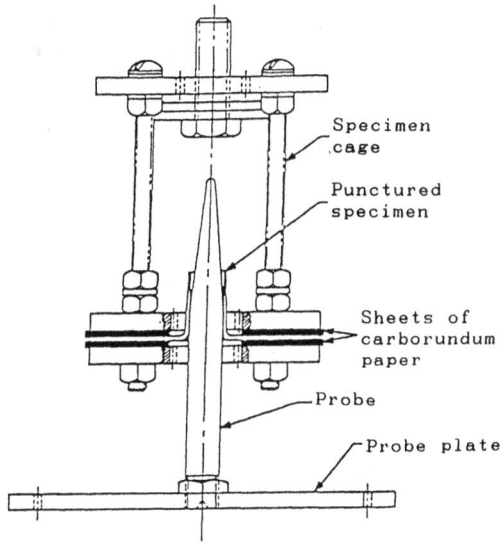

Fig. 7.6   FTMS 101 C - Method 2065 Tapered Rod Puncture
            (not to scale)

D) CBR Punger puncture test
   This DIN   standard (DIN   54307-1982) is   very   commonly
   used to characterize both geomembranes and geotextiles.
   A test specimen is clamped without tension between   two
   circular plates attached   to a   CBR mold   secured in   a
   tensile/compression test   machine. A   force is   exerted
   against the   center   of the   specimen   by a   steel   CBR
   plunger 50 mm in diameter with a flat end and 45 degree
   chamfered edges. The opening in the plates is 150 mm in
   diameter. The maximum   force required   to puncture   the
   specimen at a speed of 50 mm per minute is recorded.
   Fig. 7.7   gives some details of the test device.

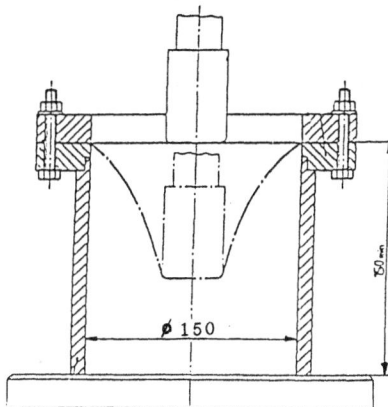

Fig. 7.7   DIN 43507 CBR Plunger test

In geotextile testing, there is a direct relationship between the puncture resistance value obtained from the rod puncture and the tensile strength of the fabric. This is because the fabric between the inner edge of the specimen holder and the outer edge of the puncture rod is in a state of axisymmetric tension (Koerner, 1986). This same relationship should be true of geomembrane testing as both product types are in fact planar sheet materials and generally axisymmetric within reasonable values.

Correlation of CBR Puncture with strip tensile was made by Cazzuffi on geotextiles at ENEL (Cazzuffi, 1986). His research claimed a very good comparison between strip tensile tests on 500 mm wide speciments and standard CBR puncture. For isotropic nonwoven geotextiles a conversion was proposed as follows :

$$\alpha_f \ (Kn/m) = 2 \ \pi \ F_p \ (Kn) \qquad\qquad (3)$$

where
α_f = force per unit width tensile
F_p^f = breaking force for CBR puncture.

A similar conversion may be appropriate for geomembranes but using a smaller width tensile specimen (i.e. 200 mm). Moritz and Murry (1982) made similar observations but again on geotextiles.

According to them, the more isotropic the material the better the correlation, especially with wider width specimens (greater than 200 mm width).

E) Modified CBR Plunger Puncture Test
In an effort to simulate the effect of sharp aggregates or waste debris being pushed into a geomembrane, Werner (1986) and Puhringer (1990) developed a modified CBR test using a pyramid plunger and a very rigid subsurface. Failure detection was by electrical current. Fig. 7.8 illustrates the device and dimensions of the pyramid point.

Lhôte and Rigo (1988) used the CBR mold with clay as the base substrate. The clay was changed as to its cohesive properties by changing the moisture content. In testing the puncture resistance of geotextiles over the clay, it was found that significant changes in puncture resistance resulted as a function of the changes in soil cohesive properties.

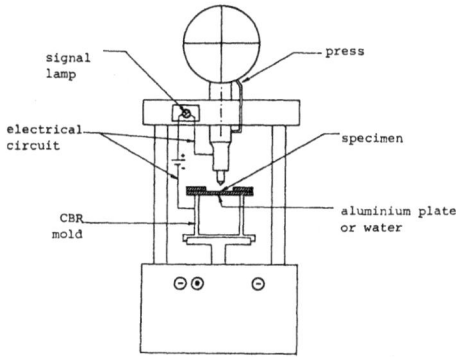

(a) Modified CBR mold and press

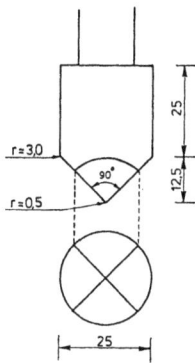

(b) Pyramid point (dimensions in mm)

Fig. 7.8   Modified CBR Plunger Test with Pyramid Point (Puhringer, 1989)

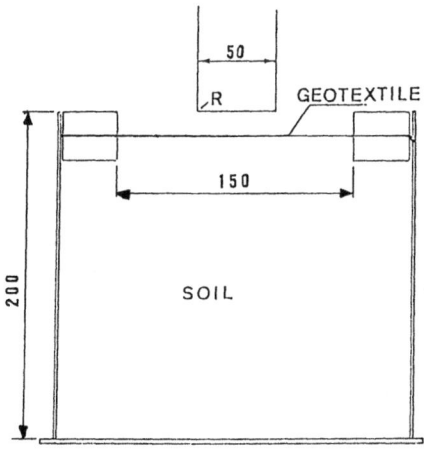

Fig. 7.9   CBR Puncture Test taking the presence of a soil into account (Lhôte & Rigo, 1988)

7.2.1.2 Dynamic puncture tests
A) Drop cone method - Australian standards
   Draft method 5 - 1989
   This method makes use of the modified CBR mold.
   Evaluating the resistance to tear initiation and
   puncture, this test is particularly relevant in
   situations where coarse aggregate or riprap is dropped
   or otherwise dynamically forced against the geotextile
   or geomembrane.
   A circular specimen is gripped around its entire
   circumference by an upper and lower clamping ring. A
   cone of specified mass is dropped into the surface of
   the material from a specified drop height. The diameter
   of the punctured hole, in combination with the drop
   height, gives a measure for puncture resistance. The
   relationship between the drop height and the diameter
   of the hole has been found, from testing a wide range
   of geotextiles, to be

$$d_2 = d_1 \; (\frac{h_2}{h_1})^{0,68}$$

   or

$$h_2 = h_1 \; (\frac{d_2}{d_1})^{1.47}$$

   where :
   h_1 = drop height (first value), in millimetres
   h_2 = drop height (second value), in millimetres
   d_1 = diameter of hole corresponding to a drop height h_1
        in millimeters
   d_2 = diameter of hole corresponding to a drop height h_2
        in millimeters

   Fig. 7.10 illustrates the drop cone apparatus.

   Notes :
   1) The exponent applying to the equation giving d_2 was
      found to be generally in the range between 0.55 and
      0.7. The value of 0.68 was established as the best
      approximation.
   2) The actual diameter of the hole for a given drop
      height is an inverse measure of the resistance
      against penetration; the smaller the hole, the
      greater the resistance.

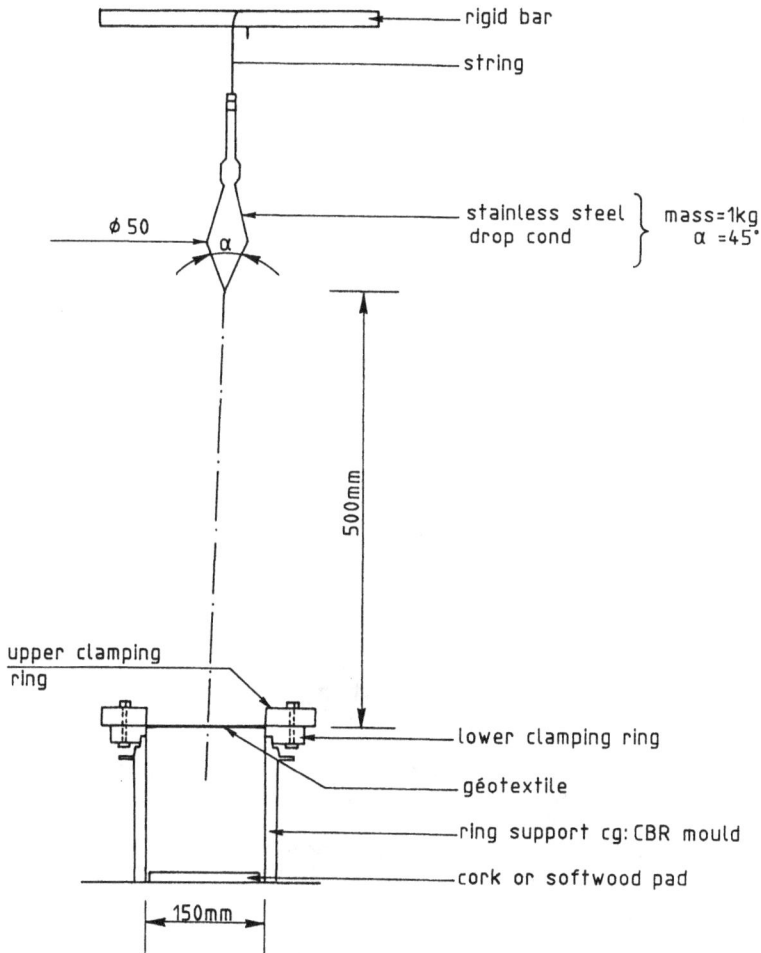

Fig. 7.10   Drop Cone Test Apparatus
Australian draft mehtod 5

B) **Drop cone method - Swiss (EMPA) standards; Nordtest Method NT Build 243; British Standards Draft BS 6906 : Part 6.**
This test method was developed to characterize geotextile resistance to damage resulting from dropping sharp-pointed stones and is a measure of resistance to installation damage. The test procedure uses the same test equipment as in the previous section A with the drop height again fixed at 500 mm. The diameter of the hole, however, is measured by placing a 600 gm cone in the hole and reading the hole diameter in mm. The resulting diameter is simply the measure of geotextile puncture resistance. This same type of method can be used for geomembranes as well.

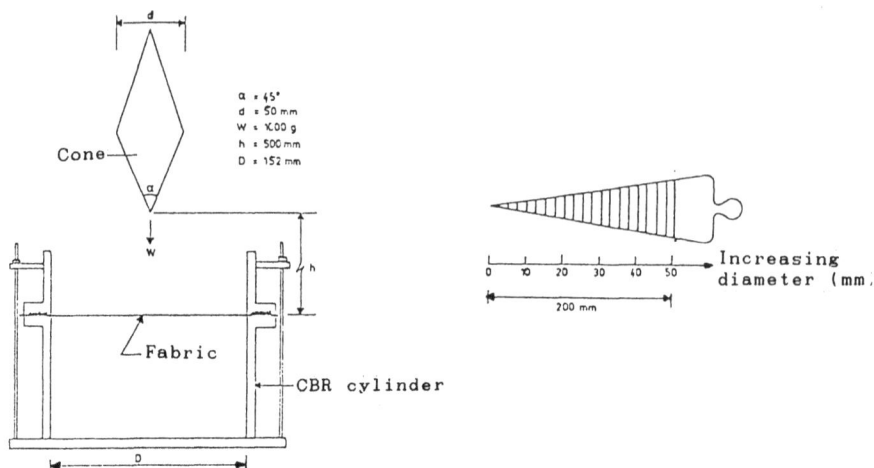

Fig. 7.11   Drop cone Method - Swiss (EMPA) standards;
            NT Built 243; BSI 6906 Part 6

7.2.2.2   Performance tests
A) ASTM Draft standard D 35.10.88.01, 1989.
   This test method is based on the principle of the
   multiaxial stress-strain testing.
   A large test specimen (at least 600 mm in diameter) is
   placed over the base of a large hydrostatic pressure
   vessel containing a grouping of standard puncture
   points. The top of the vessel is placed over the
   specimen, bolted and sealed in place, filled to a
   shallow depth with water and then pressurized gradually
   until rupture of the geomembrane occurs. Rate of
   pressure increase (strain rate) and puncture points can
   be adjusted to actual or simulated site conditions or
   loadings. The vessel must be designed for accurate
   pressure control and measurement as well as immediate
   failure detection.
B) Large scale hydrostatic pressure testing (point
   loading)
   Large diameter multiaxial stress-strain testing on
   geomembranes and geotextiles has been carried out on a
   research basis for over 13 years. Several noted
   researchers including Rigo (1977), Frobel (1981, 1983),
   Loudiere et al. (1983), Fayoux (1984), Frobel et al.
   (1987), Laine et al. (1989) have demonstrated the
   various testing devices developed to simulate an
   in-service quantitative puncture performance test of
   primarily geomembrane or geomembrane/geotextile
   combinations. Burst testing can also be accomplished
   using these devices and is covered in detail in
   Chapter 6.

168

In principle, each of the devices simulate over burden or water pressure acting on a geomembrane when placed on a subgrade containing either natural aggregate or machined puncture points, to evaluate the geomembranes ability to conform to subgrade (or cover) irregularities such as aggregate, rock out croppings, waste debris, etc.. Fig. 7.11 is a conceptual drawing illustrating an aggregate and a simulated aggregate shape (Rigo, 1977) taken from Rigo's studies wherein he attempts to develop an "ideal" shape which can be duplicated and which can provide repetitive puncture results.

Fig. 7.13 illustrates the mechanism of deformation according to Rigo (Rigo, 1977). There are two critical zones in the geomembrane - zone 1 extends over the geomembrane - aggregate contact and zone 2 comprises the void areas formed by the geomembrane between aggregates. Thus, there are two types of failure possible - rupture by puncturing (zone 1) or rupture by bursting (zone 2).

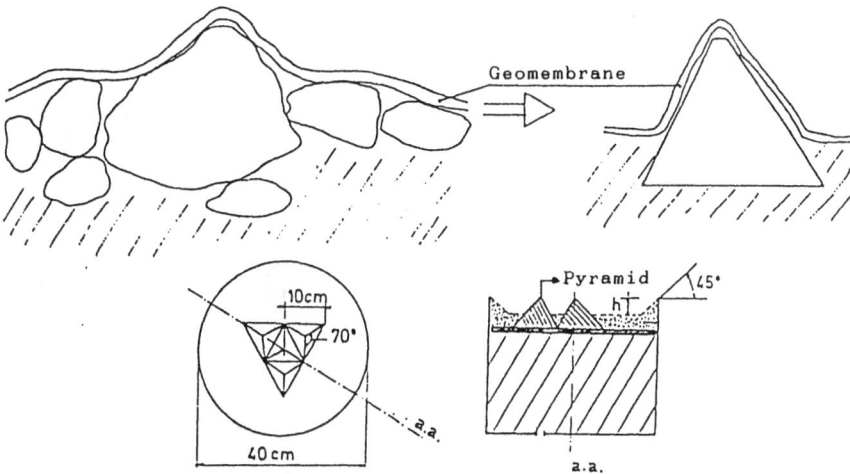

Fig. 7.12 Aggregate Shape and Machined shape for hydrostatic puncture (Rigo, 1977)

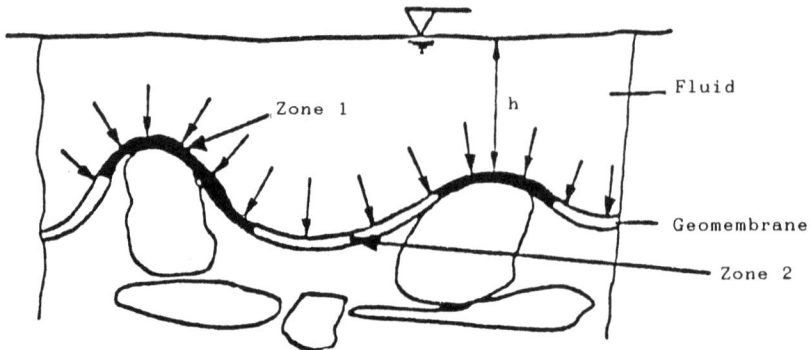

Fig. 7.13   Mechanism of geomembrane deformation over
             point loadings (Rigo, 1977)

An example of   the type of   apparatus used   for
hydrostatic puncture testing is the vessel developed by
Frobel shown in Fig.   7.14   (Frobel, 1981, 1983).
Loading rates   were   simulated   by   a   compressed
air-on-water system to provide a wide range of test
pressures and the ability to use a moisture detection
system for leakage monitoring. Maximum design pressure
was 1034 kPa (100 m head). The vessel top and bottom
sections were fabricated from 500 mm diameter pipe cap
sections with bolted flanges and double "O" ring seals.
This provided a test area of approximately 0.2 sq.
meters. The vessels were also designed to rapidly
change subgrade conditions by using interchangeable
subgrade pans. Pressurization and monitoring   was
controlled by computer   for continuous   operation.
Extensive testing with this type of device showed that
results are reproductible using the same unaltered type
of subgrade   or   puncture points   and   that   the
semicrystalline materials   failed predominately   by
puncture whereas many of the thermoplastics failed by
bursting over voids. Figures 7.15 and 7.16 illustrate
the amount of deformation that can take place in a
geomembrane over a rough surface after loading in a
hydrostatic vessel.
Fayoux (1984) also performed hydrostatic tests on a
similar device using an aggregate subgrade of 20-40 mm
quartzite stones. Fayoux observed that the thinner PVC
(thermoplastic) geomembranes   generally failed   by
bursting in voids,   whereas the more rigid, thick
geomembranes were able to span voids but fail in
puncture over sharp aggregate.
Laine et al. (1989) performed hydrostatic testing on
three types of geomembranes and geomembrane/geotextile
combinations over longer time periods and at two

170

different temperatures to determine the affect of creep over time at relatively low pressure as well as the effect of higher temperatures (50 degrees C). The simulated subgrade in this case was truncated rigid epoxy cones which were varied in height by varying sand height (similar to Rigo's and Frobel's procedures). Premature failure of the semicrystalline HDPE was also observed in this testing due to puncture over the truncated cone tips.

7.3 Conclusions : recommendations and future trends

Considering the obvious importance of puncture resistance properties on the integrity of a geomembrane system, the geosynthetics community has not seen a concerted effort at developing a special puncture test that is relevant to geomembranes and their unique applications. To date, test methods have largely been "borrowed" from existing standards.
The development of the large scale hydrostatic test as a standard for performance testing in ASTM is the first real development in the study of puncture resistance. This type of performance test must be adopted by other standards organizations and especially ISO as well. Large scale performance testing of all geosynthetics will no doubt be the next generation of test methods.
In additions to an accepted performance test method, a minimum of two index test methods for puncture must also be developed. These two index test methods should evaluate the slower puncture propagation (i.e. similar to ASTM D 4833, DIN 54307) and the rapid or tear initiation and subsequent puncture (i.e. similar to SA draft method 5). Until these universal test methods are developed, geomembrane puncture testing will continue to be nonstandard.
In utilizing any of the existing standard methods or draft methods, the user must first determine the applicability of the method to the type and thickness of material and what the test results will be used for. Due to the current variability in test methods (specimen dimension, deformation rate, probe size and shape) the end user must be somewhat knowledgable in the method that a material is specified under as the actual puncture result will be directly affected by the test method chosen.

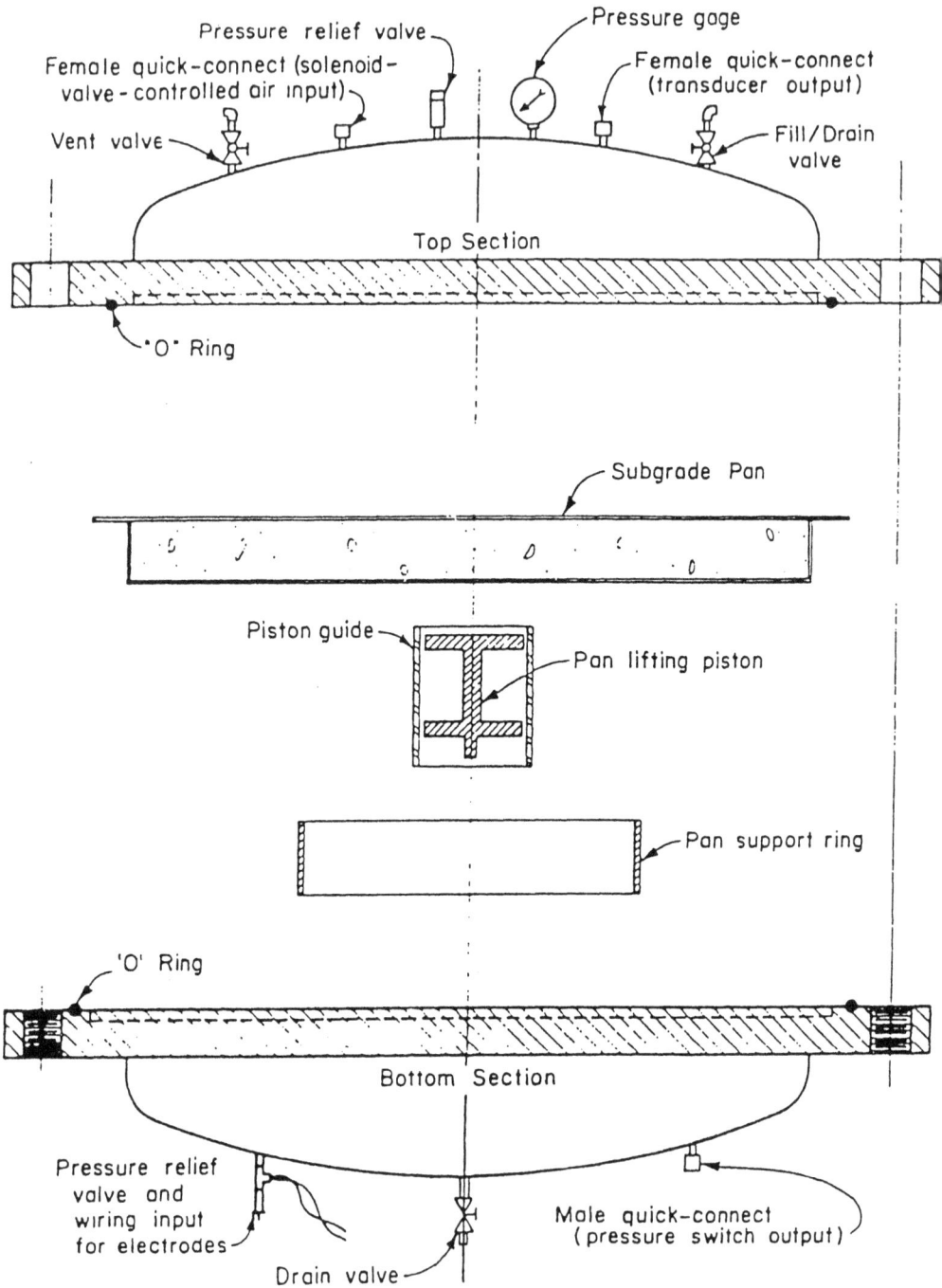

Fig. 7.14   Example of a Large Scale Hydrostatic
Test Chamber (Frobel, 1981, 1983)

Fig. 7.15 Base of a hydrostatic vessel with
           aggregate subgrade (Frobel, 1981)

Fig. 7.16 Example of Distorsion in a geomembrane
           after testing over an aggregate subgrade

7.4 References

American Society for Testing and Materials (1988), **Test method for Index Puncture Resistance of Geotextiles, Geomembranes and Related Products.** ASTM D 4833-88, ASTM, Philadephia, PA, 1988.

American Society for Testing and Materials (1989), **Recommended Standard Practice for Large Scale Hydrostatic Pressure Testing of Geosynthetics.** Draft Method D 35.10.88.01, ASTM Committee D35, ASTM, 1987.

Association suisse des professionnels de géotextiles (ASPG). **Le manuel des géotextiles - Résistance à la perforation diamètre du trou Od.** 1988.

British Standards Institute (1989). **Determination of Resistance to Perforation (Cone Drop Test).** Draft Method BS 6906 : Part 6, 1989.

Cazzuffi, D., Venesia, S., Rinaldi, M. and Zocca A. **The Mechanical Properties of Geotextiles : Italian Standard and Inter-laboratory test comparison.** Proceedings Third International Conference on Geotextiles, Vienna, IFAI Publishers, 1986.

Deutsche Norm (1982). **CBR Plunger Test.** DIN 54307, DIN, Berlin, 1982.

Fayoux D. and Loudiere D. (1984). **The Behavior of Geomembranes in relation to the soil.** Proceedings International Conference on Geomembranes, Denver, IFAI Publishers, 1984.

Federal Test Method Standards (1980). **Puncture Resistance.** Method 2031, FTMS 101C, FTMS, Washington, D.C., 1980.

Federal Test Method Standards (1980). **Puncture Resistance and Elongation Test.** Method 2065, FTMS 101C, FTMS, Washington, D.C., 1980.

Frobel, R.K. (1983). **A micro Computer-based Test Facility for Hydrostatic Stress Testing of Flexible Membrane Linings.** Proceedings Colloque sur l'Etanchéité Superficielle des Bassins, Barrages et Canaux, Paris, February 1983.

Frobel, R.K. (1981). **Design and Development of an Automated Hydrostatic Flexible Membranes Test Facility.** REC-ERC-80-9, U.S., Bureau of Reclamation, Denver, Colorado, 1981.

Frobel, R.K., Youngblood, W. and Vanderoort, J. (1987). **The composite Advantage in the Mechanical Protection of Polyethylene Geomembranes.** Proceedings Geosynthetics 87, New Orleans, IFAI Publishers, February, 1987.

Geosynthetic Research Institute (1986). **CBR Puncture Strength.** GRI Test Method GS-1-86, GRI, Drexel University, Philadelphia, PA, 1986.

Koerner, R.M. (1986). **Designing with Geosynthetics.** Prentice Hall, Englewood Cliffs, N.J. 1986.

Laine, D.L., Miklas, M.P. and Parr, C.H. (1989). **Loading Point Puncturability Analysis of Geosynthetic Liner Materials.** Proceedings Geosynthetics 89, San Diego, IFAI Publishers, 1989.

Lhôte, F., RIGO, J.M., THOMAS, J.M. **Modelisation and design of geotextiles submitted to puncture loading.** Kyushu International Symposium on Theory and Practice of Earth Reinforcement, Japan. 1988.

Loudiere, D. and Pignon N. (1983). **Puncture Resistance of Geomembrane.** Proceedings Colloque sur l'Etanchéité Superficielle des Bassins, Barrages et Canaux, Paris, February 1983.

Mortiz, K. and Murray H. (1982). **Comparison between Different Tensile Tests and the Plunger Puncture Test (CBR Test).** Proceedings Second International Conference on Geotextile, Las Vegas, 1982.

Puhringher, G. **Geotextile/Geomembrane Composite - Pyramid Testing.** Proceedings ASTM Symposium on Geosynthetic Testing for Waste Containment Applications. Las Vegas, Nevada, January 1990.

Rigo, J.M. (1977). **Correlation of Puncture Resistance over Ballast and the Mechanical Properties of Impermeable Membranes.** Matériaux et Constructions, vol. II, n° 65, RILEM, 1977.

Standards Australia (1989). **Determination of Puncture Resistance-Drop Cone Method.** Draft Method CE/20/4/88/17, 1989.

Werner (1986). **Design criteria for separation function of geotextiles on the basis of mechanical test procedures.** Third International Conference on Geotextiles, Vienna, 1986.

8 FRICTION TESTING

T. S. INGOLD
Consultant, United Kingdom

8.1 Introduction

In many applications such as the lining of canals or waste containments, a geomembrane will be applied to an inclined surface. This will lead to a component of gravitational force acting in the plane of the geomembrane which can cause it to slide down the inclined surface. Consequently it is important to be able to assess the bond properties of the interface between the geomembrane and the inclined surface. The bond strength which can be made available may be frictional, cohesive or a combination of the two. As will be considered in following sections, the inclined surface on which the geomembrane rests may be soil, or it may be some other material such as a geotextile underliner or drain, or a granular drainage layer. That the geomembrane may be underlain by a composite made up of a series of layers of different materials leads to the need to determine which is the most critical sliding surface. For most applications there is a need to give some protection to the upper surface of the geomembrane to reduce the effects of the environment. These effects would include direct mechanical damage such as abrasion, tearing and puncture, as well as photochemical effects such as ultra-violet light attack and oxidation. The form of this surface protection may be simply a layer of soil, however, in canal linings this surface protection may be a revetment comprising rip-rap rock fill, a soil layer and a geotextile underliner. Obviously consideration must be given to the frictional properties at the various interfaces involved. These are considered in more detail in the following sections.

8.2 Composite geomembrane systems

Geomembranes are rarely installed directly on the slope of a canal or waste containment and left without surface protection, Figure 1(a). Consequently there may be one or several layers of material below the geomembrane. These can be termed "under-layers". Similarly the upper surface of the geomembrane may be covered with a protective layer or a composite layer including a second geomembrane. These can be termed "over-layers".

Figure 1 shows five types of geomembrane lining systems, the most simple of which is a single geomembrane placed directly on the formation soil of the slope as shown in Figure 1(a). A single geomembrane may be placed over a low permeability soil, such as clay, to form a composite geomembrane-clay liner. In more demanding situations, such as containments for toxic leachate or certain revetment works, a double liner system might be employed with a drainage layer, or leakage detection system, incorporated between the two geomembrane layers, Figures 1(c) to 1(e). As shown in Figure 2, the drain or leachate collection system might use a

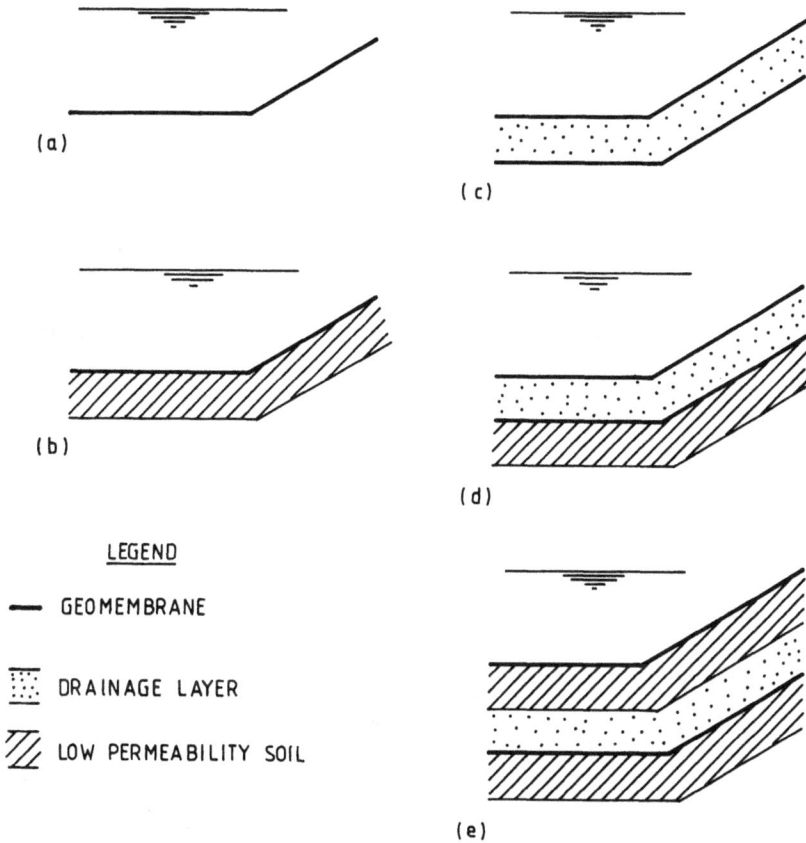

LEGEND

— GEOMEMBRANE

⬚ DRAINAGE LAYER

▨ LOW PERMEABILITY SOIL

Figure 1   Five types of lining systems: (a) Single liner: geomembrane liner; (b) Single liner: composite liner; (c) Double liner: double geomembrane liner; (d) Double liner: upper geomembrane liner and lower composite liner; (e) Double liner: double composite liner.

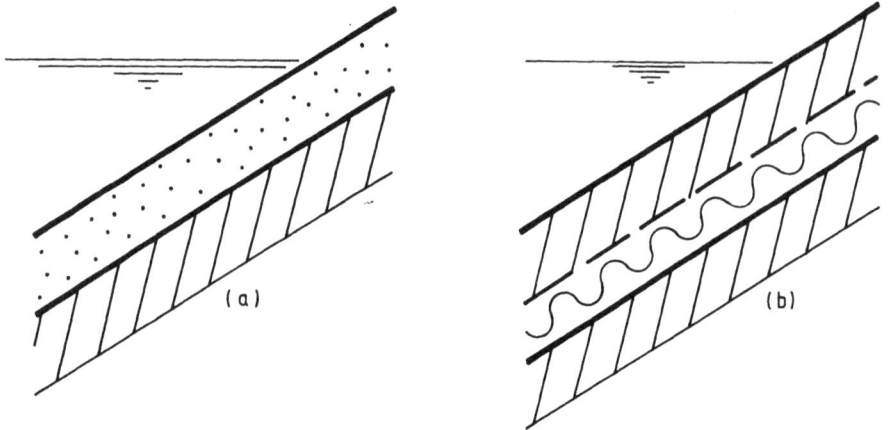

Figure 2   Examples of double liner systems using two geomembranes:
(a) the top liner is a geomembrane, the leakage collection
layer is a granlular material, and the bottom liner
is a geomembrane-low-permeability soil composite liner;
(b) both liners are geomembrane-low-permeability soil
composite liners, the leakage collection layer is
a geomesh separated from the overlying low-permeability
soil by a geotextile acting as a separator and a filter.

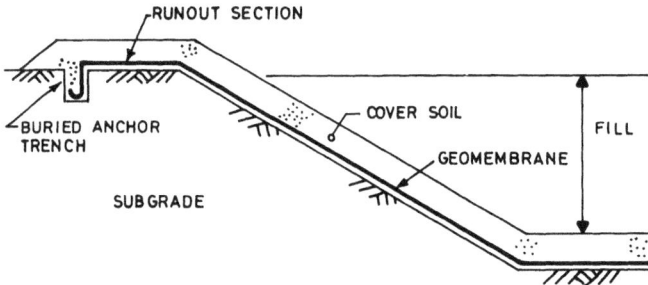

Figure 3   Typical layout of geomembrane of lined enclosed side
slope.

graded granular drainage medium, Figure 2(a), or a synthetic mesh as a water conducting core, Figure 2(b). In the latter case, it is not unusual to place a geotextile over the mesh, to act as a separator/filter, before placing the geomembrane. Where single geomembrane liners, Figure 1(a), or the lower liner of double liner system, Figure 1(c), are placed directly on the slope formation, a thick needle punched geotextile may be used as a cushion layer or as a vent drain for gases developed in the formation.

In reviewing the above composite systems, no consideration has been given to protection to the upper layer. A typical layout of a single liner system with a cover soil is shown in Figure 3. The cover soil may rest·directly on top of the geomembrane or may be separated from the geomembrane by a geotextile. Martin and Koerner (1984) have suggested that a geotextile overlay might be beneficial in mobilizing higher friction than the geomembrane and, therefore, is more effective in holding the cover soil in place. Where the geomembrane is used as a canal liner, the cover soil might be replaced by a revetment system. In this case, it would almost certainly be necessary to use a geotextile overlay to protect the geomembrane.

From the above review, it is apparent that a geomembrane might form a sliding surface with a wide range of material types, either above or below the geomembrane. For example:

> geomembrane - formation soil
>
> geomembrane - clay liner
>
> geomembrane - geotextile
>
> geomembrane - geomesh drain
>
> geomembrane - granular fill drain
>
> geomembrane - cover soil
>
> geomembrane - revetment

All of these combinations are likely to give rise to different friction angles or bond strengths for a given geomembrane and this must be taken into account in defining test methods. In a composite system comprising several layers, the stability of the system is likely to be governed by the weakest interface which would exhibit the lowest friction angle. This aspect of behaviour must be given careful consideration at design stage.

8.3 Failure modes

Geomembranes may fail through a variety of mechanisms such as chemical attack, ultra-violet light attack, oxidation and mechanical abrasion or puncture. Although none of these are directly related to the frictional properties of the geomembrane, they must be

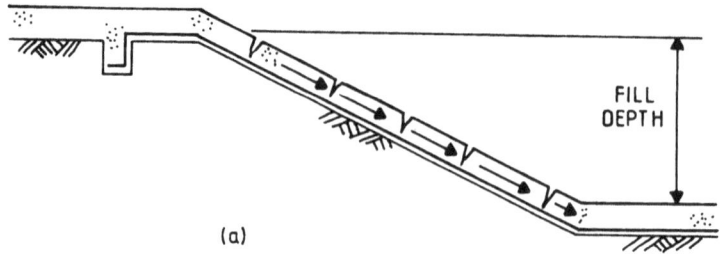

(a)

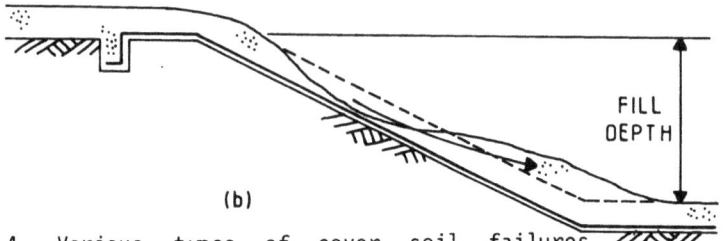

(b)

Figure 4   Various types of cover soil failures.
(a) Tensile splitting failures. (b) slumping failure.
(After Martin and Koerner, 1985)

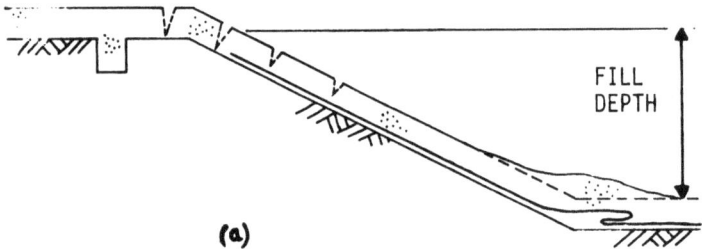

(a)

(b)

Figure 5   Various types of anchor trench failures. (a) Pull-out
of lines;   (b) localised trench failure.
(After Martin and Koerner, 1985)

taken into account during design. Other aspects to be taken into account include the stability of the slope-geomembrane system. Such stability will include consideration of the effects of soil-geomembrane friction. Several related failure modes are illustrated in Figures 4 and 5. Figure 4 deals with failures in the cover soil. Two modes are shown. The first, Figure 4(a), shows the development of tensile cracks in the cover soil which can be caused by too low a factor of safety on soil-geomembrane friction, or creep extension in the geomembrane. The latter can be caused by the shear stress imparted from the cover soil which may induce axial tension in the geomembrane. Another method of failure, Figure 4(b), involves slumping of the cover soil. This may be particularly acute if there is low soil-geomembrane friction. Figure 5 shows various types of anchor trench failure. The pull-out mode shown in Figure 5(a) will be controlled in part by the friction which can be developed between soil and geomembrane.

8.4 Existing methods of testing

Methods of testing to determine the bond strength at a geomembrane-soil interface are similar to those employed to investigate the bond properties of geotextiles. Test methods fall into the two broad categories of pull-out tests or direct shear tests. Direct shear testing is dominant since, among other things, there can be difficulty in interpreting results from pull-out tests on extensible materials. The many researchers who have studied geomembrane-soil friction have made almost exclusive use of direct shear tests. In essence there are five basic methods which can be used for direct shear testing. The principles of each type of apparatus are illustrated in Figure 6.

   i) Fixed Shear Box: This employs a standard direct shear box in which the geotextile or geomembrane is mounted on a rigid block which is placed in the lower half of the shear box. The upper half of the shear box is filled with soil which is sheared over the geotextile or geomembrane below. (Ingold, 1982).

  ii) Partially Fixed Shear Box: In which the geotextile or geomembrane is laid over soil filling the lower half of the shear box. One end of the geotextile or geomembrane is clamped to the lower half of the shear box and soil in the upper half of the shear box is sheared over the interface. (Ingold, 1982).

 iii) Free Shear Box: Is similar to the partially fixed shear box except the geotextile or geomembrane is free at both ends. (Richards & Scott, 1985).

  iv) Large Base Shear Box: Is similar to the fixed shear box test in (i), except the lower half of the shear box has a larger plan dimension than the upper half of the shear box containing the soil. This method has the advantage

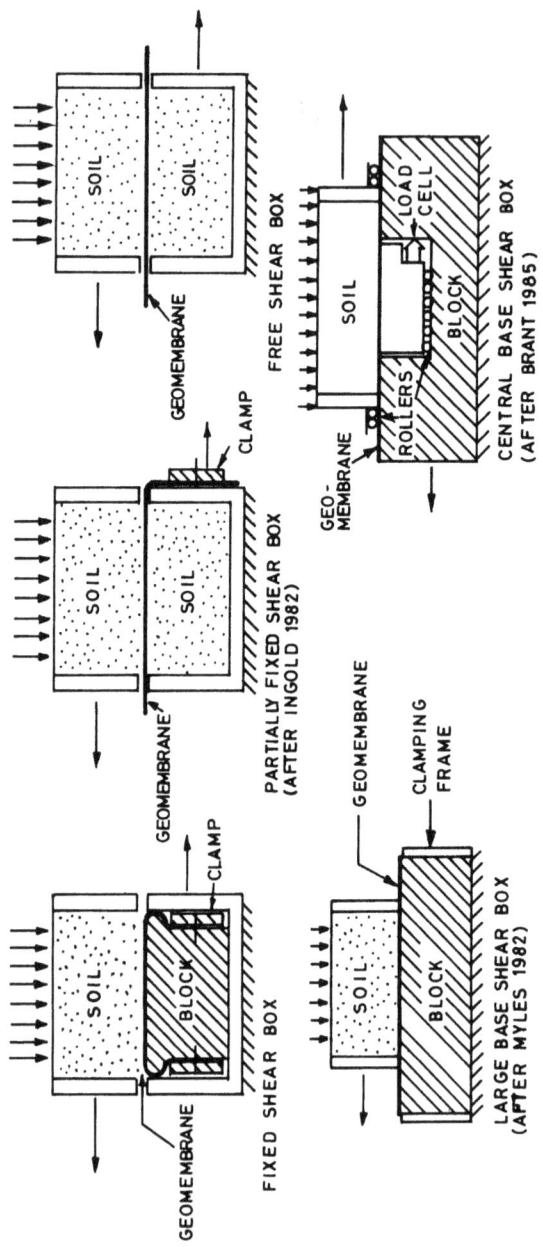

Figure 6   Types of Direct Shear Apparatus

that there is a constant contact area between soil and geotextile or geomembrane. (Myles, 1982).

This is not the case in the fixed shear box test (i) where the contact area reduces as the upper half of the shear box moves relative to the lower half. This is not likely to introduce important inaccuracy unless the strain required to fully mobilise the peak frictional force is large.

v) Central Base Shear Box: This operates on a similar principle to the large base shear box (iv), except in this case the frictional force generated in the geotextile or geomembrane is measured over a plan area smaller than the plan area of the soil in the upper half of the shear box. (Brandt, 1985).

The test methods in (i) to (iii) above are likely to prevail since they employ standard soil mechanics testing apparatus. The test methods in (iv) and (v) again employ direct shear techniques, however, they do require special apparatus.

8.5 Published bond strength values

Many researchers have investigated the bond strength properties at soil-geotextile, soil-geomembrane or geotextile-geomembrane interfaces. The frictional shear strength of soil is defined by the conventional parameters c', ϕ', where c' is the effective cohesion intercept, in kPa, and ϕ' is the effective internal angle of shearing resistance in degrees. The corresponding parameters for soil-geomembrane, or soil-geotextile, bond strength are $\alpha c'$ and δ'. The coefficient α is an adhesion factor similar to that used in pile design with the product $\alpha c'$ defining the soil-geomembrane, or soil-geotextile, adhesion. The frictional bond between soil and geomembrane is defined by δ' the angle of bond stress. The following seven tables give bond strength parameters derived by various researchers using direct sheat testing methods. It is left to the reader to compare the various values obtained, however, certain factors are worthy of note. For example comparison of Tables 1 and 5 show an order of magnitude difference in reported values of adhesion. For a multilayer system using geomembranes in contact with geotextiles Table 4 illustrates some low angles of geotextile-geomembrane friction compared to soil-geomembrane or soil-geotextile friction. Table 6 gives an indication of the effects of saturation on the angle of bond stress.

TABLE 1

Soil-Geomembrane Bond Strength Values
(Based on Williams & Houlihan,1987)

Geomembrane ▶	HDPE		PVC	
Soil Type ▼	δ	$\alpha c'$	δ	$\alpha c'$
Ottowa Sand	19	0.8	26	0.3
Concrete Sand	27	0.7	33	0.6
Sand & 5% Clay	14	0.6	19	0.8
Sand & 10% Clay	17	0.6	19	0.7
Saprollite	21	0.4	28	0.5
Gulf Coast Clay	25	1.0	23	1.6
Glacial Till	22	0.7	25	1.0
Note: δ in degrees, $\alpha c'$ in kPa				

TABLE 2

Soil-Geomembrane Friction Values (in degrees)
(Based on Akbert et al, 1985)

Soil / Geomembrane	$\phi = 42$	$\phi = 41$	$\phi = 40$	$\phi = 40$
PVC - 2	3	34	31	30
PVC - 4	31	20	29	25
EPDM	34	31	31	33
HDPE	25	20	22	22
Bitumen (sanded)	34	35	36	38
Bitumen (smooth)	41	40	39	36
Bitumen (rough)	43	41	38	35
NOTES: Test Temperature 23°C, Normal Stress 120 kPa.				

TABLE 3

Geomembrane - Geotextile
Friction Values
(Based on Akbert et al, 1985)

Geomembrane	Friction Angle (°)
Bitumen (rough)	30
Bitumen (smooth)	13
EPDM	31
Butyl	31
CIM	21
PVC	22
CSPE (Hypalon)	28

TABLE 4

Friction Angles at Various Interfaces
(Based on Martin et al, 1984)

(a) Soil to Geomembrane Friction Angles (Degrees)

Soil / Geomembrane	Concrete Sand ($\emptyset = 30°$	Ottawa Sand ($\emptyset = 28°$	Mica Schist ($\emptyset = 26°$)
EPDM	24	20	24
PVC (Rough)	27	-	25
PVC (Smooth)	25	-	21
CSPE	25	21	23
HDPE	18	18	17

(b) Soil to Geotextile Friction Angles (Degrees)

Soil / Geomembrane	Concrete Sand ($\emptyset = 30°$)	Ottawa Sand ($\emptyset = 28°$)	Mica Schist ($\emptyset = 26°$)
CZ 600	30	26	25
Typar 3401	26	-	-
Polyfilter X	26	-	-
500 X	24	24	23

(c) Geomembrane to Geotextile Friction Angles (Degrees)

Geomembrane / Geotextile	EPDM	PVC (R)	PVC (S)	CSPE	HDPE
CZ 600	23	23	21	15	8
Typar 3401	18	20	18	21	11
Polyfilter X	17	11	10	9	6
500 X	21	28	24	13	10

(R): Rough face,   (S): Smooth face.

TABLE 5

Geomembrane to Fine-Grained Soil Bond Strengths
(Based on Koerner et al, 1986)

SOIL TYPE	Soil No.1 ML-CL c=9 Ø=38		Soil No.2 CL-ML c=12 Ø=34		Soil No.3 CL c=20 Ø=30		Soil No.4 SP-CH c=25 Ø=24		Soil No.5 CH-SP c=28 Ø=22	
Soil-Geomembrane Adhesion/Friction	αc	δ	αc	δ	αc	δ	αc	δ	αc	δ
PVC	8.5	39	3.7	23	14.0	16	7.0	24	12.0	17
CPE	8.0	40	3.2	24	13.0	17	8.0	23	10.0	19
EPDM	5.0	33	5.0	23	8.0	23	7.5	20	9.0	18
HDPE	5.0	26	2.0	23	14.0	15	3.0	21	14.0	15
Embossed HDPE	9.0	35	11.0	29	18.0	27	15.0	26	16.0	25

Note: αc and c are in units of kPa, Ø and δ are in degrees.

TABLE 6

Friction Angles Between Smooth Polyethylene Film
and Different Construction Materials
(Based on Weiss & Batereau,1987)

Material in contact with PE Film	Friction Angle (degrees)	
	Dry	Saturated
Geotextile (non woven)	11 - 14	9
Sand	14 - 24	11 - 22
Prefabricated concrete slab	16 - 22	14 - 22

TABLE 7

Friction Angles Between Smooth PVC Film
and Different Construction Materials
Based on Weiss & Batereau,1987)

Material in contact with PVC Film	Friction Angle (degrees)	
	Dry	Saturated
Non woven	16 - 24	14 - 24
Sand	27 - 31	22 - 27
Prefabricated concrete slab	27 - 45	27 - 45

8.6 Factors affecting results and design considerations

In soil-geomembrane friction testing the fundamentals of soil mechanics must be applied rigorously. Many of the bogus results obtained stem from a lack of understanding of how soils behave rather than some complex interaction between soil and geomembrane. As a first step the engineer must decide on drainage conditions. Is the sample to be sheared drained or undrained? Drained testing applies to both short and long term analysis of cohesionless soils, e.g., sands and gravels, and long term analysis of cohesionless soils, e.g., clays and clay bound soils. Under these conditions soil strengths are represented by the effective cohesion, $\alpha c'$, and the effective internal angle of shearing resistance, ϕ. The corresponding values of drained bond strength are $\alpha c'$ and $\delta '$, where α is the drained adhesion factor. By definition drained tests are sheared at a rate of strain low enough to prevent the generation of excess porewater pressures. Undrained testing applies to the short term analysis of saturated cohesive soils where soil strength is represented by the undrained shear strength of the soil cu. For fully saturated soils the undrained shear strength, cu, will be independent of normal stress level, e.g., the undrained friction angle ϕ_u will be zero. The corresponding value of undrained bond strength is βcu, where β is the undrained adhesion factor. In applying these bond strengths to design, the engineer must carry out two analyses; one in terms of drained strength and the other in terms of undrained strength, to assess which is the most critical. It is vital to carry out these two types of analysis for cohesive soils since bond strength will vary with time as drainage takes place.

When testing cohesionless soils, such as sand, great care must be taken to compact samples to the dry density likely to be achieved on site. Similarly the normal stress levels used in the laboratory tests must cover the range of normal effective stresses likely to prevail on site. Any departure from these basic requirements will lead to bogus results. The interaction of dry density and normal stress level is illustrated below for a sand.

The internal angle of shearing resistance ϕ for a cohesionless soil is made up of two components:

$$\phi = \phi_{cv} + \phi_{dc}$$

The ϕ_{cv} component is the residual angle of shearing resistance obtained when the soil is shearing at constant volume. This value could be regarded as the basic frictional strength. The second component ϕ_{dc} is related to the capacity of the soil to dilate, or increase volume, during shearing. This capacity relates to the initial porosity of the soil as placed in the shear box apparatus. A low initial porosity equates to a high dry density and vice-versa. Where a soil is free to dilate during shear the initial porosity will have a huge effect on ϕ and, therefore, $\delta '$. This can be gauged from Figure 7 which shows $\phi_{cv}=32.5°$ for the

189

sand placed at high porosity (low dry density) increasing to $\phi'=42°$ at low porosity (high dry density). Clearly if all samples are not compacted at the same initial density then for the sand in Figure 7 there is a potential for a 10° difference in the results obtained.

Where cohesionless soils are well compacted to the same dry density there is the potential to achieve a high value of ϕ, or δ', through the mechanism of dilation, i.e., expansion of the soil during shearing. However, where the soil is tested at high normal stresses the ability to dilate and, therefore, generate a high ϕ, may be suppressed. This phenomenon and the problems it can lead to are demonstrated below.

Figure 8 shows the results from shear box testing carried out at normal stresses of 50, 100 and 200 kPa. A straight line failure envelope gives $\alpha c' = 20$ kPa, $\delta' = 30°$. These results are applied to a project where there is a 500mm cover of cohesionless soil giving a normal stress level of 10 kPa. The engineer concludes that the available bond strength between the soil cover and the geomembrane is (20 + 10 x tan 30°) = 25.8 kPa. Using this for his particular slope he calculates a factor of safety of 2 against the cover soil sliding, however, during construction the cover soil slips. To investigate the failure further testing is carried out at normal stresses of 50, 100, 200 and 10 kPa. Due to the effects of dilatancy the true failure envelope at low normal stress level is found to curve down through the origin, Figure 9. The true bond strength is found to have no adhesion component, e.g., $\alpha c' = 0$, and the angle of bond stress, δ', is found to be a function of normal stress level. At a normal stress level of 10 kPa, corresponding to the 500mm soil cover, the value of δ' is 45°, and the available bond strength between the soil cover and the geomembrane is found to be only (10 x tan 45°) = 10 kPa. This relates to a factor of safety of 0.8 against the cover soil sliding.

As shown above it is important to relate laboratory test soil densities and normal stress levels to those likely to prevail on site. It is equally important to define the correct drainage conditions for the laboratory tests, particularly where cohesive, e.g., clay, soils are being used. Failing this some extremely misleading results can be obtained. To illustrate the problems which can arise it is useful to briefly review some of the major factors affecting the measured strengths of cohesive soils. Clearly these factors will also affect measured values of bond strength.

As with cohesionless soils the strength of a clay will be affected by its density and consequently the densities used for laboratory samples must reflect those achieved on site. The strength of clays will also be strongly affected by moisture content. For a saturated clay there is approximately a linear log-log relationship between moisture content and undrained shear strength.

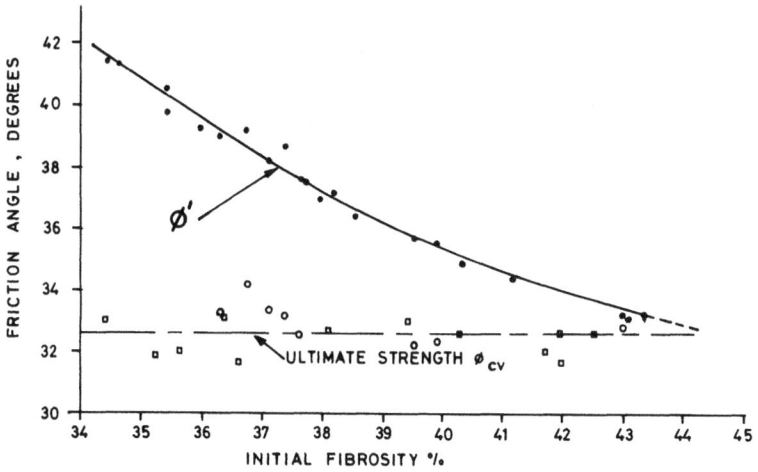

Figure 7   Strength-density relationships, Brasted sand.

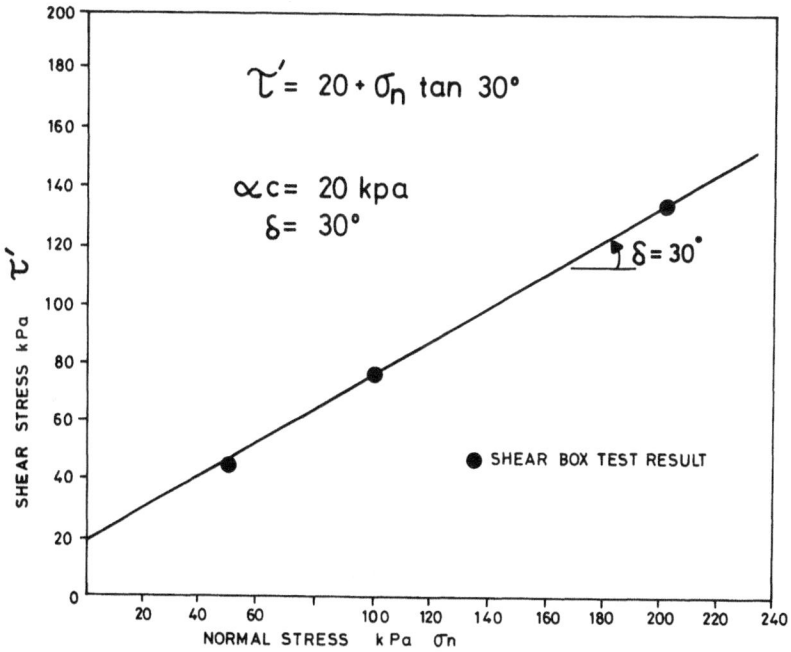

$$\tau' = 20 + \sigma_n \tan 30°$$

$$\alpha c = 20 \text{ kpa}$$
$$\delta = 30°$$

$$\delta = 30°$$

● SHEAR BOX TEST RESULT

Figure 8   Bogus Interpretation - Cohesionless Soil.

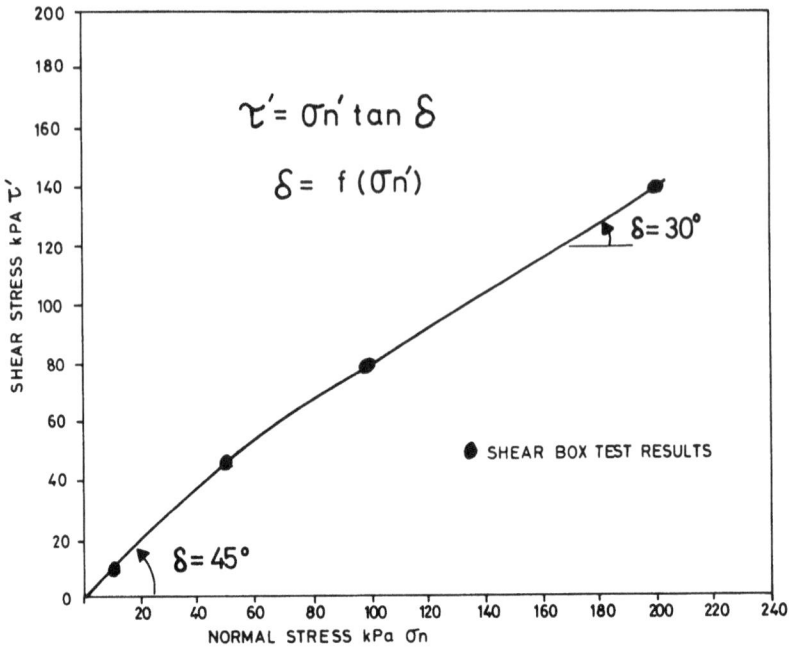

Figure 9   Correct Interpretation - Cohensionless Soil.

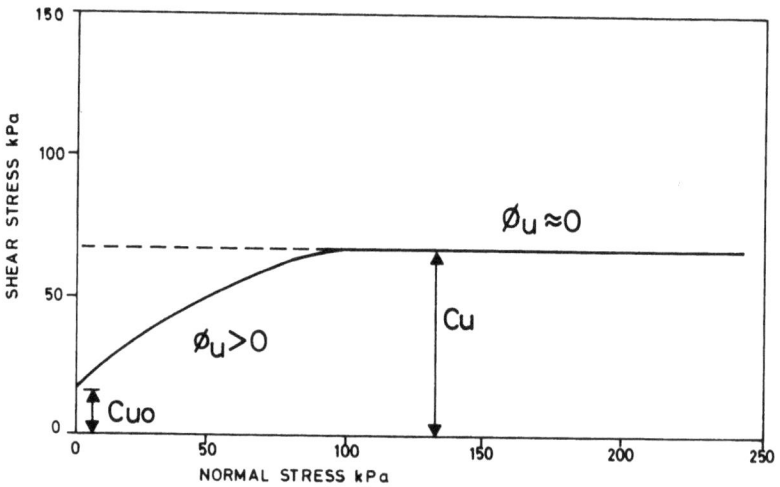

Figure 10   Undrained Strength Parameters - Cohesive Soil.

Moisture contents giving degrees of saturation less than 100% will have an important effect on measured undrained shear strength. Other important factors are the geological history of the soil. Intact normally consolidated clays will behave differently to fissured overconsolidated clays.

The total stress shear strength parameters for a cohesive soil are determined using the so called "quick-undrained" tests. In this test samples are sheared rapidly, without prior consolidation, in the shear box. A typical rate of strain for this test would be 2% per minute. If the soil is fully saturated the measured undrained shear strength, cu, should be found to be independent of normal stress level. The failure envelope is a horizontal straight line and consequently the undrained angle of shearing resistance, ϕu, is zero, Figure 10. For partially saturated soils, or heavily overconsolidated clays, the failure envelope at low normal stress levels may be curved, Figure 10. In this case ϕu $\neq$ ϕ and the undrained shear strength will be a function of normal stress level. If such a soil is tested at a high enough normal stress it will become saturated. At this point the failure envelope reverts to a horizontal straight line, where ϕu = 0.

The effective shear strength parameters for a cohesive soil, c' and ϕ', are determined using consolidate-drained tests. It is vital that the rate of strain applied during the shear stage is low enough to prevent the generation of excess porewater pressures within the soil sample. The requisite rate of shear can be calculated from the time a sample takes to consolidate under the applied normal stress. Rates of strain will vary with soil type but will be much lower than those used in the undrained test. Whereas an undrained test may have a shear stage lasting less than half-an-hour, the shear stage on a drained test may run to several days if not weeks. The failure envelope obtained will be affected by the geological history of the soil and, in the case of compacted fills, the compactive effort. For normally consolidated soils the effective cohesion intercept, c', will be zero or very small. The failure envelope is usually linear and so defines a unique value of ϕ'. For heavily overconsolidated soils c' will have a non-zero value and the failure envelope may be curved so giving a ϕ' value which is a function of normal effective stress.

Drained and undrained tests are distinctly different. For clays they give two distinctly different sets of shear strength parameters. The undrained test defines cu, and ϕu if the soil is unsaturated. The drained test defines c' and ϕ'. Where soil-geomembrane bond is being measured the corresponding values are βcu, δu for undrained tests, and αc', δ' for drained tests. In a design problem stability must be checked twice. Once to determine short term stability using the undrained parameters and once to determine long term stability using the drained parameters. Whichever gives the lowest factor of safety will be the most

critical.   It does not follow that the long term will always
be the most critical and this is often the explanation for failures
occurring during construction.

Extremely misleading test results can be obtained when the test
regime   is   neither   strictly   unconsolidated-undrained   or
consolidated-drained.   Intermediate conditions often arise where
a test has been run at a rate of shear too low for the undrained
condition yet not low enough to give a fully drained condition.
Although values of adhesion and friction angle can be obtained
they are meaningless.   Design application of the values obtained
can give completely bogus factors of safety, especially if the
intermediate values are wrongly assumed to be the true drained
values.   This is illustrated in Figure 12 where it can be seen
that intermediate values of cohesion and friction angle fall
between those derived for the truly drained and truly undrained
tests.

In specifying design testing using the shear box there will be
many other factors which will affect the measured soil-geomembrane
bond strength.   Care must be taken to control sample size effects
by choosing a shear box size compatible with the maximum particle
size of the soil being tested.   As a rule of thumb the maximum
soil particle size should not exceed one-eighth of the maximum
depth of the soil sample.   For a 300mm x 300mm shear box this
puts an upper limit of about 10mm on the maximum particle size
which can be tested with a good degree of reproducibility.   Clearly
the results will also be affected by the manner in which the
geomembrane sample is mounted in the shear box, Figure 6.   Since
geomembranes are made from visco-elastic materials, test temperature
and test duration will also affect the results.

No clear cut advice can be given on these parameters since they
will vary from site to site and application to application.
At all times the fundamental principles of soil mechanics must
be applied to both testing and design.   Over and above this the
test must reflect as closely as possible conditions applying
on site, particularly soil densities, normal stress levels,
temperatures, rates of loading and, of course, the nature and
form of material in contact with the geomembrane.

All of the above comments have been made in reference to direct
shear testing using the conventional shear box as opposed to
pull-out testing.   One huge advantage of the shear box test is
that the soil moves relative to the geomembrane over the whole
contact area.   This means that the area of the geomembrane resisting
the applied shear force is acurately known and, therefore, average
bond stresses can be calculated with a degree of confidence.

The same argument does not apply to pull-out tests on extensible
materials such as geomembranes.   When a geomembrane is buried
in soil and a pull-out force is applied to the free end there

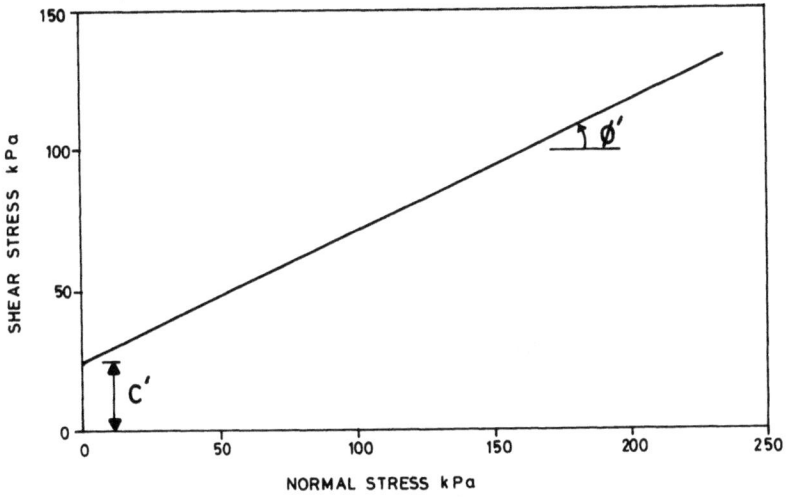

Figure 11   Drained Strength Parameters - Cohesive Soil.

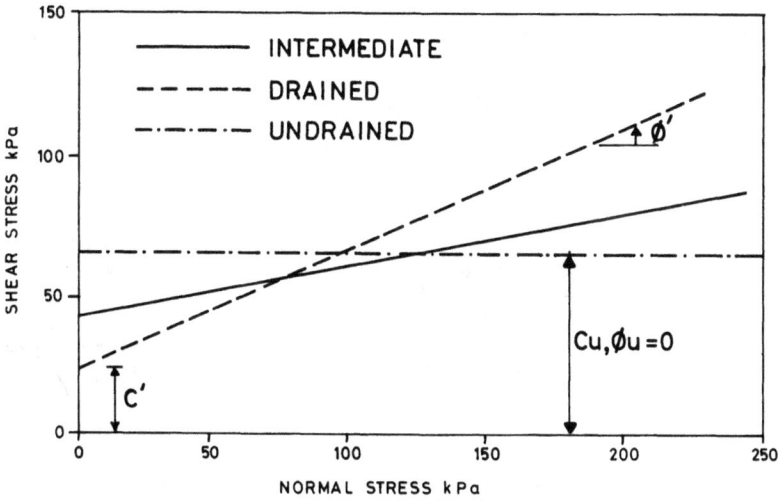

Figure 12   Influence of Drainage Conditions - Cohesive Soil.

is no indication of how that pull-out force is being resisted along the embedded length of the geomembrane. Consequently there is no basis on which to analyse such a test <u>unless</u> the embedded length is instrumented. A good analogy is to a railway engine coupled to several carriages by extensible springs. If the springs are very flexible the railway engine could have left the station before the last carriage even moves!

Although the pull-out test is difficult to analyse it does give a good qualitative check on how materials might behave under pull-out. One approach to using this test is to first calculate the theoretical pull-out resistance of a sample, based on bond strength parameters measured in the shear box, embedded area and normal stress level. This theoretical value can then be compared with that actually obtained in the pull-out test.

8.7   Index Testing

The purpose of index testing is not to produce results or data to be used in design. Rather it is very simple testing in which many of the test variables are closely controlled in order to give a high degree of reproducibility in the results. If this is achieved then a meaningful comparison can be made between the results derived using different geomembranes.

A draft of a suitable index test specification is given in Appendix 1. The test is based on use of the direct shear apparatus with the geomembrane placed in the lower half of the shear box. To ensure that the upper surface of the geomembrane is flat, and aligned with the shear plane defined by the two halves of the shear box, the geomembrane sample is mounted on a rigid substrate which can be placed in the lower half of the shear box.

The geomembrane is sheared against a dry uniformly graded sand which is used to fill the upper half of the shear box. The sand is used dry to circumvent problems which might arise from different laboratories using different compaction moisture contents. The grading of the sand is chosen so that it is difficult to compact the sand to too high a density. The reason for this is to encourage the sand to shear at constant volume, or at least low rates of dilatancy. The objective of this is to produce pairs of values of shear stress and normal stress which define a linear failure envelope passing through the origin. This is an attempt to supress a curved failure envelope which can lead to difficulties in interpretation.

To give a check on the absolute values of angle of bond stress, δ', obtained the test specification also requires measurement of the internal angle of shearing resistance, ϕ', of the standard sand. No value of δ' can ever exceed ϕ', e.g., $\delta' \not> \phi'$. This equality proves useful in testing drainage grids or meshes where careless testing or interpretation can lead to apparent angles of bond stress higher than the internal angle of shearing resistance of the sand, e.g., $\delta'>\phi$.

8.8 Conclusions

In the context of geomembranes, and related products, there are two broad reasons for testing. The first is to allow a meaningful comparison of different products. The second is to determine parameters to be used in design. Almost by definition this requires two different test specifications since in one case the prime variable is the geomembrane and in the other it is the site specific parameters which are varied. This is the essence of the difference between index testing and design testing. Whilst index testing may be limited to laboratory testing, design testing should not. Usually the limiting factors in design testing are budget and programme. Where these permit then design testing should not be limited to the laboratory. More decisive results are usually obtained by adhoc testing on site. This is systems testing rather than laboratory design testing, however, it does allow fuller modelling of certain site specific conditions such as normal stress levels, nature of formation and, of course, the multi-layers which may be used in a geomembrane lining system.

Although this Chapter is not dedicated to design, it must be remembered that in complex systems the art of design, or systems, testing is to determine minimum bond strength parameters at the most critical interface.

8.9 References

Akbert, S.Z., Hammamji, Y., & Lafleur, J. 1985
    "Frictional characteristics of geomembranes, geotextiles and geomembrane-geotextile composites". Proc.Second Canadian Symp.on Geotextiles and Geomembranes, Edmonton, Alberta. pp.209-217.

Andrawes, K.Z., McGowan, A., & Wilson-Fahing, R.F. (Undated).
    "Measurement of the soil friction properties of geotechnical fabrics". Report issued by Dept.Civ.Engrg., Strathclyde University.

Brandt, J.R.T. 1985.
    "Behaviour of soil-concrete interface". Ph.D. Thesis, Dept. of Civil Engrg. University of Alberta.

Ingold, T.S. 1982.
    "Some observations on the laboratory measurement of soil-geotextile bond". ASTM Geotechnical Testing Journal, Vol.5, No.3/4, pp.57-67.

Koerner, R.M., Martin, J.P., & Koerner, G.R. 1986.
    "Shear strength parameters between geomembranes and cohesive

soils". Int.Jrnl. Geotextiles & Geomembranes, Vol.4, No.1, pp.21-30.

Martin, J.P., & Koerner, R.M. 1985.
    "Geotechnical design considerations for geomembrane lined slopes : Slope Stability". Int.Jrnl. Geotextiles & Geomembranes, Vol.2, No.4, pp.299-321.

Martin, J.P., Koerner, R.M., & Whitty, J.E. 1984.
    "Experimental friction evaluation of slippage between geomembranes, geotextiles and soils". Proc.Int.Conf. on Geomembranes, Denver, Vol.1, pp.191-196.

Myles, B. 1982.
    "Assessment of soil-fabric friction by means of shear". Proc.Second Int.Conf. on Geotextiles, Las Vegas, Vol.3, pp.787-791.

Richards, E.A., & Scott, J.D. 1985.
    "Soil geotextile frictional properties". Proc.Second Canadian Sypm. on Geotextiles & Geomembranes, Edmonton, Alberta, pp.13-24.

Weiss, W. & Batereau, C. 1987.
    "A note on Planar Shear between geosynthetics and construction materials". Int.Jrnl. Geotextiles & Geomembranes, Vol.5, No.1, pp.63-67.

Williams, N.D., & Houlihan, M.F. 1987.
    "Evaluation of interface friction properties between geosynthetics and soils". Proc.Geosynthetics 87, IFAI, New Orleans, pp.616-627.

A8.1 Scope

This specification describes an index test method to be used to determine the frictional properties of geomembranes in contact with a standard sand under standard loading conditions. The results obtained define the index properties of geomembranes which are intended to be used to make initial comparisons between the frictional characteristics of different geomembranes. Specialised testing is required to determine frictional, or cohesive, parameters to be used in design.

A8.2 Introduction

The magnitude of the frictional shear stress developed between a geomembrane and a noncohesive soil is defined by the product of the normal effective shear stress, σ_n', acting on the system and the peak angle of bond stress, δ'. The magnitude of δ' will be a function of many variables including the loading regime, soil type and geomembrane type. The following specification fixes all these variables save the geomembrane type and consequently any variation in the measured value of δ' is related to the geomembrane type. The laboratory test method prescribed uses direct shear box testing to determine the magnitude of the angle of bond stress, δ', for a given geomembrane, under standard conditions, Values of δ' obtained for different geomembranes are index properties which may be used to make a direct comparison of the frictional properties of these geomembranes. Index properties are not likely to be appropriate to design.

A8.3 Definitions and Symbols

N   applied normal force (kN)

F   applied shear force (kN)

A_o initial area of slip surface (m²)

A   corrected area of slip surface (m²)

σ_n' applied normal effective stress (kPa)

τ_f maximum stress at $_n'$ (kPa)

ϕ' effective internal angle of shearing resistance for standard sand (°)

δ' effective angle of sand/geomembrane bond stress (°)

A8.4 Test Method

8.4.1 Preparation of geomembrane samples
       (a)   Number of tests
Tests are carried out on a standard sand alone and on the geomembrane in contact with the standard sand. A minimum of three tests are carried out on the standard sand alone with each test undertaken at the specified normal stresses. For tests with the geomembrane in contact with the standard sand the tests are carried out according to the geomembrane structure with a minimum of three tests each carried out at the specified normal stresses. The number of testing directions required will depend on the type of geomembrane. For a material with an isotropic surface on both sides it is sufficient to test with shear applied in the machine direction on the obverse side of the geomembrane. Conversely if these are different surface treatments on the obverse and reverse sides of the geomembrane and these differ in the machine and transverse-machine directions, it is necessary to test in four directions.

(b)   Sample selection
Geomembrane samples are to be taken in compliance with Rilem

Recommendation TC.103 MHG. A sufficient number of samples shall be taken to serve the requirements of 8.4.1(a).

(c)   Conditioning
Geomembrane samples shall be tested dry at a temperature and humidity specified in Rilem Recommendation TC 103 MHG.

8.4.2 Preparation of the Sand
The sand used for index testing shall be uniformly graded sand complying with Rilem Recommendation TC 103 MHG. It shall be dry and not have been used for more than twenty successive tests.

8.4.3 Apparatus
The geomembrane and standard sand shall be tested in direct shear using a conventional shear box apparatus.

8.4.4
(a) Number of tests
A minimum of two series of tests shall be carried out with each series of tests comprising three tests. Each of these three tests is carried out at the prescribed normal load intensities. The first series of three tests is used to determine the frictional properties of the standard sand alone. Dry sand is poured into the upper and lower halves of the shear box and compacted in the manner described below. To determine the frictional properties of the geomembrane in contact with the standard sand the geomembrane in contact with the standard sand, the geomembrane is mounted on a rigid block which is placed in the lower half of the shear box. The upper half of the shear box is filled with standard sand in the prescribed manner. One series of three tests at the prescribed normal load intensities is to be carried out for each test direction described in 8.4.1(a).

(b) Placing of geomembrane sample
For each test direction three samples are cut to the plan dimensions of the shear box to give a sliding fit. Each sample is then mounted on a rigid block as shown in Figure A.1. For thick geomembranes or geomembranes with a high flexural stiffness, the method shown in Figure A1(a) is preferred. This involves bonding the geomembrane sample to a disposable rigid wooden block using an adhesive. The adhesive employed should be chemically compatible with the material of the geomembrane and set rigid. The edges and upper surface of the geomembrane shall be free of any adhesive. The alternative method shown in Figure A1(b) may be employed to mount thin or flexible geomembranes. The total thickness of the geomembrane and mounting block shall be less than the depth of the lower half of the shear box. The mounted sample is placed in the lower half of the shear box and packed from beneath with metal shims such that the top surface of the geomembrane is securely supported to be flush with

the top edges of the lower half of the shear box. The upper
half of the shear box is placed into position and secured
by the locating screws which must subsequently be removed
before commencement of the shear stage.

(c) Placing of standard sand
Where tests are carried out to determine the frictional
properties of the sand alone, both the upper and lower halves
of the shear box are filled with sand. Where tests are
carried out to determine the frictional properties of the
geomembrane/sand only, the upper half of the shear box is
filled with sand. A known weight of dry sand is poured
into the shear box and lightly rodded. The surface of the
sand shall be made level and the thickness of the sand
determined. A normal stress of 100 kPa is applied and
maintained until any vertical compression ceases. The
magnitude of any such compression is recorded and used to
determine the thickness of the sand and consequently its
dry density.

(d) Normal Load
Fresh samples of sand and geomembrane shall be prepared
in the manner described above for each normal load. Normal
loads shall be applied to produce normal stress levels of
25 kPa, 50 kPa and 100 kPa. The shear box locating screws
should be removed and a zero reading taken on the vertical
movement indicating device in readiness for the shear stage.

(e) Shear Stage
The shear force shall be applied at a constant rate of
displacement equal to 1% per minute of the shear box size,
e.g., 0.5mm/min for a 50mm square shear box or 1.00m/min
for a 100mm square shear box. Readings of vertical
displacement and horizontal shear force shall be taken at
regular time intervals not exceeding 1 minute after
commencement of the shearing stage. Shearing shall continue
until there is no further increase in the applied horizontal
shear force. At the end of each shear stage the geomembrane
sample shall be recovered and a note made of any damage
to the upper surface of the geomembrane. Any failure in
the bond between the geomembrane and the mounting block
renders the test invalid and requires repetition of the
test.

A8.5 Calculation

The effective normal stress, σ_n', shall be obtained by dividing
the normal applied force, N, by the initial area of the slip
surface which may be taken as the area of the sample in the shear
box.

$$\sigma_n' = \frac{N}{A_0} \text{ (kPa)}$$

The maximum shear stress, τ_f, is calculated by dividing the applied shear force, F, by the corrected area of the slip surface.

$$\tau_f = \frac{F}{A} \text{ (kPa)}$$

For each set of three tests a plot is prepared of pairs of maximum shear stress, τ_f, on the vertical axis against effective normal stress, σ_n', on the horizontal axis using the same scale for each axis. A best fit linear failure envelope passing through the origin is drawn through the points obtained. The slope of each failure envelope is measured, or calculated, in degrees from the horizontal. For tests involving sand alone, this result is recorded as ϕ' the effective internal angle of shearing resistance for the standard sand. For tests involving geomembrane samples in contact with sand, this result is recorded as δ' the effective angle of sand/geomembrane bond stress.

A8.6 Report

The report shall include:

8.6.1.   A description of the apparatus used, including dimensions and the method adopted for mounting the geomembrane sample.

8.6.2.   A description of the geomembrane including structure, thickness and polymer type, together with a note of test direction and surface finish in this direction.

8.6.3.   The dry density of the sand for each test.

8.6.4.   The rate of displacement (mm/min).

8.6.5.   Plots of shear stress and vertical displacement against horizontal displacement for each normal stress level and each test. Plots of τ_f -v- σ_n' for each series of tests.

8.6.6.   A description of each geomembrane sample at the end of each test.

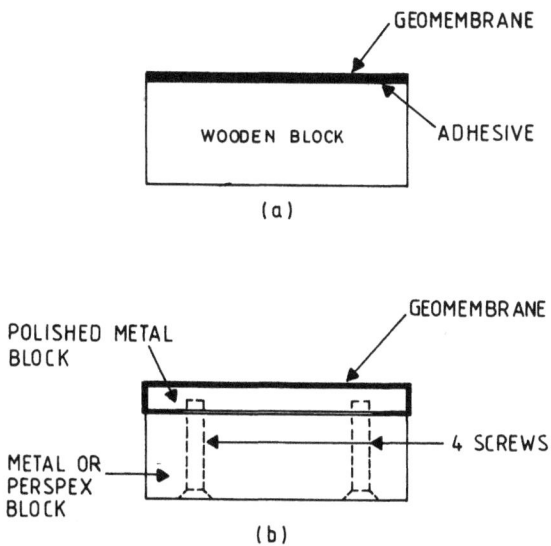

Figure A1 : Approved methods for maintaining goemembrane sample.

9 GEOMEMBRANE ANCHORAGE BEHAVIOR USING A LARGE-SCALE PULLOUT APPARATUS

R. M. KOERNER and M. H. WAYNE
Geosynthetic Research Institute, Drexel University,
Philadelphia, PA, USA

9-1 Introduction

For the subsurface storage and containment of various solid and liquid materials a geomembrane liner generally extends out of an excavation and into an anchorage extending around the periphery of the site. A designer or installer usually chooses from one of the following types of configurations, see Figure 1.

- Horizontal anchor
- Shallow "V" anchor
- Horizontal and vertical (trench) anchor
- Horizontal and vertical (trench) anchor with return
- Concrete anchor block

In the selection and dimensioning of these different configurations a number of variables are involved. They include, berm and cover soil (surcharge) type and related index properties, cover soil (surcharge) thickness, slope angle (β) of exiting geomembrane, geomembrane type, thickness and related properties, and the anchorage geometry and dimensions. Since several of these items are determined well in advance of the anchorage design, it is essential that the remaining items, e.g., surcharge thickness and anchorage geometry, be evaluated by design rather than by estimation or empirical relationships. This chapter focuses on the use of a large scale anchorage apparatus to determine anchorage behavior of geomembranes and then illustrates the resulting design implications.

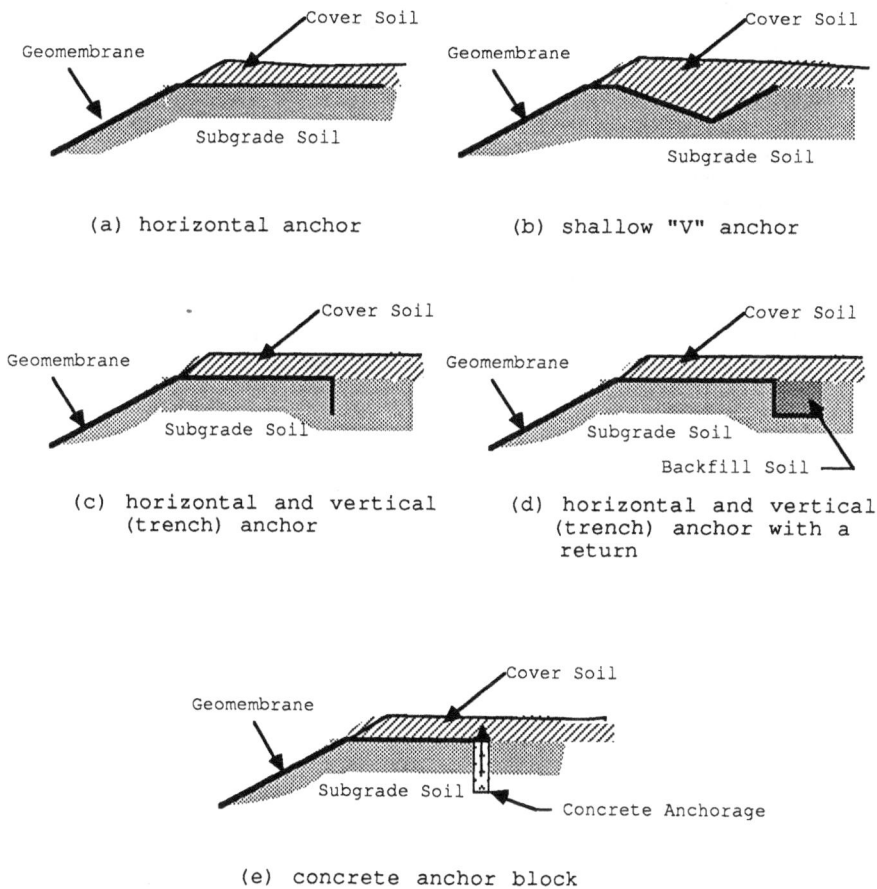

(a) horizontal anchor

(b) shallow "V" anchor

(c) horizontal and vertical
(trench) anchor

(d) horizontal and vertical
(trench) anchor with a
return

(e) concrete anchor block

Fig. 1 Various types of geomembrane anchorage configurations.
After Richardson and Koerner (1987) and Koerner (1990)

9-2 Background and related studies

Well established in the geosynthetics literature is the
anchorage testing and evaluation of geogrids and geotextiles. With these materials there are two limiting modes
of anchorage resistance, i.e. shear and pullout, Milligan
(1988). For geogrids, a limited amount of shear resistance
is gained along the rib surfaces, but the large apertures
allow for soil strike-through and a significant bearing
capacity against the transverse ribs. For geotextiles, the
surface shear against the fabric is much more of a contri-
butor since soil particle interaction can only occur to a
limited extent, and only with the fine soil fraction. In
contrast to geogrids and geotextiles, geomembranes (having
no open voids or rough surfaces) will provide anchorage
resistance completely from shear along the upper and lower
surfaces.
   Relavent information gained from the anchorage testing of
geotextiles and geogrids performed by Ingold (1982, 1983)
Tzong and Cheng-Kuang (1987) and Swan (1987) indicates that
both types of materials exhibit a nonuniform strain distribution pattern. These tests clearly indicate that during
pullout, frictional resistance is not fully mobilized along
the entire length of the embedded material. As will be
presented in the sections which follow, it will be shown
that geomembranes in an anchorage mode respond to external
stresses in a similar manner. Factors which influence the
resulting stress mobilization pattern include the soil type,
physical properties of the geomembrane considered (i.e.,
thickness, dimensions, hardness, surface texture, etc.),
loading conditions (i.e., normal stress), and exit angle of
the geomembrane. The full details are found in the thesis
of Wayne (1989).

9-3 Anchorage test setup and related details

By their very nature geosynthetic anchorage tests are
performance tests and require very large modeling facilities. Thus to avoid scale effects the test containers are
typically more than one meter in size in each direction.
Geomembrane anchorage testing should be no exception. The
test container used in these experiments measured 91 cm by

107 cm by 189 cm long. It was made as a heavy wooden box with a steel framework surrounding it so as to provide adequate reinforcement. As seen in Figure 2, the geo-membrane exited the box at a controlled angle (β) near to the center of the smallest side. The geomembrane under test extended into the container where it was confined with soil on both sides and could be further loaded via an air bag located inside the upper box surface. Thus varying exit angles, normal (surcharge) stresses, and geometries could be evaluated as pullout stress was mobilized and monitored from outside of the box.

To determine how strains are mobilized within the con-fined geomembrane itself, 50 mm long strain gages were bonded to the geomembrane under test at the locations shown in Figure 3. The attached gages were protected from damage by the adjacent soil by using a small piece of needle-punched nonwoven fabric and a small piece of HDPE geo-membrane as a cushion and slip-layer respectively. Separate wide-width tensile tests on the same type of geomembranes were made to obtain a calibration curve to convert the measured strains into stresses.

The geomembrane types evaluated in this study were as follows:

• 1.5 mm thick high density polyethylene (HDPE)
• 0.9 mm thick reinforced chlorosulfonated polyethylene (CSPE-R)
• 1.0 mm thick very low density polyethylene (VLDPE)
• 0.5 mm thick polyvinyl chloride (PVC)

All tests were performed with a subgrade and cover soil consisting of a well graded coarse to fine concrete sand which was placed dry at a relative density of 92 ± 1%. Each of the above listed geomembranes were then tested under separate surcharge pressures of 5, 17 and 36 kPa and at an inclination angle (β) of 45° to the horizontal in a con-figuration as shown in Figure 1(c). To determine the influence of inclination angle on anchorage behavior, sub-sequent tests were performed using the HDPE geomembranes at β angles of 30°, 15°, and 0°.

The external stress was generated by means of a jack attached to a load frame which was capable of stressing the geomembrane at a controlled slope (or "β" angle) ranging from 0° through 45°. The test specimens were stressed until they failed outside of the box or until the geomembrane slid out of the container.

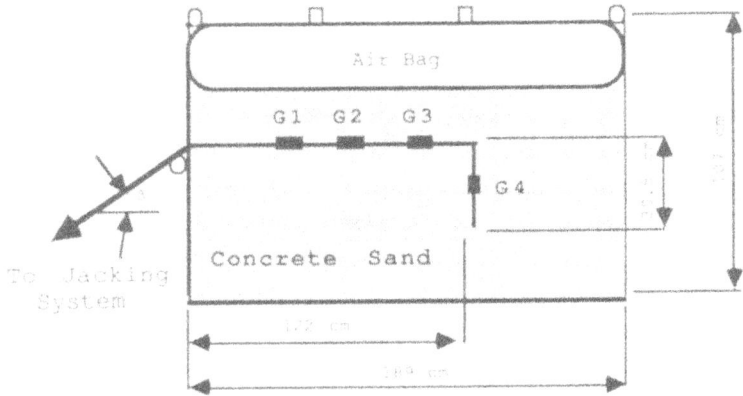

Fig. 2 Photograph and schematic diagram of geomembrane
anchorage box used in this study

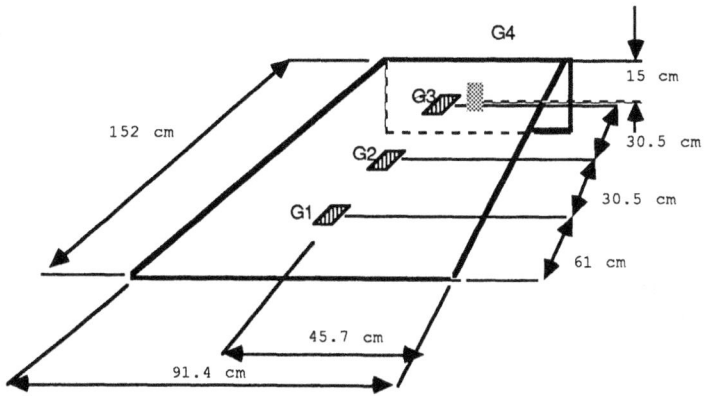

Fig.3 Typical strain gage installation on geomembrane test
       sample and wide width calibration for strain-to-stress
       conversion

9-4 Test results

Information obtained from the various tests includes the external force in the geomembrane (T) which is mobilized by the jack and strain gage readings for each of the gages installed at the various locations indicated in Figure 3. Table 1 summarizes the results obtained from the fifteen separate tests which were performed, and lists relevant comments regarding each experiment.

Table 1   Maximum External Stress and Corresponding Values of Strain Along Geomembrane

Test No. and Geo-membrane	Surcharge Pressure (kPa)	β (deg)	T (kN/m)	Strain (%) At Location				Comments on test conclusion
				G1	G2	G3	G4	
1 CSPE-R	5	45	8.1	4.71	3.50	1.65	0.50	slippage
2 CSPE-R	17	45	19.4	n/a	11.50	5.28	0.00	break
3 CSPE-R	36	45	25.0	n/a	5.54	0.73	0.00	break
4 VLDPE	5	45	5.6	5.29	0.94	0.09	0.00	ultimate
5 VLDPE	17	45	4.9	4.13	0.79	0.11	0.00	ultimate
6 VLDPE	36	45	9.6	1.34	0.15	0.04	0.00	ultimate
7 PVC	5	45	.9	6.79	1.40	0.14	0.00	ultimate
8 PVC	17	45	3.2	6.79	0.33	0.00	0.00	ultimate
9 PVC	36	45	3.2	2.00	0.05	0.00	0.00	ultimate
10 HDPE	5	45	10.5	1.98	n/a	1.25	0.36	slippage
11 HDPE	17	45	18.6	2.72	2.02	1.15	0.32	slippage
12 HDPE	36	45	27.5	8.42	4.38	2.27	0.22	yield
13 HDPE	36	30	27.5	9.65	5.04	2.42	0.09	yield
14 HDPE	36	15	26.6	9.86	5.15	2.23	0.11	yield
15 HDPE	36	0	25.0	10.90	5.06	1.87	0.16	yield

Note: n/a = not available

The values of strain in Table 1 were converted to stress via the wide width calibration tests and then made into a ratio of the strength of the particular geomembrane involved. Note that the scrim reinforcement breaking strength value was used for CSPE-R, the ultimate strength was used for VLDPE and PVC, and the yield strength was used for HDPE. Curves relating this mobilization of stress as a function of the distance of geomembrane embedment at each applied surcharge pressure are shown in Figures 4 through 8.

The CSPE-R response shown in Figure 4 mobilizes gradually diminishing stress from the front of the box throughout the geomembrane when adequately confined. For the lowest confinement pressure however, the geomembrane actually slid out of the box after an initial, but low, value of tensile stress was mobilized.

The VLDPE responses of Figure 5 and the PVC responses of Figure 6 are quite similar to one another. Both geomembrane

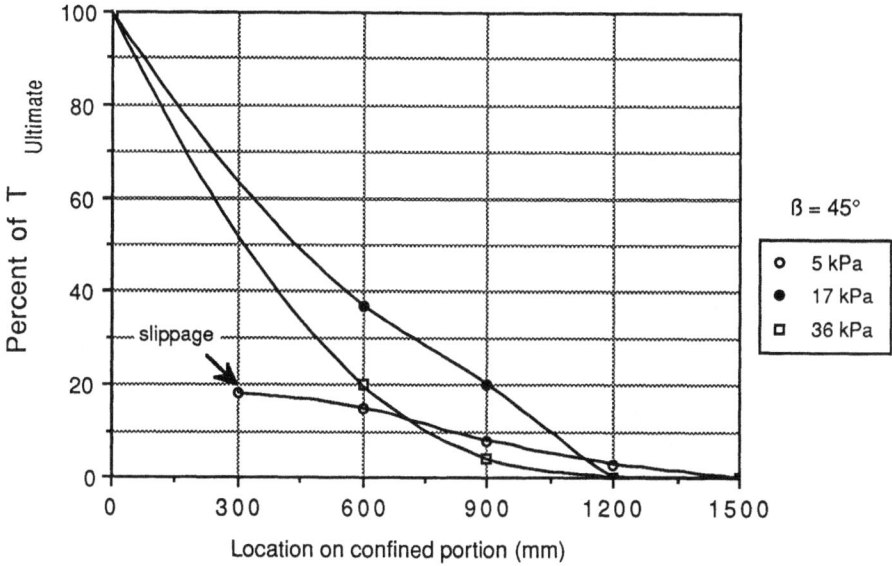

Fig. 4 Percent of ultimate versus location for 0.9 mm CSPE-R

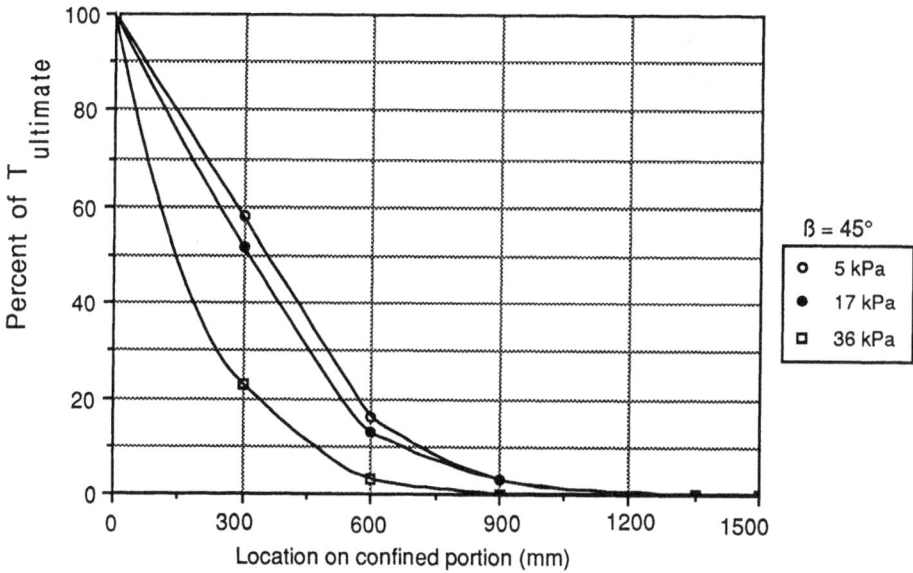

Fig. 5 Percent of ultimate versus location for 1.0 mm VLDPE

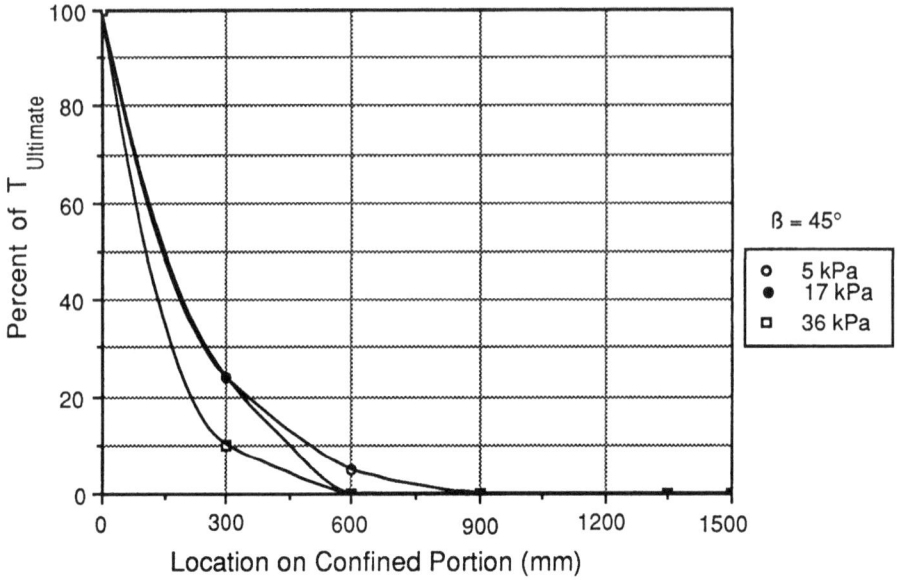

Fig. 6 Percent of ultimate versus location for 0.5 mm PVC

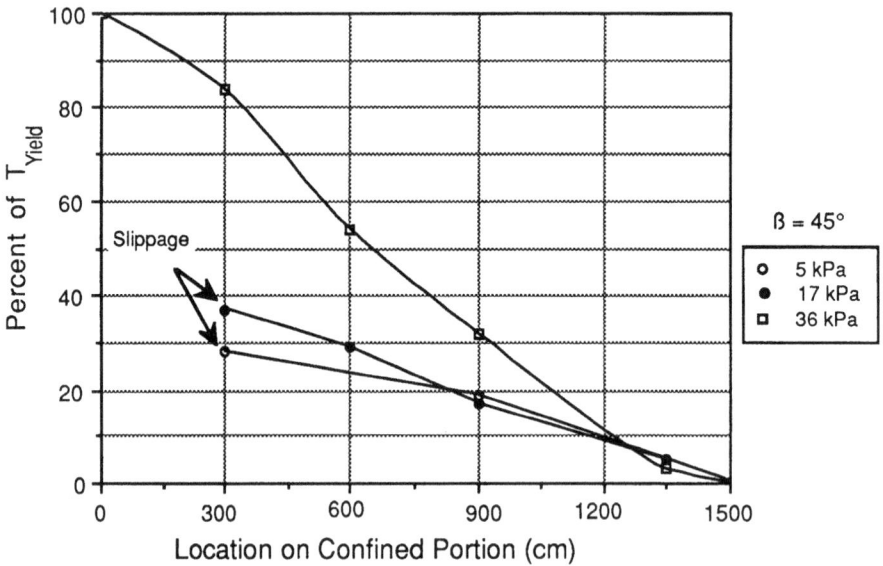

Fig. 7 Percent of yield versus location for 1.5 mm HDPE

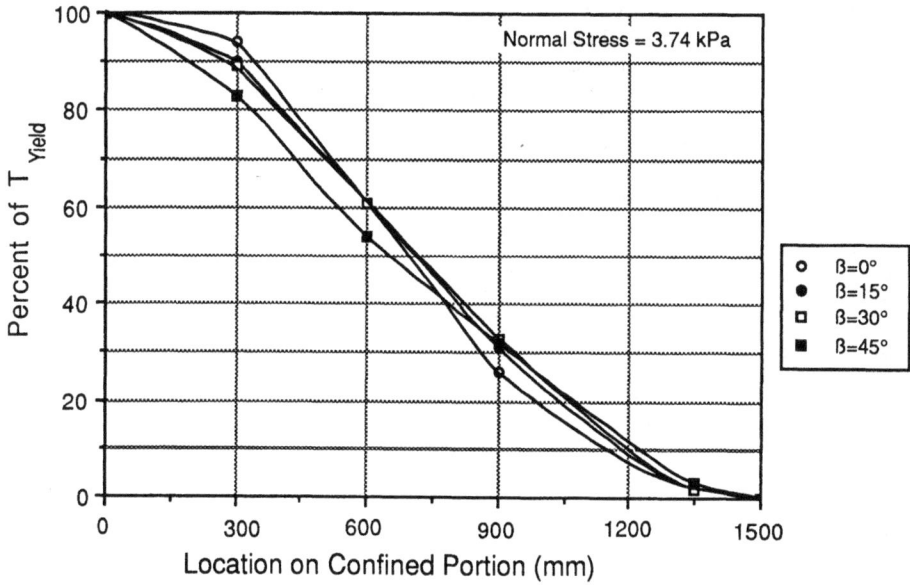

Fig. 8 Percent of yield versus location for 1.5 mm HDPE

types take up stress very rapidly near the front of the box.
The stress dissipates rapidly with only minor propagation
back into the box as the geomembrane deforms under relatively low tensile stresses. High values of normal pressure
accentuates the phenomenon considerably.

The HDPE responses are shown in Figure 7 in which
slippage is seen to occur at the lower two values of normal
pressure. At the highest normal pressure, however, the
response is felt completely through the geomembrane in an
approximately linear fashion.

Figure 8 reflects stress mobilization as a function of
geomembrane exit inclination, or "β" angle for the HDPE
geomembranes. An indicated by this figure, these results
reveal that there is a greater stress transfer at the front
of the box for an angle of 0° than at 45°. However, the
variation between different "β" angles is not significant.

9-5 Design implications

Given the information obtained through the anchorage testing
of the four different geomembrane types, it is possible to
determine factor of safety (FS) values as they apply to
current anchorage design. For the experimental setup, shear
forces are mobilized both above and below the geomembrane
because of the applied surcharge pressure and the vertical
component of external pullout force. Using sum of the
forces in the horizontal direction from the free body
diagram of Figure 9 leads to the following relationship:

$$T \cos \beta = T_{Uq} + T_{Lq} + T_{LT} \qquad\qquad (1)$$

where

T = geomembrane stress external to the box
β = geomembrane exit angle
T_{Uq} = anchor trench force above geomembrane due to
surcharge pressure
T_{Tq} = anchor trench force below geomembrane due to
surcharge pressure
T_{LT} = anchor trench force below geomembrane due to
external force component in vertical direction

Since these horizontal forces within the anchor trench area
are mobilized by friction, equation (1) results in the following:

$$T \cos \beta = q \tan \delta_U \,(L_{RO}) + q \tan \delta_L \,(L_{RO}) + \frac{1}{2}\,T \sin\beta \tan \delta_L \qquad (2)$$

where

T = geomembrane stress external to the box
β = geomembrane exit angle
q = surcharge pressure (above and below geomembrane)

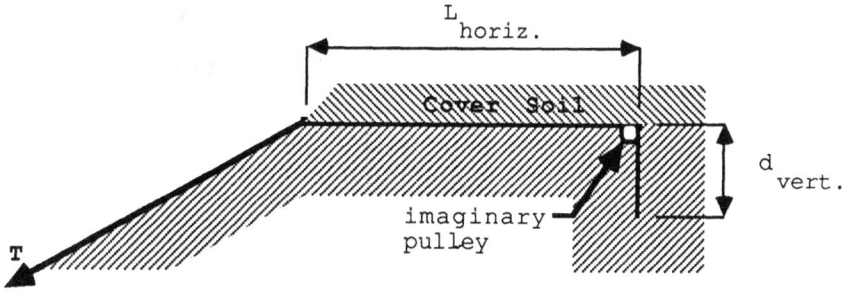

(a)　anchorage geometry

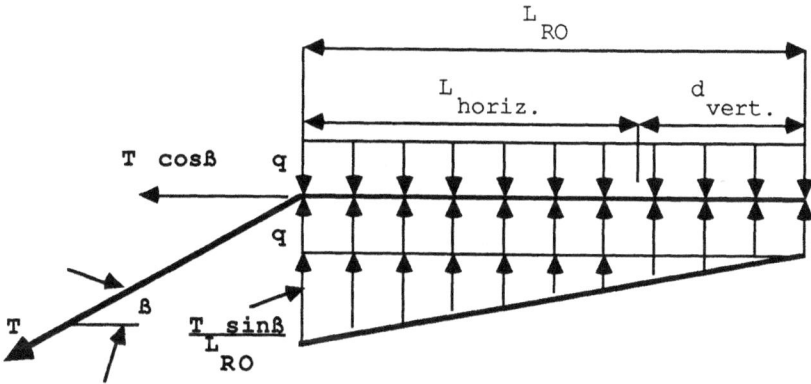

(b)　equivalent anchorage geometry and associated forces

Fig. 9　Anchorage geometry and associated force diagram for theoretical development and design purposes.

δ_U = soil-to-geomembrane friction angle on the upper surface of the geomembrane

δ_L = soil-to-geomembrane friction angle on the lower surface of the geomembrane

L_{RO} = runout length (including the anchor trench depth)

Since the same soil type was used above and below the geomembrane $\delta_U = \delta_L = \delta$. The actual value for the different geomembrane types was found by direct shear testing as per ASTM D-3080. The above equation can be reduced to the following:

$$L_{RO(req'd)} = \frac{R\ (\cos\beta - 0.5\ \sin\beta\ \tan\delta)}{2q\ \tan\ \delta} \qquad (3)$$

By comparing $L_{RO(req'd)}$ in the above equation to $L_{RO(act)}$ in the experimental device, a factor of safety can be calculated for these particular experiments and the configuration used therein, i.e.,

$$FS = \frac{L_{RO(act)}}{L_{RO(req'd)}} \qquad (4)$$

Note that this factor of safety is on the geomembrane's ultimate strength and not on its ability to pullout of the anchorage system. Using the above equations versus the actual runout lengths in the experimental device one arrives at Table 2.

Table 2   Factor of safety values for laboratory anchorage simulation tests.

Material	t (mm)	T (kN/m)	Surcharge (kPa)	δ (deg)	β (deg)	$L_{RO(req'd)}$ (mm)	$L_{RO(act)}$ (mm)	FS Value
CSPE-R	0.9	19.4	17	27.5	45	580	1520	2.6
	0.9	25.0	36	27.5	45	350	1520	4.3
VLDPE	1.0	4.9	17	26	45	160	1520	9.5
	1.0	9.8	36	26	45	150	1520	10
PVC	0.5	0.9	5	28	45	90	1520	17
	0.5	3.2	17	28	45	90	1520	17
	0.5	3.2	36	28	45	40	1520	38
HDPE	1.5	25.0	36	25	0	750	1520	2.0
	1.5	26.6	36	25	15	720	1520	2.1
	1.5	27.5	36	25	30	620	1520	2.5
	1.5	27.5	36	25	45	450	1520	3.4

Here it can be seen that the factor of safety values exactly replicate the test results in that very large FS values of 9.5 to 38 are obtained for geomembranes that easily deform and take stress rapidly, i.e. VLDPE and PVC. In other words, the anchorage length in these experiments was very excessive. Conversely, geomembranes with high modulus and tensile strength distribute the pullout stress back into the anchorage system and result in low FS values, i.e., 2.0 to 4.3. Therefore, the anchorage length in these experiments was reasonable and resulted in appropriate factor of safety values.

9-6 Summary and conclusions

This chapter has presented details of a geomembrane anchorage test and in so doing has illustrated the behavior of four types of polymeric geomembranes, CSPE-R, VLDPE, PVC, and HDPE. Each were evaluated under a number of surcharge pressures in a large scale laboratory pullout box. Strain gages on the geomembranes allowed for determination of the stress distribution as the geomembrane extended into the confining facility.

It was clearly evident that the low modulus- low strength geomembranes (like VLDPE and PVC) mobilize their full tensile strength over very short distances with respect to the applied pullout stress. Distances as small as 40 to 160 mm are adequate to mobilize full geomembrane tensile strength. Conversely, high modulus- high strength geomembranes (like CSPE-R and HDPE) mobilize their full tensile strength over large distances with respect to the applied pullout stress. Distances in these cases are from 350 to 750 mm.

In conclusion, the tests have shown that for the usual case of a meter long runout distance with an additional 300 mm of anchor trench, all of the geomembranes evaluated in this study will have their ultimate strength exceeded before pulling out of their anchorages. However, they do so at significantly different distances. The test device, while burdensome due to its size and complexity, seems appropriate for a performance test to evaluate the anchorage behavior of geomembranes. It is felt that a wide variety of geomembrane types, geometric configurations, surcharge pressures, soil conditions, etc., can be evaluated using this type of test device, procedure and analysis methodology.

9-7 Acknowledgements

This project is funded by the U.S. Enviromental Protection Agency under Project No. CR-813953-02. Our sincere appreciation is extended to the Agency and in particular to the Project Officer, Robert E. Landreth.

References

Ingold, T.S., "Some Observations on the Laboratory Measurement of Soil-Geotextile Bond," Geotechnical Testing

Journal, Vol. 5, No. 3/4, 1982, pp. 57-67.

Ingold, T.S., "Laboratory Pull-Out Testing of Grid Reinforcement in Sand," Geotechnical Testing Journal, American Society of Testing and Materials, Vol. 6, No. 3, Sept. 1983.

Koerner, R. M, Designing with Geosynthetics, Second Edition, Prentice-Hall, Englewood Cliffs, NJ, 1990.

Milligan, G.W.E., "Evaluation of Material Properties - Discussion," The Application of Polymeric Reinforcement in Soil Retaining Structures, P.M. Jarrett and A. McGown Eds., NATO ASI Series, Vol. 147, Kluwer Academic Publications, Dordrecht, The Netherlands, 1988, pp. 359-361.

Richardson, G.N. and Koerner, R. M., Geosynthetic Design Guidance for Hazardous Waste Landfill Cells and Surface Impoundments, Contract No. 68-03-3338, U.S. EPA, Cincinnati, OH, 1987.

Swan, R.H., "Anchorage and Shear Behavior of Various Geosynthetics," Master Thesis, Drexel University, Philadelphia, PA, June 1987.

Tzong, W.H. and Cheng-Kuang, S., "Soil-Geotextile Interaction Mechanism in Pullout Test," Proc. of Geosynthetics '87, Vol 1, New Orleans, Louisiana, 1987, pp. 250-259.

Wayne, M. H., "A Large Scale Laboratory Study on the Anchorage Behavior of Geomembranes," Master Thesis, Philadelphia, PA, June 1989.

10 PERMEABILITY TESTING

H. E. HAXO Jr
Matrecon (USA)
P. PIERSON
Université Joseph Fourier, France

10-1 Introduction

A geomembrane, even if it is used as a barrier to movement
of various fluids, cannot be completely impermeable to all
gases, vapors, liquids, and dissolved species. Tests are
necessary to characterize the permeability of the membrane
to the constituents of a fluid with which it is in contact.
    Before studying the different permeability tests used, it
is important to know the mechanism which leads to a fluid
flow rate through a geomembrane. This is the objective of
the first section. In the second section, examples of tests
for the permeability of geomembranes to various species are
discussed and the results that have been obtained in
performing the respective tests are presented.

10-2 Permeation through Geomembranes

Geomembranes have to be considered as non-porous media
(Matrecon, 1988), where permeation proceeds by diffusion,
with dissolution of the permeant into the membrane. The
driving force is then the chemical potential gradient $d\mu_i/dx$
(N. mol^{-1} or N. kg^{-1}) of the permeant i in the membrane. The
steady state flow rate J_i ($mol.m^{-2}.s^{-1}$ or $kg.m^{-2}.s^{-1}$) of
permeant i per membrane area unit due to $d\mu_i/dx$ is then
(Park, 1986 and Yasuda et Al., 1971) :

$$J_i = - c_i / \left(\widetilde{N} . f_i \right) . d\mu_i / dx \qquad (1)$$

where c_i is the molar concentration of the permeant i in the
membrane ($mol.m^{-3}$ or $kg.m^{-3}$), $\widetilde{N}$ the Avogadro number and f_i
the friction coefficient ($N.m^{-1}.s.mol^{-1}$ or $N.m^{-1}.s.kg^{-1}$). The
sign - in equation (1) is due to the μ_i decreasing in the
direction x of the permeant flow.
    The expression of μ_i depends on the fluid nature :

10-2.1 The permeant is a gas

In this case, μ_i is a function of the temperature T (K), which is supposed to be constant here and of the partial pressure p_i (Pa) of the gas in solution in the membrane. For an ideal gas (Morlaes, 1982 and Conture et Al., 1980) :

$$\mu_i\left(T, p_i\right) = \mu_{io}(T) + R.T.Ln(p_i/p_o) \qquad (2)$$

where μ_{io} is the reference chemical potential of the gas at temperature T and at the reference pressure p_o (generally 1 atm i.e. 10^5 Pa). R represents the ideal gases constant.

In a gas, a modification of the partial pressure p_i is accompanied by a modification of the molar concentration c_i ; p_i and c_i are related by Henry's law (Van Krevelen and Hoftyzer, 1976) :

$$c_i = S_i . p_i \qquad (3)$$

where S_i is the solubility coefficient (m^3 gas.Pa^{-1}.m^{-3} memb.). If S_i is constant, eq. (1) combined with eq. (2) and (3) gives :

$$J_i = -\overline{P}_i . \rho_i . dp_i/dx = -D_i . dc_i/dx \qquad (4a)$$

$$\qquad (4)$$

$$\text{with} : \left\{ \begin{array}{l} D_i = R.T/\left(\tilde{N}.f_i\right) \qquad (4.b) \\ \overline{P}_i = S_i . D_i \qquad (4.c) \end{array} \right.$$

Eq. (4.a) is usually known as the Fick's law where D_i is the coefficient of diffusion (m^2.s^{-1}). ρ_i is the mass of gas per volume unit (kg.m^{-3}) and $\overline{P}_i$, which can be calculated from (4.c), as a function of D_i and S_i, is the permeation coefficient or permeability coefficient ([m^3 gas.m^{-2} memb. s^{-1}.Pa^{-1}.m memb.thick)] or [mL gas.cm^{-2} memb.s^{-1}. cm Hg^{-1}.cm memb.thick.] as used in section 10-2.1). It can be noticed that permeability coefficients can also be expressed in [kg gas.m^{-2} memb.s^{-1}.Pa^{-1}.m memb.thick.] in SI or [metric perm.cm^{-1} memb.thick.] as used in 10-2.2 and 10-2.3 where the metric perm is the unity of permeance $\overline{P}_i$.ρ_i/t_g (t_g= membrane thickness) : [metric perm] = [g gas.m^{-2} memb.day^{-1}.mm Hg^{-1}]. These permeability coefficients P_i represent then the quantity : $P_i = \overline{P}_i$.ρ_i.

If Di is a constant, eq. (4.a) can be written :

$$J_i = -D_i . (c_2 - c_1) / t_g \qquad (5)$$

where c_1 and c_2 are the molar concentrations in the membrane respectively at the upstream and downstream sides. t_g represents the membrane thickness.

10-2.2 The permeant is a liquid
The hydrostatic pressure p (Pa) of the liquid has to be taken into account when $p \gg p_o$; in this case, for an ideal solution (Morlaes, 1982) :

$$\mu_i(T, p) = \mu_{io}(T) + p . V_i + R . T . Ln(c_i / c_o) \qquad (6)$$

Where μ_{io} is the reference chemical potential of the liquid at the reference concentration c_o, at temperature T and pressure p_o. The molar volume V_i is supposed to be constant during the tests.

The general expression (6) has to be used for instance in the case of a membrane used for the tightness of a water reservoir, where the water pressure gradient between the up and downstream surfaces of the membrane can be important. (It can be such as the diffusion flow due to the water concentration gradient is annuled : this pressure is known as the osmotic pressure). The pressure gradient creates a chemical potential gradient which has to be added to the one generated by the concentration gradient dc_i/dx. Here, c_i and p are two separate variables (this was not the case of c_i and p_i for gases), which lead to express J_i, derived from (1), as a function of the gradients dc_i/dx and dp/dx.

This appears in the expression (7) (derived from eq. 1 and 6) of the flow J_i, which is the sum of two flowrates : the first one due to the concentration gradient and the second one due to the pressure gradient :

$$J_i = -D_i . dc_i / dx - K_i . \rho_i . dp/dx \qquad (7)$$

$$\text{with}: \begin{cases} D_i = R . T / (\tilde{N} . f_i) & (m^2 . s^{-1}) \\ K_i = c_i . V_i / (\tilde{N} . f_i . \rho_i) \end{cases}$$

(K_i is expressed in m^3 permeant.s^{-1}.Pa^{-1}.m^{-1} memb.or m memb.s^{-1} if p is expresed in permeant height, often used for water).

D_i and K_i are respectively the diffusion and permeability coefficients. The coefficients $\overline{P}_i$ (permeation coefficient also named permeability coefficient defined for gases : cf section 10-2.1) and K_i (permeability coefficient defined for liquids) look very similar but it has to be noted that :

$\overline{P_i}$ characterizes the flow due to a partial pressure gradient (i.e. a molar concentration gradient) of gas in the membrane ;

K_i characterizes the flow due to a liquid pressure gradient (the liquid is obviously supposed to be noncompressible).

Characterizing the permeability of a geomembrane to a permeant i means determining D_i and (or) K_i. It is therefore necessary to conduct the test in the presence of only one gradient : dc_i/dx or dp/dx ; if this is not the case, the test will lead to the determination of a coefficient which will be neither D_i nor P_i :

- If $dp/dx \cong 0$ (low pressure gradient in the membrane) eq. 7 becomes similar to eq. 4 (Fick's law). D_i is then a specific characteristic of the membrane ; it can be calculed if the concentration gradient is known and if the diffusive flowrate is measured. In fact, for real solutions, Fick's law is not perfectly verified and D_i might depend on the concentration c_i. D_i is then determined in a range of concentrations. Measuring D_i gives interesting informations on the permselectivity of the membrane, i.e. its propriety to select the diffusion flow of the species in function of the species themselves. Such informations are necessary in waste disposal facilities applications for instance.

- If $dc_i/dx \cong 0$ (low concentration gradient in the membrane) eq.7 becomes very similar to Darcy's equation used in porous media :

$$Q = k.dh/dx \qquad (8)$$

where Q is the flowrate per membrane area unit ($m^3.s^{-1}.m^{-2}$). dh/dx is the hydraulic gradient (m water.m^{-1}) and k, Darcy's coefficient of permeability ($m.s^{-1}$).

But in a porous medium, contaminated water flows with all the dissoved species from pore to pore. It is not the case in a geomembrane where water molecules or dissolved species molecules diffuse individually. In eq.(8), k is a specific characteristic of the porous medium considered. In eq.(7), the permeability coefficient K_i of the non porous membrane is a function of the molar volume V_i, of the mass per volume unit ρ_i, and of the friction coefficient f_i which depend on the pressure p. Furthermore, the permeation can be accompanied by a swelling of the membrane, which modifies the diffusion rate and which also depends on the pressure p.

However, in the particular case of a membrane used for the tightness of a reservoir, it is interesting to compare the water tightness of different geomembranes (coefficient K_i) and of different soils (coefficient k of porous media) (Giroud, 1984). K_i then has to be determined by specifying the water pressure used during the test. Such tests are described in 10-3.5.

10-3 Testing of the Permeability of Geomembranes

A variety of test methods are available for assessing the permeability of polymeric membranes and films to the different chemical species that may be encountered in the specific application in which it is to be used (Crank and Park, 1968 ; Yasuda et al. 1968). However, it is essential to select test methods that simulate as closely as possible the conditions of the intended use. A variety of the methods that are used in measuring the permeability of geomembranes and films to gases, vapors, dissolved species, and liquids are described in the following sections and data obtained by the respective methods are presented.

10-3.1 Permeability of Geomembranes to Gases
Knowledge of the permeability of geomembranes to gases is important in such applications as in lining materials for waste disposal facilities, methane barriers in construction, and in the lining of tunnels through ground that may contain methane. A wide range of test procedures is used in assessing the gas permeability of polymeric membranes and various films (Crank and Park, 1968). In the case of geomembranes, two methods are commonly used to measure the permeability of a geomembrane to a single gas :
    1) measurement of the volume of the selected gas passing through the membrane under specific test pressure or concentration gradient,
    2) the measurement of the increase in pressure with a manometer on the downstream side from an initial vacuum. Both of these methods are described in ASTM D1434. The volumetric method, which uses the equipment illustrated in figure 1, has been used by Haxo et al.(1985) to measure the permeability of a wide range of geomembranes to methane, carbon dioxide, and nitrogene.
    In the apparatus shown in Fig.10.1, the membrane is in contact on the two sides with the selected gas. The downstream side is at atmospheric pressure (P_{atm}), and the upstream side is at pressure greater than atmospheric. Thus, a pressure gradient exists across the membrane and, in the case of gases, leads to a concentration gradient in the membrane and to a permeation and diffusion of the gas from higher to lower pressure.
    Results for the permeability of geomembrane based on 9 different polymers to CO_2, CH_4, and N_2 obtained by Haxo et al.(1984) and Haxo et al.(1985) are presented in Table 10.1.

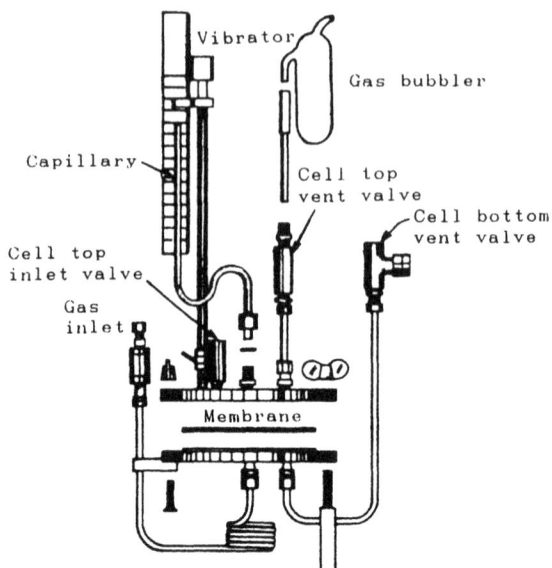

Fig.10.1 Gas permeability apparatus in ASTM D1434,
Procedure V - Volumetric.

The data show measurements on 15 different commercial
geomembranes and a standard test film based on polyethylene
terephthalate. They include data on the gas transmission
rates for the specific geomembranes and the normalized data
for the permeability coefficients in "barrers", a unit which
is equal to 10^{-10} mL (STP).cm/cm^2.s.cm Hg. The data show the
great range in permeability coefficients among the
geomembranes as well as among the three gases.

Data reported as gas transmission rates indicate the
performance of a specific geomembrane which varies inversely
with thickness ; the permeability coefficients are material
properties and reflect the permeabilities of the material
compound. Gas transmission rates (GTR) in mL (STP)/m^2·d·atm
can be converted into permeability coefficients $\boxed{\mathbf{P}}$ in
barrers [10^{-10} mL(STP).cm/cm^2.s.cm Hg.] using the following
equations (ASTM D1434) :

$$\overline{P} = 0.01532 \text{ x (thickness in mm) x GTR.} \qquad (9)$$

A polymeric geomembrane was tested by Haxo et al (1984) and
Haxo et al (1985) for permeability to carbon dioxide,
methane, and nitrogen at three different temperatures (10°,
23°, and 33°C). Data are reported as GTR for a 0.158 mm
thick specimen under a pressure difference across the
geomembrane of one atmosphere. These results show that

TABLE 10.1 Permeability of Polymeric Geomembranes to Selected Gases at 23°C, Determined in Accordance with ASTM D1434, Procedure V

Geomembrane description				Gas transmission rate $mL(STP^{[b]})/m^2 \cdot d \cdot atm$			Gas permeability coefficient (P), barrer[c]		
Base Polymer	Density	Thickness[a], mm	Compound type[d]	CO_2	CH_4	N_2	CO_2	CH_4	N_2
Butyl rubber	...	1.60	XL	512	120	19.7	12.5	2.92	0.480
Chlorosulfonated polyethylene	...	0.82R	TP	122	21.6	26.2	1.52	0.27	0.33
	...	0.86	TP	418	124	27.1	5.47	1.62	0.36
Elasticized polyolefin	...	0.58	CX	1450	280	125	12.8	2.47	1.10
Ethylene propylene rubber	...	0.89R	TP	2720e	...	...	36.8e	...	...
	...	0.90	XL	5260	1400	314	72.0	19.2	4.30
Neoprene	...	0.90	XL	716	80.9	31.1	9.81	1.11	0.43
Polybutylene	...	0.71	CX	818	248	62.3	8.84	2.68	0.67
Polyethylene[f] (low density)	0.921	0.25	CX	6180e	1340e	...	23.5e	5.10e	...
Polyethylene (linear low density)	0.923	0.46	CX	1370	322	...	9.59	2.25	...
Polyethylene (high density)	0.945	0.61	CX	729	138	...	6.77	1.28	...
	0.945	0.86	CX	467	104	...	6.11	1.36	...
Polyvinyl chloride (plasticized)	...	0.25	TP	7730e	1150e	108	29.4e	4.38e	0.81
	...	0.49	TP	3010	446	...	22.4	3.32	...
	...	0.81	TP	2840e	285e	...	35.0e	3.51	...
Polyethylene terephthalate[g]	...	0.022	CX	357	...	...	0.119	...	...

a R = fabric-reinforced.

b STP = standard temperature and pressure.

c One barrer = 10^{-10} mL(STP)·cm/cm²·s·cm Hg.

d XL = crosslinked; TP = thermoplastic; CX = semicrystalline.

e Measured at 30°C.

f Natural resin (no carbon black).

g NBS Standard Material 1470. The determination was made at 15.0 psi, under which condition the NBS Certified CO_2 transmission rate was calculated to be 338 mL(STP)/m²·d·atm.

permeability of a given geomembrane to gases increases with the temperature in accordance with Arrhenius's equation (Matrecon, 1988).

If the geomembrane is to be used as a barrier to a mixture of gases, the permeability can be measured by using an apparatus with two compartments separated by a specimen of the geomembrane and following the transmission and the composition of the mixture by GC or GCMS (Yasuda et al, 1968). Due to the permselectivity of the polymeric geomembranes, the rates of permeation will differ from geomembrane to geomembrane, though the ratio of the rates tends to remain relatively constant.

10-3.2 Permeability of Geomembranes to Moisture Vapor

Another important permeability characteristic often needed in assessing geomembranes for specific applications is that of permeability to moisture vapor. This is important in such applications as reservoir covers and moisture barriers in buildings. Moisture vapor transmission is commonly determined by measuring the loss in weight of a small cup containing a small amount of water, the mouth of which is sealed with a specimen of the geomembrane. In this type of test, the cup is placed in an atmosphere of a constant relative humidity less than 100 % at a constant temperature, either with water resting on the geomembrane or inverted with the water on the bottom of the cup and exposing the geomembrane to vapor only. Various procedures for performing such a test are described in ASTM Test Method E96.

Some results obtained by Haxo et al (1984) and Haxo et al (1985), in accordance with Procedure BW of the above method are presented in Table 10.2, which includes data on geomembranes based on 17 different polymers and 2 films of very low permeability, e.g. polyvinylidene chloride (Saran) and polytetrafluoroethylene (Teflon). The number of geomembranes tested of each type and the ranges of thickness as well as the average values and ranges of test values are shown.

In these tests, an aluminium cup was partially filled with deionized water and a specimen of the geomembrane was sealed with wax at the open end. The cup was placed in an inverted position (so that water was in contact with the specimen) in a controlled environment where the relative humidity (H) was 50 % and the air flow across the specimen surface was controlled. Water vapor permeated the membrane from the higher concentration surface (H = 100 % in the cup) to the lower concentration surface (H = 50 % on the outside of the cup). The water vapor flow J_2 (or WVT) can be measured during the test by periodically weighing the cup.

The permeation coefficient $\overline{P}_i$ is then derived from Fick's equation (4.a), where the partial pressure gradient dp_i/dx can be derived from the relative humidity gradient

TABLE 10.2 Moisture Vapor Permeability of Geomembranes and Films Based on Different Polymers Measured in Accordance with ASTM E96[a], Procedure BW

Base polymer	Type of compound[b]	Number of membranes in average	Range of thickness, mm	Moisture vapor permeability (10^{-2} metric perm·cm)	
				Average	Range of values
Butyl rubber	XL	3	0.85-1.18	0.081	0.016-0.170
Chlorinated polyethylene	TP	11	0.53-0.97	0.485	0.213-1.27
Chlorosulfonated polyethylene	TP	10	0.74-1.07	0.460	0.234-0.845
Elasticized polyolefin	CX	2	0.61-0.72	0.090	0.083-0.097
Epichlorohydrin rubber	XL	2	1.16-1.65	22.3	22.2-22.4
Ethylene propylene rubber	XL	6	0.51-1.70	0.260	0.131-0.367
Neoprene	XL	6	0.51-1.59	0.354	0.147-0.517
Nitrile rubber	TP	1	0.76	3.98	...
Polybutylene	CX	2	0.19-0.69	0.064	0.055-0.072
Polyester elastomer	CX	2	0.20-0.25	1.99-10.6	1.99-10.6
Polyethylene (low density)	CX	1	0.76	0.041	...
Polyethylene (high density)	CX	2	0.80-2.44	0.014	0.013-0.014
Polyethylene (high density) – alloy	CX	1	0.86	0.039	...
Poly (ethylene vinyl acetate)	TP	1	0.53	0.76	...
Polyvinyl chloride	TP	8	0.28-0.79	1.30	0.77-1.45
Polyvinyl chloride (elasticized)	TP	2	0.91-0.97	2.10	1.79-2.40
Polyvinyl chloride (oil-resistant)	TP	3	0.79-0.84	3.08	2.60-3.35
Polyvinylidene chloride (Saran)	TP	1	0.013	0.0070	...
Polytetrafluoroethylene (Teflon)	CX	1	0.10	0.0021	...

[a]ASTM E96-80, Procedure BW: Inverted water method at 23°C; 50% humidity on downstream side. Permeance in metric perms = g m^{-2} d^{-1} $(mm\ Hg)^{-1}$ = WVT/Δp in mm Hg, where Δp = the vapor pressure difference = 10.53 mm Hg (at 23°C and 50% humidity on downstream side). Permeability in metric perms·cm = permeance x thickness of geomembrane in cm.

[b]XL = crosslinked; TP = thermoplastic; CX = semicrystalline.

dH/dx i.e.(100-50)%/t, where t is the specimen thickness) as
follows :

$$dp_i/dx = p_s.dH/dx \qquad (10)$$

where p_s represents the water vapor pressure at saturation
(H = 100 %) : $p_s = p_i/H$.

The results in Table 10.2 are reported as permeability
coefficients, i.e. metric perm·cm. These data can be
converted into water vapor permeance by dividing the
permeability coefficient by thickness in centimeters. The
results can be converted into water vapor transmission rates
(grams per square meter per day, $g/m^2/d$) for the specific
conditions of the test (23°C and 100 % relative humidity, H,
in the cup and 50 % H on the outside of the cup) by
multiplying the permeance by vapor pressure difference
across the membrane at 23°C, i.e. 10.53 mm Hg (Matrecon,
1988).

10-3.3 Permeability of Geomembranes to Organic Vapors
A procedure similar to that described above for measuring
the moisture vapor permeability can be used for measuring
solvent vapor transmission (SVT). For testing solvents, an
all metal cup is used which is not inverted ; thus, the
solvent vapor contacts the geomembrane surface ($H_{solvent}$ =
100 %), while, outside the cup, $H_{solvent}$ is 0 % ; this
difference of vapor pressure of 100 % is the driving force
of the permeation.

Results obtained using this test for five different
volatile organic solvents with six geomembranes and
a 0.10 mm film of tetrafluoroethylene (Teflon) are presented
in Table 10.3 (Haxo et al., 1985 ; Matrecon, 1988), in
terms of solvent transmission rates and permeability
coefficients, which correct for thickness and reflect the
permeabilities of the respective polymeric compositions.
The data on the two HDPE geomembranes of different
thicknesses show : 1) the effect of thickness on solvent
transmission rates, and 2) the equal values for the
permeability coefficients. The results also show
significant differences among the geomembranes and among the
organics. The low permeation of the more polar solvents,
i.e. methyl alcohol and acetone, compared with the less
polar solvents should be noted.

August and Tatsky (1984) measured the permeation of the
six individual organics in an equivolume mixture through an
HDPE geomembrane. The results are presented in Table 10.4.
They used an apparatus consisting of two compartments
separated by the geomembrane being tested. The upper
compartment contained the solvent mixture, the composition
of which was held constant, and the lower compartment was
partially evacuated. The permeating vapors were analyzed by

TABLE 10.3 Permeability of Polymeric Geomembranes to
Various Solvents, Measured in Accordance with ASTM E96,
Procedure B (Modified to Test Solvents)

Polymer[a]	ELPO	HDPE	HDPE	HDPE-A	LDPE	PB	Teflon
Average thickness, mm	0.57	0.80	2.62	0.87	0.75	0.69	0.10
Type of compound	CX	CX	CX	CX	CX	CX	CX
SVT, $g\ m^{-2}\ d^{-1}$							
Methyl alcohol	2.10	0.16	...	0.50	0.74	0.35	0.34
Acetone	8.62	0.56	...	2.19	2.83	1.23	1.27
Cyclohexane	7.60	11.7	...	151	161	616	0.026
Xylene	359	21.6	6.86	212	116	178	0.16
Chloroform	3230	54.8	15.8	506	570	2120	20.6
Solvent vapor permeability[b], 10^{-2} metric perms·cm							
Methyl alcohol	0.11	0.01	...	0.04	0.05	0.02	0.003
Acetone	0.23	0.02	...	0.09	0.10	0.04	0.006
Cyclohexane	0.49	1.05	...	14.7	13.6	47.8	0.0003
Xylene[c]	292	24.6	25.6	262	124	175	0.002
Chloroform	103	2.46	2.32	24.6	24.0	82.2	0.12

[a]ELPO = elasticized polyolefin; HDPE = polyethylene (high density);
HDPE-A = polyethylene (high density) – alloy; LDPE = polyethylene
(low density); PB = polybutylene.

[b]The median thickness value was used to calculate the permeability.

TABLE 10.4 Permeation Rates of the
Components of a Mixture of Organics Through
a 1.0 mm HDPE Geomembrane

Organic	Permeation rate, $g\ m^{-2}\ d^{-1}$
Trichloroethylene	9.4
Tetrachoroethylene	8.1
Xylene	3.0
Isooctane	0.8
Acetone	1.4
Methanol	0.7
Total	23.4

Source: August and Tatsky (1984).

GC. The resulting data show the great difference in the transmission rates among the solvents, demonstrating the permselectivity of this geomembrane.

10-3.4 Pouch Test for Assessing Permeability of a Geomembrane to Dissolved Species
When a geomembrane is to be used as a lining material for waste disposal facilities where a complex liquid or leachate may be encountered, it is desirable to know its permeability to the various constituents of the liquid to be contained. The pouch test, which simulates some of the in situ conditions in waste disposal facilities, can be used for characterizing simultaneously the permeabilities of a geomembrane to ions, water, and the organic constituents of a leachate or other waste liquid.

A small pouch made of the membrane material is filled with the test liquid and immersed in a liquid of known composition, generally dionized water (DI water). The initial pouch contents are analyzed, i.e. composition, pH, and electrical conductivity (EC) measurements, to know what is permeating the membrane and at which direction and rate. At the end of the exposure, the pouch is dismantled, and the pouch wall and contents are weighed, analyzed, and tested for changes in composition and physical properties. The movement of the various constituents in the solution or waste liquid is illustrated schematically in Fig.10.2.

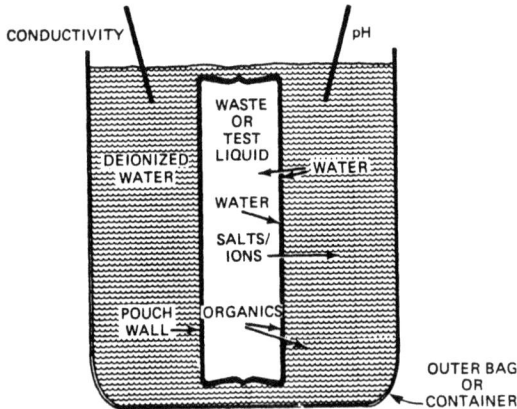

Fig.10.2 Pouch assembly showing the movement of constituents during the pouch test. In the case illustrated by this drawing, the pouch is filled with an aqueous waste or test liquid and immersed in DI water in a closed container. Arrows indicate the movement of specific constituents.

To measure the permeability of geomembranes to ions and to water, the pouch can be filled with an aqueous salt solution, e.g. 5% NaCl, and immersed in DI water in an outer pouch or a jar. The permeation of the ions through a membrane can be assessed by measuring the EC, pH, and specific ions of the salts as a function of time. The permeation of water through the pouch wall can be determined by measuring the weight of the pouch as a function of time. If the liquid in the pouch contains volatile organics, the outside container should be closed and sealed and a port with a septum should be incorporated in the top through which the outer liquid could be sampled and analyzed by GC.

It is to be noted that a pouch containing a high concentration of ions will increase in weight during the course of the test and, consequently, the water outside the pouch will move into the pouch with time due to the lower chemical potential of the water in the pouch compared with that in the outer water. The chemical potential of the water is reduced by the presence of the ions.

As an example of the measurement of the permeation of ions and water, pouches of PVC were filled with 5 and 10% solutions of lithium chloride (LiCl) and placed in outer plastic bags containing DI water. The electrical conductivity (EC) of the outer water exhibited almost no change during exposures of up to 600 days. However, the pouches gained in weight, as is shown in Fig.10.2. The fact there was very little change in EC of the outer water indicates that the ions did not permeate the pouch walls ; however, the gain in weight of the pouches show that the water permeated into the pouch from the outer water. Comparing the results for the pouches containing different concentrations of LiCl demonstrate the importance of concentration on the rate and direction of the transmission of water that is present on both sides of a geomembrane. Similar results were obtained with pouches of oil-resistant PVC containing LiCl and with pouches made of other geomembranes and filled with high ion concentration waters. Because the lithium and chloride ions do not permeate a geomembrane, the results indicate the potential use of lithium ions as a tracer for the detection of leaks in a geomembrane.

Pouch tests can also be used in assessing the permeability of geomembranes to the constituents of aqueous solutions containing organics, such as leachates generated in landfills. This was illustrated by a series of tests in which pouches made of an HDPE alloy (HDPE-A) were filled with xylene and placed in xylene or in water. In this test, a soluble, nonvolatile organic dye which might be used as a leak-detection tracer was placed in the xylene to assess its movement through a geomembrane. These results are illustrated in Figure 10.3, which shows the changes in the weight of the pouches and the movement of the xylene in the

pouches as well as the dye. The dye, which is not soluble
in water, permeated the pouch wall and deposited on the
outer surface of the pouch that was placed in water. For
the pouch that was placed in xylene the dye also permeated
the wall and, in this case dissolved outside into the
xylene. The pouch that was placed in xylene gained weight,
whereas the pouch that was placed in water lost weight due
to the permeation of the xylene out of the pouch. The
permeation or the transmission of the constituents in the
pouches can be determined by measurements of the weight
changes and the amount of solution that has been absorbed by
the pouch walls when the pouches are dismantled.

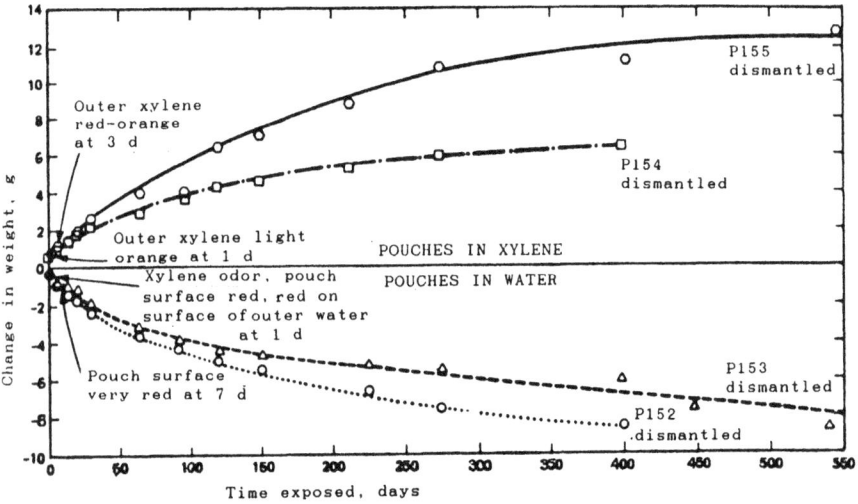

Fig.10.3 Weight changes of HDPE-A pouches filled with
xylene immersed in xylene and in DI water. Some of the
xylene that permeated into the water and rose to the surface
of the water was lost to the air by evaporation.

    The accuracy of the pouch test depends upon the
preparation of durable, leak-free pouches, the seams of
which would not allow liquids to by-pass the pouch wall to
yield high transmission values. The experimental work on
the pouch test was restricted to thermoplastic geomembranes
which could be heat-sealed or welded to yield non-leaking
seams.
    The test should also apply to vulcanized geomembranes if
the pouch can be fabricated to yield no leaks in the seams.
This seaming procedure would probably require heat
vulcanization at the edges.

232

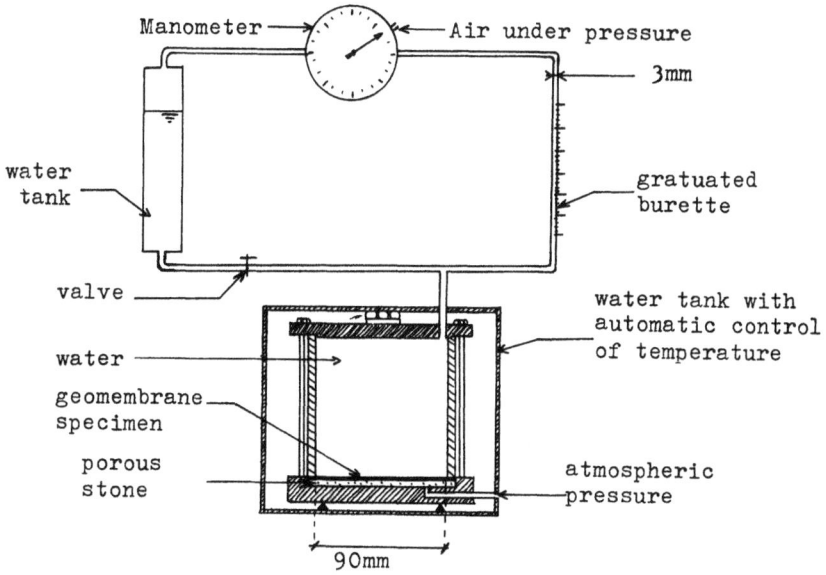

Fig.10.4   Apparatus for the measure of the permeability coefficient of geomembranes to water under pressure (Faure et Al, 1989

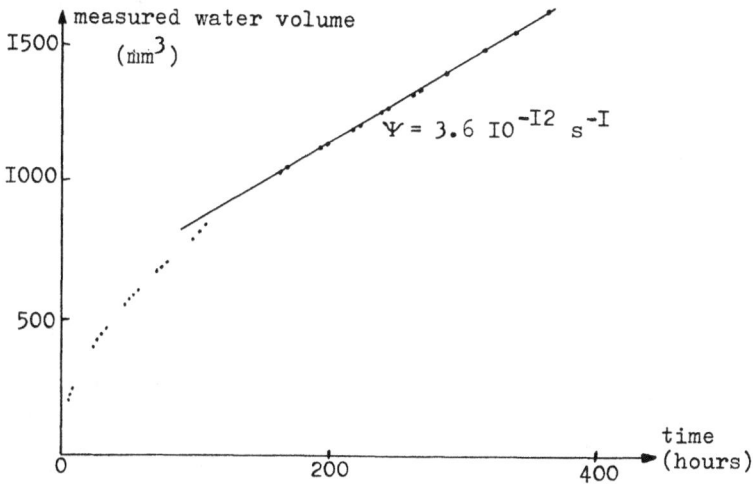

Fig.10.5   Results obtained with the apparatus of fig. 4 on a bituminous sample (t = 4.2 mm, p = 350 kPa, T = 25°c). (Faure et al, 1989).

10-3.5 Permeability of Geomembranes to Water under Pressure
The usefulness of tests conducted under water pressure has
already been mentionned in section 10-2.2. These can be
conducted on stressed or nonstressed samples (Bernhard,
1988). Here follows an example of geomembrane permeameter
as well as the results obtained with the influence of the
main parameters on the coefficient of permeability.

10-3.5.1 Description of a permeameter cell with the
apparatus to measure the coefficient of permeability.
Fig.10.4 shows the apparatus used at IRIGM (Grenoble-
France) to measure the permeability coefficient K_i of a
membrane to water (Faure et Al, 1989) : the geomembrane
specimen (Φ=90mm) is in contact with water above (under
variable pressure) and with a porous stone below.
    The measure of the water flowrate through the membrane is
made on the upstream side of the sample : the fluid below
the specimen may contain some air from water after
decompression, which can cause errors in the value of K_i, if
the flowrate is measured on the downstream side. However,
the upstream side measure takes into account the water
leakage through the edges of the sample, which have to be
correctly controlled : the real permeability coefficient is
then less than the one measured, which gives a security
margin.
    It is also important to control the surrounding
temperature , the variations of which can be responsible
for :
- variations of volume of the fluid and of the specimen
- modifications of the membrane permeability (see paragraph
10-3.5.2).

10-3.5.2 Result obtained by Faure et Al (1989)
Fig.10.5 shows the results obtained with the apparatus of
Fig.10.4 used for testing bituminous membranes permeability
to water, characterized by the permittivity Ψ (s^{-1}) :

$$\Psi = K/t = J/[\rho \cdot (p_{upstream} - p_{downstream})] \qquad (11)$$

which is derived from (7) and where t is the specimen
thickness. It can be observed that :
    1) 200 hours are required to reach the steady state
    2) the straight line J (time) which gives Ψ can be drawn
with better precision than in Fig.10.6, which corresponds
to results obtained without any control over the surrounding
temperature (difference : 30%).
Fig.10.7 shows the results of a test conducted on foam
rubber.
    The influence of the two main parameters, pressure p and
temperature T, has been studied as follows :
    Influence of p :

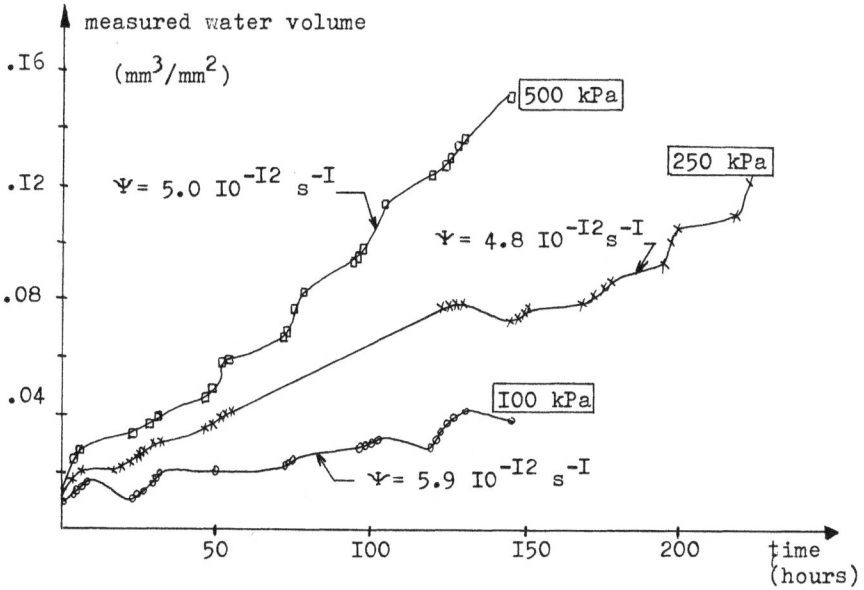

Fig.10.6   Results of tests on a bituminous membrane
(t = 2.7 mm) (Faure et Al, 1989)

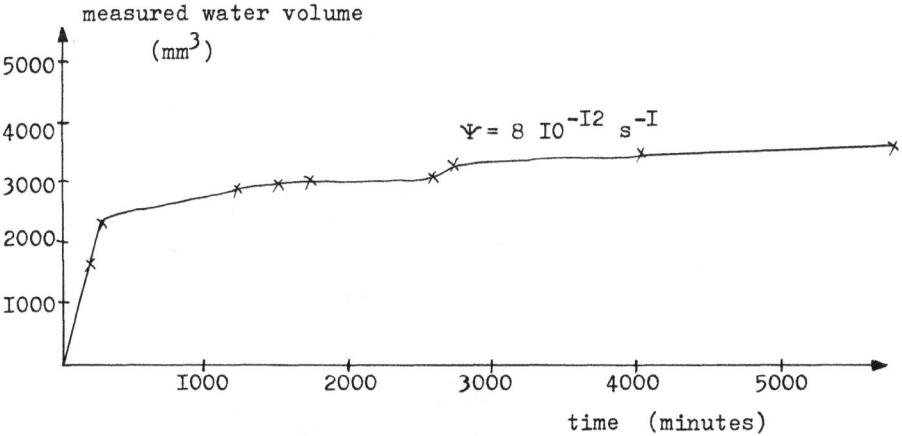

Fig.10.7   Results obtained on a sample of foam rubber
(t = 2.25 mm, p = 500 kPa).(Faure et Al, 1989)

Contrary to Darcy's coefficient of permeability, it has already been observed (section 10-2.2) that K_i depends on p; this is illustrated in Fig.10.6, in which one finds the results of three tests conducted on the same bituminous membrane under different water pressures. Here, the variations of Ψ (or K_i) are small when p is modified (from 100 kPa to 500 kPa) ; but the results obtained can be very different from one membrane to another one, because of the variable modifications of the membrane geometry (see section 10-2) : Giroud(1984) worked on butyl rubber samples subjected to elongations in both directions. The results are given in Fig.10.8 : the function K (p) is characterized by a peak for elongations less than 10 % at about 300 kPa.
   Influence of T :
Fig.10.9 shows the variations Ψ (T) for a bituminous membrane. These variations which are given by Arrhenius' Law (exponential function of (-1/T), are quite different from those observed in a porous medium : see Fig.10.9, in which the variations Ψ (T) due to water viscosity variations in a fictitious porous medium (characterized by the same permittivity as the one of the membrane studied at 25°C) are also plotted.

10-4   Conclusion

All the tests now used in laboratories to measure diffusion or permeability coefficients have not been considered here. Only the general principles and some examples of tests have been described.
   Some important remarks can be pointed out concerning the permeability tests on geomembranes :

10-4.1 About the mechanism of diffusion
   1 The transport or permeation of a chemical species through a polymeric membrane is a three-steep process :
   Step 1 - Dissolution of the permeating species into the surface of the membrane on the upstream side.
   Step 2 - Diffusion of the permeating species through the membrane from higher to lower chemical potential.
   Step 3 - The escape of the permeating species on the downstream side by evaporation or dissolution in the liquid on the downstream side.
   2 Transport of permeation species through a polymeric membrane is on a molecular scale by diffusion.
   3 The driving force for permeation or transport through a membrane is chemical potential as indicated by gradients in concentration of the species, vapor pressure of the species, or temperature and hydrostatic pressure across the membrane.
   4 Geomembrane are permselective, i.e. the rates of permeation of different species differ, depending upon

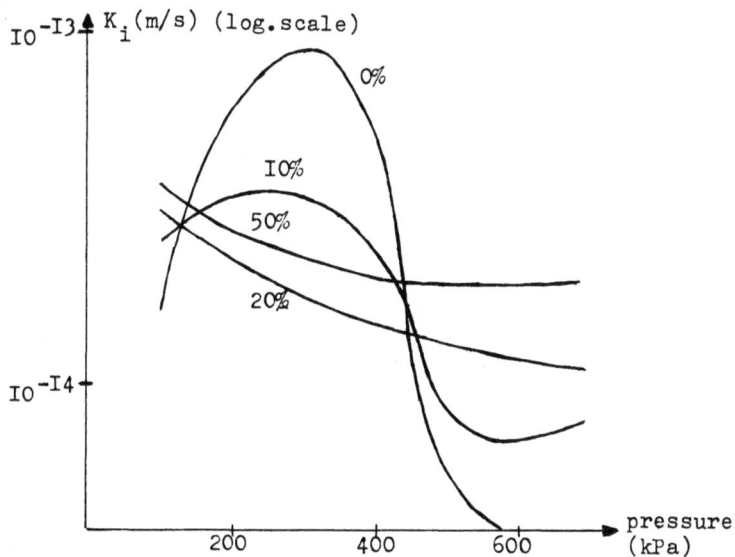

Fig.10.8   Coefficient of permeability of a Butyl rubber
geomembrane subjected to elongation.(Percentages are
increases in area of the Butyl rubber geomembrane. [Giroud,
1984]

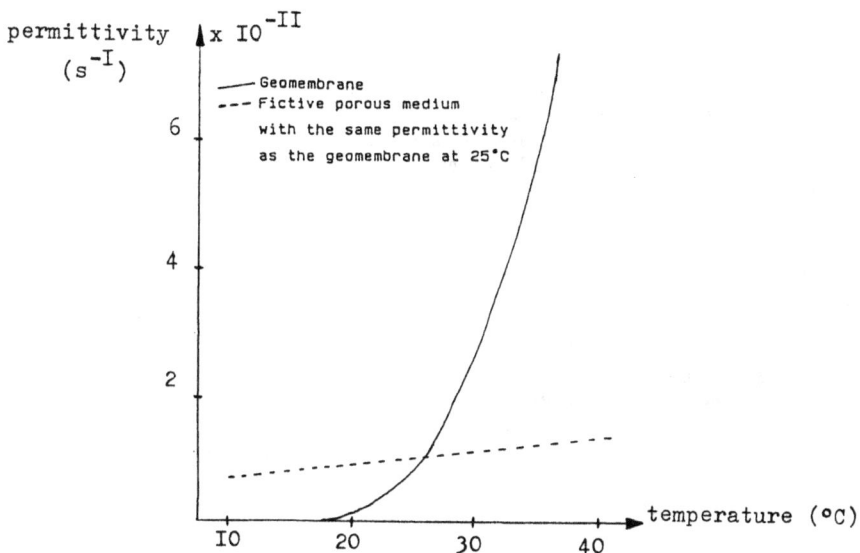

Fig.10.9   Influence of temperature on the permittivity of a
bituminous geomembrane (t = 4.2 mm, p = 250 KPa). [Faure et
Al, 1989].

237

solubility of the species in the membrane and diffusivity of the species within the membrane.

5 Inorganic species, e.g. dissolved metal salts, do not permeate geomembranes.

6 Water can permeate a membrane under pressure. However, the less contaminated water on the downstream side of a geomembrane liner can permeate the liner against the hydrostatic pressure of the water on the upstream side which is more contaminated. Such a condition may be encountered in a waste disposal facility.

10-4.2 About the tests of permeability

1 It has to be known what is really measured during the test : a coefficient of diffusion (corresponding to a concentration gradient in the membrane) or a coefficient of permeability ; for a gas, this coefficient P_i can be related to the coefficient of diffusion, but not for a liquid : the corresponding coefficient K_i is then defined for an hydrostatic pressure p.

2 According to the very small amount of fluid diffusing through the membrane, the measurement of the flowrate must be very accurate and must take into account the leakage due to the apparatus ;

3 Other characteristics of the membrane have to be known to deal correctly with the calculation of the permeability and diffusion coefficients : particularly, the thickness and specific weight of the membrane must be known as functions of the fluid absorption, temperature and time.

Diffusion and permeability coefficients are important characteristics of geomembranes but they would be more useful for designers if they were also performed on seams, on elongated membranes and on membranes submitted to different leachates : permeability tests are often recognized as the most sensible tests to characterize the aging of geomembranes. The problem of such tests is the difficulty to get specimens in situ : it is then necessary to design landfill simulators such as Haxo et Al's (1985) and to test the mechanic and permeability characteristics of geomembranes before and after several years of exposure to typical waste landfill leachates in the simulator.

Other permeability tests would be interesting to characterize the fluid leakage through flaws in the membrane. Such a test has been developed by Jayawickrama et Al (1988), who studied the influence of a slit and seam failure in the liner on the leakage flowrate.

References

ASTM. Annual Book of ASTM Standards. Issued annually in
several parts. American Society for Testing and Materials,
Philadelphia, PA :
   D1434-84. "Test Method for Determining Gas Permeability
Characteristics of Plastic and Sheeting to Gases," Section
08.01.
   E96-80.   "Test Methods for Water Vapor Transmission of
Materials." Sections 04.06, 08.03, and 15.09.
August, H., and Tatzky, R. (1984) Permeabilities of
Commercially Available Polymeric Liners for Hazardous
   Landfill Leachate Organic Constituents, in Proceedings of
   the International Conference on Geomembranes, June 20-24,
   1984, Denver, CO., Vol. 1, Industrial Fabrics Association
   International, St. Paul, MN., pp. 163-168.
Bernhard, C. (1988) Note sur l'essai de perméabilité à l'eau
   des géomembranes, RILEM TC-103 MHG, June 7-8,
   1988,Montreal.
Couture, L., Chahine, Ch. and Zitoun, R. (1980) Thermo
   dynamique classique et propriétés de la matière, Dunod
   Ed., Paris.
Crank, J. and Park, G.S. (1968) Diffusion in Polymers,
   Academic Press Inc., NY, 452 pp.
Faure, Y.H., Pierson, P. and De La Lance, A. (1989) Study
   of a water Tightness Test for Geomembranes, in Proceedings
   of the second international Landfill Symposium, October 9-
   13, 1989, Sardinia, Italy.
Giroud, J.P. (1984) Impermeability : the Myth and a rational
   Approach, in Proceedings of the International Conference
   on Geomembranes, June 20-24, 1984, Denver, Co., Vol. 1,
   Industrial Fabrics Association International, S^T Paul,
   MN., pp. 163-168.
Haxo, H.E. (1988) Transport of Dissolved Organics from
   Dilute Aqueous Solutions Through Flexible Membrane Liners,
   in Proceedings of the Fourteenth Annual Solid Waste
   Research Symposium : Land Disposal, Remedial Action,
   Incineration and Treatment of Hazardous Waste, May 9-11,
   1988, EPA/600/9-88/021. U.S. EPA, Cincinnati, OH., pp.
   145-166.
Haxo, H.E., Miedema J.A. and Nelson, N.A. (1985)
   Permeability of Polymeric Membrane Lining Materials for
   Waste Management Facilities. Elastomerics, 117 (5), pp.
   29-26, 66.
Haxo, H.E., Miedema J.A. and Nelson, N.A. (1984)
   Permeability of Polymeric Membrane Lining Materials, in
   Proceedings of the International Conference on
   Geomembranes, June 20-24, 1984, Denver, Co., Vol. 1,
   Industrial Fabrics Association International, S^T Paul,
   MN., pp. 151-156.

Haxo, H.E., Nelson, N.A. (1984) Permeability characteristics
    of flexible Membrane Liners measured in Pouch Tests, In
    Proceedings of the 10th Annual Research Symposium : Land
    Disposal of Hazardous Waste, EPA/600/9-88/007. U.S. EPA,
    Cincinnati, OH., pp. 230-251.
Haxo, H.E., White, R.M., Haxo, P.D. and Fong, M.A. (1985)
    Liner Materials exposed to municipal solid waste Leachate,
    in Waste Management and Research 3, 1985, pp. 41-54.
Jayawickrama, P.W., Brown, K.W., Thomas, J.C. and Lytton,
    R.L. (1988) Leachate Rates through Flaws in Membrane
    Liners. J. Environ. Eng., 114 (6), pp. 1401-1420.
Matrecon, Inc. (1988) Lining of Waste Containment and other
    Impoundment Facilities, 2nd Ed., EPA/600/2-88/052. US.
    EPA, Cincinnati, OH, 991 pp.
Morlaes, J.C. and P. (1982) Thermodynamique chimique.
    Vuibert Ed., Paris.
Park, G.S. (1986) Transport Principles - Solution. Diffusion
    and Permeation in Polymer Membranes, in Synthetic
    Membranes : Science, Engineering and Applications, Bangay,
    P.M. et Al. Ed., Reidel Publishing Company, pp. 57-107.
Van Krevelen, D.W. and Hoftyzer, P.J. (1976) Properties of
    Polymers, Their Estimation and Correlation with Chemical
    Structure. Elsevier scientific publishing Company.
Yasuda, H., Lamaze, C.E. and Peterkin, A. (1971) Diffusive
    and hydraulic Permeabilities of Water in Water-swollen
    Polymer   Membranes. J. of Polymer Science, A-2, Vol.
    9, pp. 1117-1131.
Yasuda, H., Clark, H.G., and Stannett, V. (1968)
    Permeability, In Encyclopedia of Polymer Science and
    Technology, Vol. 9, Interscience NY, pp. 794-807.

ACKNOWLEDGEMENT

This chapter was written with the assistance of:

F. GOUSSÉ   CEMAGREF, 92164 Antony Cedex, France
Y.H. FAURE   IRIGM, Université de Grenoble, 38041 Cedex, FRANCE

11 STRESS CRACK TESTING OF POLYETHYLENE GEOMEMBRANES

Y. H. HALSE, R. M. KOERNER and A. E. LORD Jr
Geosynthetic Research Institute, Drexel University,
Philadelphia, PA, USA

11-1 Introduction

Stress cracking is defined by ASTM as "an external or internal rupture in a plastic caused by a tensile stress less
than its short-term mechanical strength". For geomembranes
there are a number of possible contributors to this
undesirable phenomenon:
- effects of surface wetting agents (e.g., environmental
  stress cracking)
- effects of residual stress
- stress cracks originating from scratches or other
  defects
- stress cracks originating from irregular geometry
- fatigue, or cyclic, induced cracking
- combinations of the above

While not exclusively the case, the above mechanisms often occur at geomembrane field seams. Here, liquids are in contact with the surface, residual stresses often occur due to the extreme heating and rapid cooling, grind marks are often present due to surface buffing to remove oxide layers, the geometry creates stress raising via a load eccentricity, and thermal stresses due to temperature expansion and contraction can occur.

However, one other feature is required for stress cracking to occur in plastics and that is the presence of a highly crystalline material. To understand why, one must be aware of the molecular structure of those polymers which are sensitive to stress cracking. Polyethylenes, for example, are approximately 50% to 65% crystalline, with the balance being amorphous. One common view of the molecular structure as proposed by Lustiger and Markham (1985), has the molecules folding back upon themselves numerous times forming lamellae as shown in Figure 1. These lamellae form the crystalline regions in the polymer. In the initial stages of deformation the intact lamellae move apart with the tie molecules in the amorphous region doing the unfolding thereby accommodating the separation. The three sketches of Figure 1(a) illustrate the likely phenomenon. If additional separation occurs and these tie molecules are sufficiently numerous and strong, the crystalline regions themselves will pull apart under stress and eventually result in a ductile behavior of the geomembrane, see the two sketches of Figure 1(b). This is easily observable by the very large post-yield elongation of HDPE specimens tested in a one-dimensional tension test. If, however, the tie molecules abruptly break, as shown in the sketches of Figure 1(c), the crystalline regions remain intact and failure occurs between the lamellae often in an abrupt and brittle manner. This latter situation is that of stress cracking and is the focus of this Chapter. It must be mentioned that Figure 1 is only one possible mechanism of stress cracking and is not universally accepted. It does, however, nicely illustrate the phenomenon on the basis of easily grasped concepts.

In such instances of brittle stress cracking the sheet materials on both sides of the crack maintain their original thickness and ductile elongation (along with gradual thinning of the sheet thickness) does not occur. Once initiated, the stress crack propagates vertically through the sheet's thickness and then, depending on the continuation of the applied stress magnitude and orientation, propagates along the sheet in a horizontal direction. The rate of crack propagation can be slow or rapid, Halse, et al. (1989). Slow crack growth is evidenced by a fibrous surface morphology when viewed under a low magnification in a scanning electron microscope. Crack growth rates can

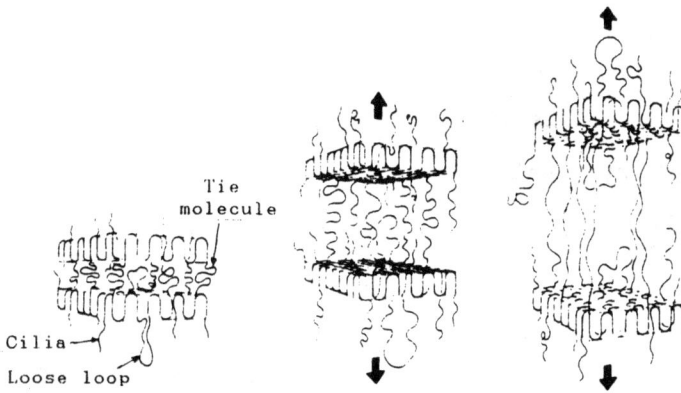

(a) Initial Stages of Lamellae Separation

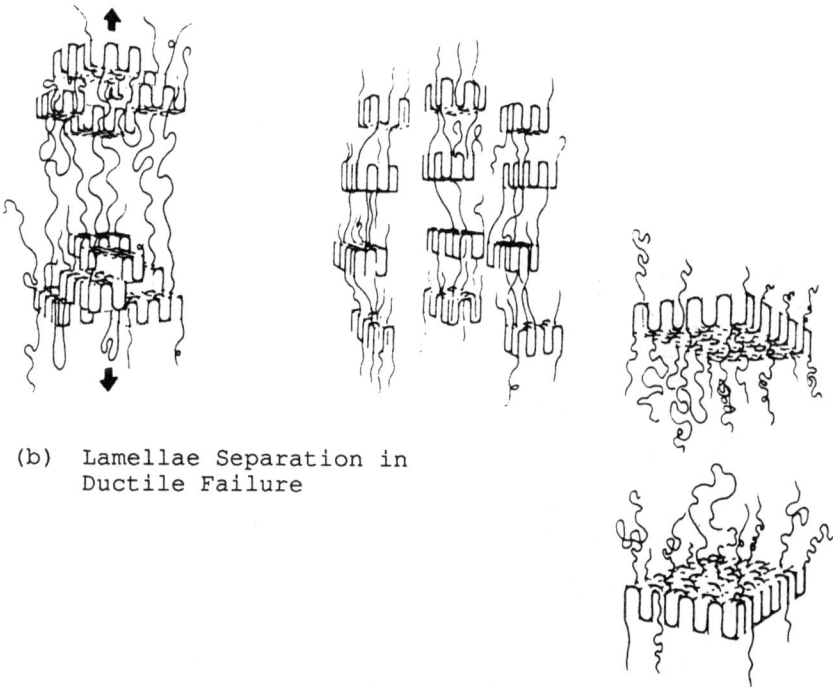

(b) Lamellae Separation in
Ductile Failure

(c) Tie Molecule Separation
in Brittle Failure

Fig.1 Conceptualization of ductile and brittle stress
cracking in polyethylene, after Lustiger and Markham, 1985.

range from minutes to months per centimeter of crack opening. Rapid crack growth is evidenced by a flaky surface morphology and can be many meters per second under severe conditions. In either case (slow or fast crack growth rates) the situation is of great concern whenever it occurs.

11-2   Laboratory evaluation

The most widely used laboratory test method to evaluate stress cracking of polyethylene geomembranes is the ASTM D-1693 test method. Its proper title is, "Environmental Stress Cracking of Ethylene Plastics" although it is widely referenced as the "Bent Strip Test". The test specimens are 38 mm long by 13 mm wide with a controlled depth, or notch, placed along one surface, see Figure 2(a). The depth of the notch is approximately 1/5 of the geomembrane's thickness. The test specimen is then bent 180° and placed within the flanges of a small channel, see Figure 2(b). Thus a constant tensile strain is exerted on the outside of the specimen along the surface where the notch is located. The entire assembly, along with a number of similarly notched test specimens, is immersed in a surface wetting agent at an elevated tempera- ture, see Figure 2(c). The test procedure allows for the user to select these conditions. The National Sanitation Foundation (1983) has recommended 100% Igepal and 100°C. If any cracks or crazes are observed on the geomembrane's surface before an arbitrary defined length of time the sheet material is considered to be sensitive to stress cracking. The cracks generally develop at the sides of the notch near the top of the arch. They generally grow towards the outer edge of the specimen at right angles to the notch.

Most commercially available geomembrane sheet materials perform quite well in this test. Manufacturers regularly list times of 500 to 2000 hours of crack free performance. If, however, the test were conducted at 10% Igepal solution and 50°C the results would be somewhat poorer since these are more severe conditions for this particular test.

Alternatively, one could evaluate stress cracking of polyethylene geomembranes by ASTM D-2552, entitled "Environmental Stress Rupture of Type III Polyethylene Under Constant Tensile Load". The test specimen used in this test is dumbbell shaped without any notches. The neck region has a continuous curvature with a radius of 13 mm. The constant load mobilizes a tensile stress which is concentrated on the narrowest section of the neck which is 3 mm in width. Specimens are loaded to a given proportion of their yield stress by static load, see Figure 3. They are then immersed in a surface wetting agent at an elevated temperature. Often 10% Igepal at 50°C is used. Twenty specimens are tested simultaneously and after 50% fail in a brittle manner the test is concluded, i.e. F_{50} is determined. Materials with high F_{50} values will have better stress cracking

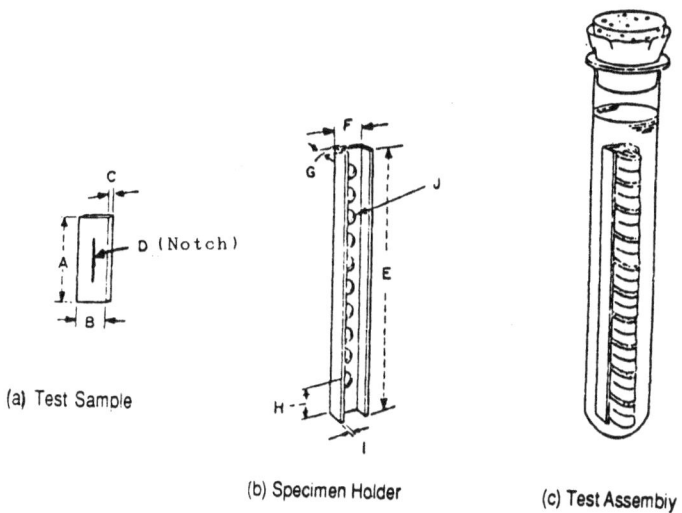

(a) Test Sample

(b) Specimen Holder

(c) Test Assembly

Fig. 2   Schematic and details of ASTM D-1693 for evaluating stress cracking of geomembrane sheet materials.

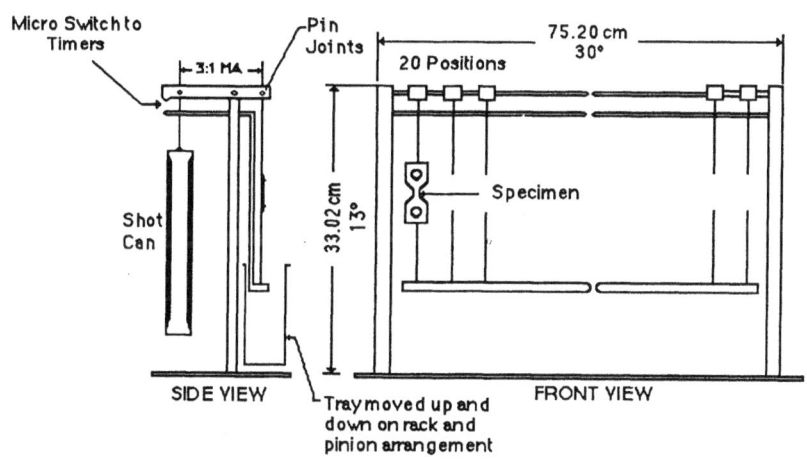

Fig. 3   Schematic and photographs of ASTM D-2552 test vice used for evaluating stress cracking of geomembrane sheet materials and seams.

245

stress cracking resistance than those with low F_{50} values.

This test can be readily adapted to evaluate stress cracking of geomembrane seams. By modifying the test specimen to include a constant length central region the system can incorporate a seam in the test. Thus the 3 mm neck section can be extended as long as desired to accommodate any length of seam. This type of modified D-2552 test method has been the focus of studies done on stress cracking at the Geosynthetic Research Institute (GRI) by Halse, et al., (1989).

Beginning in 1987, a series of D-2552 (modified) tests on a random group of laboratory and field retrieved polyethylene geomembrane seams have been conducted at various percentages of their yield stress. Table 1 gives the results for 168 hour load duration and Table 2 for 1000 hour load duration.

Table 1   Summary of stress cracking tests on geomembrane seams; 168 hours duration, after Halse, et al. (1983).

Type of Seam	Number of Specimens at % Yield Stress	Elastic	Plastic	Cracked			Percent Cracked
				Level 1	Level 2	Level 3	
Fillet*	44 @ 35	21	2	2	6	13	
Extru-	45 @ 40	21	1	4	5	14	51%
sion	45 @ 45	17	4	8	3	13	
	45 @ 50	15	6	6	4	14	
Flat	10 @ 35	10	0	0	0	0	
Extru-	10 @ 40	4	0	5	0	1	15%
sion	10 @ 45	1	9	0	0	0	
	10 @ 50	0	10	0	0	0	
Hot	15 @ 35	4	5	5	0	1	
Wedge	15 @ 40	2	6	4	1	2	27%
	15 @ 45	11	3	1	0	0	
	15 @ 50	8	5	0	0	2	
Hot Air	20 @ 35	11	0	1	1	7	
	20 @ 40	8	1	4	0	7	39%
	20 @ 45	5	8	0	0	7	
	20 @ 50	3	13	1	0	3	
Ultra-	10 @ 35	9	0	1	0	0	
sonic	10 @ 40	8	0	0	1	1	18%
	10 @ 45	6	0	1	1	2	
	10 @ 50	5	5	0	0	0	

*This is combined data from three different manufacturers.

Table 2 Summary of Stress Cracking Tests on Geomembrane Seams; 1000 Hours Duration, after Halse, et al., 1989.

Type of Seam	Number of Specimens at % Yield Stress	Elastic	Plastic	Cracked			Percent Cracked
				Level 1	Level 2	Level 3	
Fillet* Extrusion	4 @ 30	3	0	1	0	0	
	4 @ 35	4	0	0	0	0	25%
	4 @ 40	4	0	0	0	0	
	4 @ 45	1	0	2	1	0	
	4 @ 50	2	1	1	0	0	
Flat Extrusion	5 @ 20	5	0	0	0	0	
	5 @ 30	5	0	0	0	0	45%
	5 @ 40	1	0	0	0	4	
	5 @ 50	0	0	0	0	5	
Hot Wedge	4 @ 30	4	0	0	0	0	
	4 @ 35	2	0	0	0	2	60%
	4 @ 40	2	0	0	0	2	
	4 @ 45	0	0	0	0	4	
	4 @ 50	0	0	0	0	4	

*This is from one manufacturer.

The test results are divided into three classes, elastic, plastic, and cracked. In addition, the last class is subdivided into three categories according to the extent of the crack growth. These categories are listed in following:

• Elastic: nominal deformation within the elastic range of the material.
• Plastic: large deformation which results from plastic deformation in the necked region.
• Cracked: deformation which results from an abrupt stress crack in the necked region transverse to the applied stress.
Crack Level #1 refers to small cracks that can only be seen at a magnification of 40x.
Crack Level #2 refers to moderate cracks that can be seen by the unaided eye.
Crack Level #3 refers to complete fracture of the specimen.

Examples of these three levels of cracking along with an original seam for comparison purposes are shown in Figure 4.

The goal of these tests is not to compare seam types, but to evaluate the general susceptibility of a wide range of

247

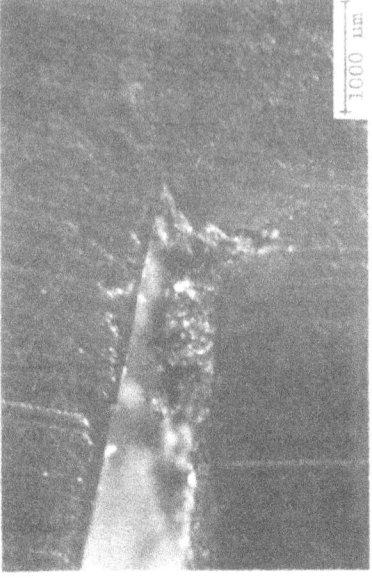

Original

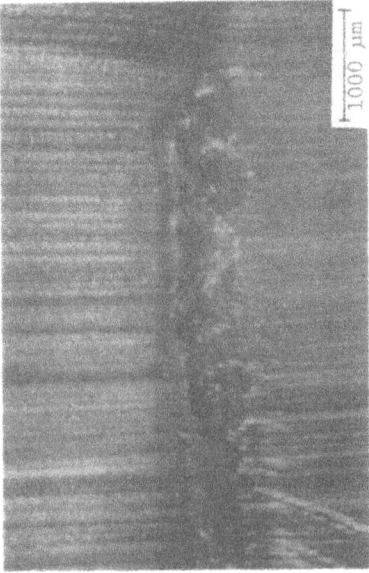

Crack level #1

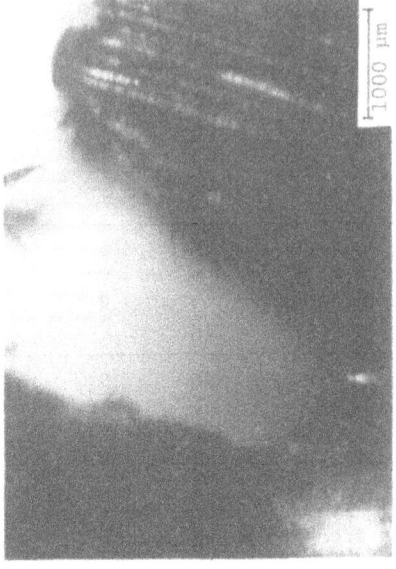

Crack level #2

Crack level #3

Fig. 4   Photographs of various levels of stress cracking.

polyethylene geomembrane seams to stress cracking. The data
from these tables clearly indicates that all types of polyethylene seams are sensitive to stress cracking but to
varying degrees and at different stress levels. It should
be noted that the original samples from which the test
specimens were formed came from a myriad of sources. Most
were from the field, made by different seaming crews under
various weather conditions.

In comparing the 168 hour test results to the 1000 hour
test results the failure percentage of flat extrusion seams
and hot wedge seams increased when the length of the
experiment increased. While this does not appear to be the
case for the fillet extrusion seams, it may be because of
the number of different manufacturers that were included in
the 168 hour test results. Only one manufacturer's material
was used in the 1000 hour tests. For the same seam, the
failure percentage value increased from 10% to 25% when the
length of the test was extended form 168 to 1000 hours.
Thus it appears that increased test time is a sensitive
parameter to evaluate the stress cracking phenomena.

After testing in the above described manner, the specimens were examined under a light microscope so that the
location, direction, and extent of the cracking could be
observed. Four different types of seams are illustrated
using the schematic diagrams shown in Figure 5. In all
cases, the cracks grew in the direction perpendicular to the
applied stress. In most of the seams, the location where
the crack is initiated is very close to the overlapping
junction of the two geomembrane sheets where the stress
concentration is the highest. The stress level at this
location depends significantly on the geometrical shape of
the junction. Generally, a pointed junction is felt to
result in a higher stress intensity than a rounded one. In
addition, scratches due to grinding can provide for crack
initiation when they occur at these locations.

For a few of the fillet extrusion seams, cracks were
observed at either end of the extrudate, as shown in Figure
5(b). However, this is not a common situation in these
tests since only one set of tests had such failures. The
amount of extrudate applied to the seam in this case was
much more than the other fillet extrusion seams, resulting
in a larger cross-sectional area across the seam. Hence,
the global stress at the overlapping junction is corre-
spondingly reduced. The maximum stress concentration
changes its location to the ends of the extrudate where the
cross-sectional area is the smallest. Note that these
locations also coincide with the area of most intense
grinding. Therefore any scratches resulting from the
grinding activity can act as crack initiation points. For
other cases of fillet extrusion seams the cracks propagated
through the extrudate itself.

While the results of Tables 1 and 2 are of concern, it
must be emphatically stated that these laboratory test

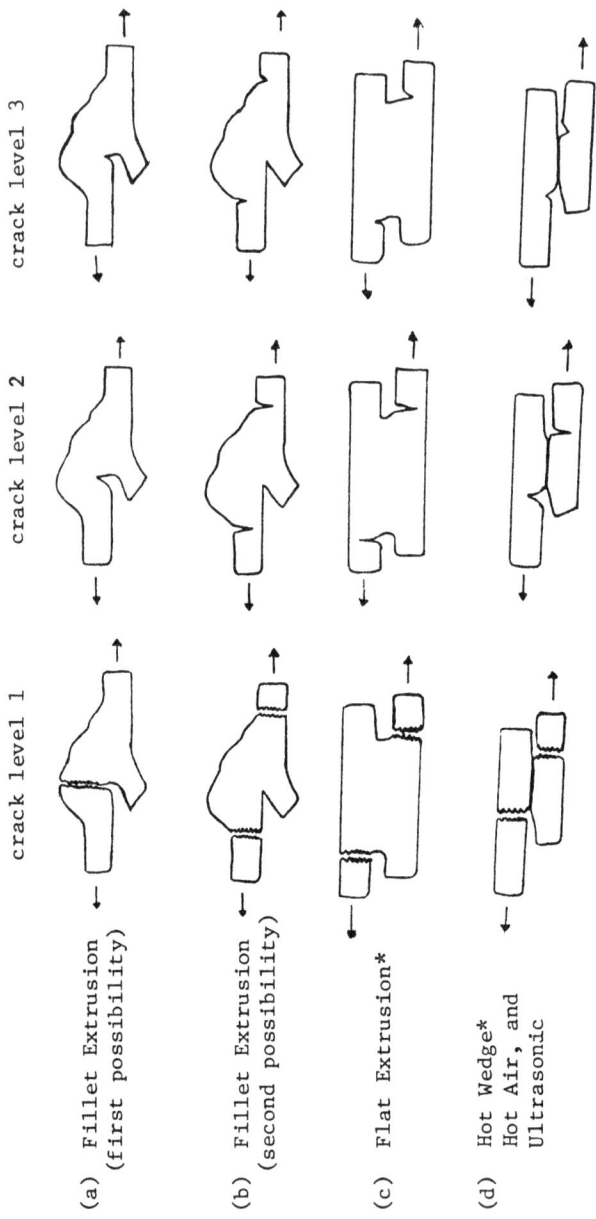

crack level 1   crack level 2   crack level 3

(a) Fillet Extrusion
    (first possibility)

(b) Fillet Extrusion
    (second possibility)

(c) Flat Extrusion*

(d) Hot Wedge*
    Hot Air, and
    Ultrasonic

Note: *Cracking only will occur on one side or
       the other but not both sides

Fig. 5   Cross section views of different seam types and locations of stress cracks.

conditions are very different from being a field simulation. For example, the following items in the laboratory test make it only an index test and clearly not a performance test.

- The test specimen width is extremely small, i.e., 3 mm.
- The test specimen has a complex geometrical shape and the stress state in the seam is very difficult to determine.
- The wetting solution can attack the test specimen on all four sides, whereas in the field it is only acting against the upper surface
- The applied stress is kept constant, thus stress relaxation in the polymer structure cannot occur as it possibly would in the field.
- A slight load eccentricity occurs in the test due to a lack of applied normal stress. This has the effect of a slight rotation with somewhat of a peel action occurring, rather than strictly a shearing action.
- Ten percent Igepal is a very aggressive wetting agent.
- The elevated temperature of 50°C is considerably higher than most field environments.

In spite of these comments, however, we do feel that the test is a good index test wherein one can compare seam types, seaming temperatures, seaming pressures, effects of surface grinding, resin types, effect of additive packages, etc., etc., so as to help optimize field seaming procedures.

11-3 Field incidence of stress cracking

There have been several field reported incidents of stress cracking. All have been on polyethylene geomembranes used for surface impoundments where the geomembranes were exposed to the atmosphere. Thus ultraviolet light and oxidation are involved, as well as expansion and contraction to the extent that the geomembrane surface varies in temperature. All of the cracks have occurred, or at least started, at field seams. Papers by Peggs (1988, 1989), Fisher (1989), and Rollin and Peggs in this Book, adequately describe the field situation and when added to the authors' case histories a reasonably well defined set of common conditions result.

1. All known stress cracking failures have occurred at surface impoundments where the geomembrane was exposed.
2. The failures almost always occur on the side slopes above the liquid level or on the horizontal runout length leading to the anchor trench.
3. The tensile stress is mobilized in large part by temperature contraction during cold winter months or during cold evenings. Inadequate slack is observed at many sites.
4. Southerly exposure side slopes where temperature extremes are the greatest, appear to have the most problems.

5. Seam failures often are accompanied by overgrinding and/or overheating.
6. Where fillet extrusion seams are placed directly over previously made flat extrusion, hot wedge or hot air seams, cracks often occur at the edge of the fillet extrusion seam.

The authors of this particular paper know of no solid waste landfill liners where stress cracking has occurred nor any situations involving covered polyethylene geomembranes. The problem appears to be focused on exposed polyethylene liners used for reservoirs, liquid impoundments and other situations where atmospheric conditions have direct impact on the geomembrane's surface.

11-4   Recommendations

At this point in time it appears that with adequate temperature compensation placed in the polyethylene geomembrane during installation and excellent workmanship in making field seams, the field documented stress cracking reports will be significantly reduced. Both of these suggestions are well within the grasp of the designer (for temperature compensation) and the installer (for proper seams). Of course, both could, and perhaps should, be verified by the owner's representative which is sometimes done via a third party engineering auditor.

Regarding the quality of individual types of field seams there is much that can be said. Most documents from engineering firms which merely ask for destructive shear and peel tests at various intervals are felt to be inadequate. In fact, to develop high shear and peel test values the installers sometimes tend to overgrind and/or overheat the seams, a process which most likely favors stress cracking at, or near, the seams. Needed is a delicate balance between surface buffing and proper temperature during seaming, and adequate seam strength after seaming. This is a difficult task, but hopefully a do-able task, see EPA (1989).

On a longer time frame than the above recommendations it is possible that more stress crack resistant PE resins may be forthcoming. Quite possibly the tie molecule bonding between lamellae can be enhanced and work is ongoing by a number of resin suppliers in this regard.

11-5 Acknowledgements

This project is funded by the U.S. Environmental Protection Agency under Project No. CR-815692-01. Our sincere appreciation is extended to the Agency and in particular to the Project Officer, David A. Carson.

252

References

ASTM D-1693, "Environmental Stress Cracking of Ethylene Plastics", Vol.08.02

ASTM D-2552, "Test Method for Environmental Stress Rupture of Type III Polyethylenes Under Constant Tensile Load", Vol.08.02 (terminated in 1987).

EPA, "Technical Resource Document: The Fabrication of Polyethylene FML Field Seams," EPA/530/SW-89/069 September, 1989.

Fisher, G. E., "Controlling Thermal Damage in Flexible Membrane Liners," Geotech. Fabrics Report, Mar./Apr., 1989, IFAI, pp. 39-41.

Halse, Y. H., Koerner, R. M. and Lord, A. E. Jr., "Stress Cracking Morphology of HDPE Geomembrane Seams," Proc. ASTM Conf. on Microstructure and the Performance of Geosynthetics, Orlando, FL, Jan. 1989, (to appear as a STP in 1989).

Halse, Y. H., Koerner, R. M. and Lord, A. E. Jr., "Laboratory Evaluation of Stress Cracking in HDPE Geomembrane Seams," Proceedings GRI-2, Durability and Aging of Geosynthetics, Jour. Geotec. and Geomemb., Elsevier Publ. Co., May, 1989, pp. 177-194.

Lustiger, A. and Markham, R. L., "Importance of Tie Molecules in Preventing Polyethylene Fracture Under Long Term Loading Conditions," Proceedings 9th Plastic Fuel Gas Pipe Symp., New Orleans, LA, Nov. 1985, pp. 132-140.

National Sanitation Foundation Standard November 54 - Flexible Membranes Liners, Ann Arbor, MI, Nov. 1983.

Peggs, I.D., "Failure and Repair of Geomembrane Lining Systems," Geotechnical Fabrics Report, Nov. 1988, IFAI, pp. 13-16.

Peggs, I. D. and Carlson, D. S., "Stress Cracking of Polyethylene Geomembrane: Field Experience," Proceedings GRI-2, Durability and Aging of Geosynthetics, Jour. Geotex. and Geomemb., Elsevier Publ. Co., May, 1989, pp. 195-211.

12 MICROANALYSIS OF POLYETHYLENE GEOMEMBRANE SEAMS

A. L. ROLLIN
Ecole Polytechnic of Montreal, Quebec, Canada
I. D. PEGGS
GeoSyntec Inc., FL, USA

12-1 Introduction

The increasing use of High Density Polyethylene (HDPE) geomembranes in geotechnical applications, the different welding techniques used to seam geomembrane sheets and the limitations of the welding apparatus have created a need to improve field quality control methods and proper welding guidelines.

The HDPE geomembrane sheets may only be assembled through polymer fusion in which basically a quantity of energy (heat) is applied at the sheets interface there by developing new molecular links between adjacent sheets. Pressure is also applied on the seam in order to ensure a homogeneous bond. The objective is to produce a homogeneous weld zone with the same microstructure and strength characteristic as the geomembrane.

The quality of an HDPE installation depends on many factors related to the membrane itself, atmospheric conditions at the time of seaming, the projet design as well as the implantation of the quality assurance program.

But the more important factors related to geomembrane welding parameters are usually identified from the type of equipment and welding technique used such as the welding speed, the welding temperature and the applied pressure (Rollin (1989b). Each of these parameters may affect the seam strength as well as the molecular structure of the seam.

The influence of these parameters on the seam bond strength, may be determined using destructive tests such as shear and peel tests. As indicated by Rollin

(1989a) and Peggs (1989a), microscopic examination of cross sections of seam samples with bond strength greater than the acceptance criteria indicated unbonded areas, poor carbon black dispersion, stress cracks and crazing. This is corroborated by recent cases reported on long term failures (Halse (1989), Peggs (1986b)).

Microscopic analysis may be performed using an optical and/or a scanning electron microscope. These techniques are used to characterize structural molecular changes, micro stress cracks within the bonded sheets, carbon black dispersion, unbonded areas, residual stress magnitude and distribution, and slow crack growth phenomena. These microscopic abnormalities in the seams (unheard of from the traditional dynamic test methods) may contribute to brittle fractures during the installation or during the projected service life of the installation.

In this paper, three microscopic analytical techniques are presented: a) light transmission microscopic analysis on microtome seam sections; b) optical microscopic analysis of seam cross sections; and c) scanning electron microscopic analysis.

12.2 Microscopic analysis of microtome seam sections

For many years, microtomes have been used to prepare very thin slices of biological materials that are translucent when observed through a light transmission microscope. Such microtome sections reveal the internal microstructure and defect distribution within the materials. The technique is equally applicable to polymers such as HDPE geomembranes (Bell (1979)) and hence to geosynthetic components, particularly those manufactured from semi-crystalline materials. This method was applied by Peggs (1989a) to geomembranes.

12.2.1 Specimen preparation

The basic microtome consists simply of a clamp capable of firmly holding the seam specimen that can be raised controllably in increments of approximately 0.1μm such that a very sharp, rigid knife can cut slices that are 8 to 20 microns thick. The microtome apparatus is shown in Figure 12.1.

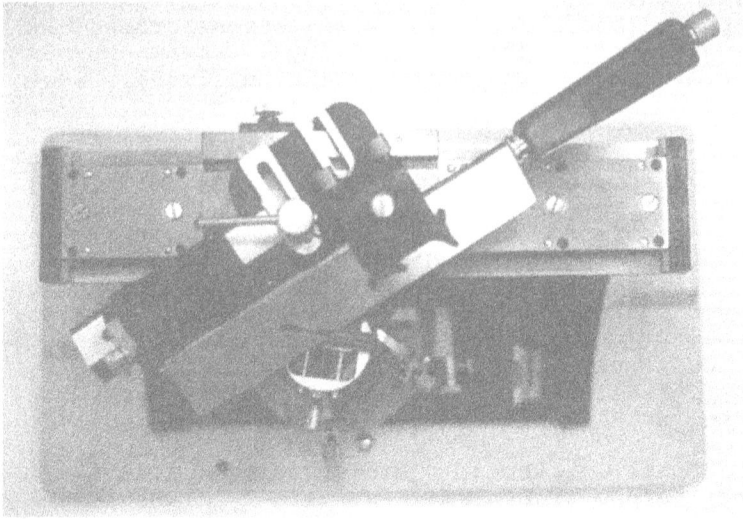

Figure 12.1   Microtome apparatus

It is still much of an art to prepare a large, thin
section since each grade of HDPE requires fine tuning
ajustments to the knife angle. As a general guide,
the knife should be at a vertical angle of
approximately 15° to the horizontal plane of the
surface being cut. The longitudinal axis of the
sample itself should be held at an angle of
approximately 8° to the direction of the motion of
the knife. This is done to ensure that the markings
caused by the knife imperfections will not be oriented
in the same direction as the extrusion pattern
of the geomembrane. The knife, in turn, is best
held at an angle of approximately 55° from normal to
the direction of motion. The direction of the knife
with respect to the specimen is illustrated in
Figure 12.2.

As the microsection is cut, it may curl. It can be
gently uncurled into a layer of balsa cement
spread on the surface of a glass slide. The
microsection must be completely covered with cement
before a transparent glass cover slide is placed on top
of it. The two slides are slightly compressed to
flatten the microsection and to expel air bubbles from
the balsa cement. Varnishes can be an alternative to
the balsa cement. A microtome specimen is shown in
Figure 12.3.

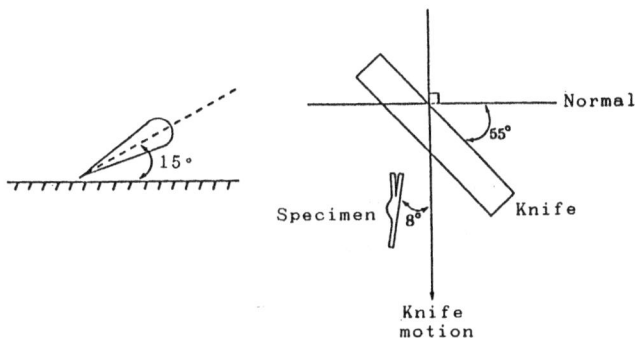

Figure 12.2   Microtome knife orientation

12-2.2   Specimen examination

The slide assembly is placed on the stage of a light transmission microscope so that rotating polarizing filters may be placed on each side of the specimen. When crossed or partially crossed polarizing filters are placed on each side of the section, the contrasts between many of the features are highlighted and improved and most significantly, the distribution and relative magnitude of residual stresses are identified shown by brilliant blue, yellow and orange colors (birefringence). It is advisable to use at least one thin cover slide so that high magnification objective lenses may be used.

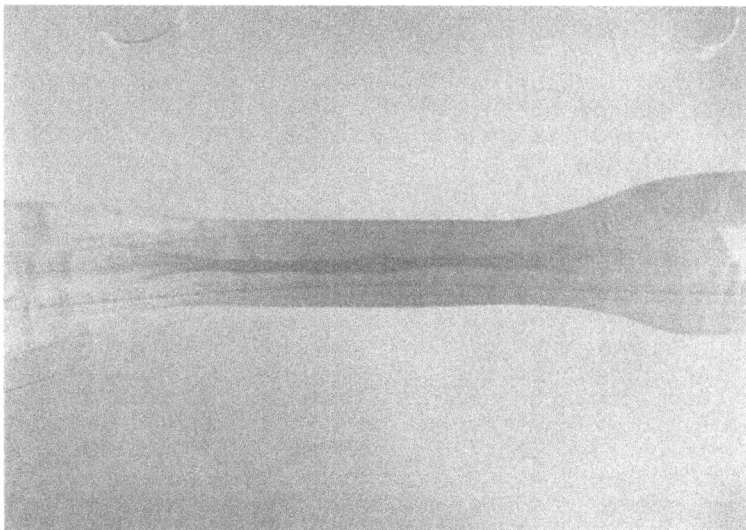

Figure 12.3   Microsection of an HDPE geomembrane seam

This method permits to identify the symmetry and homogeneity of the seam, the density matching of extrudate and parent resin, the effectiveness of extrudate/sheet mixing, the elimination of distinct interface, the presence of unbonded areas, voids and dirt particles, the distribution of carbon aggregates and the presence of crazes and cracks. It can also be used to measure the stress crack growth rates in the specimen used with constant tensile load tests described in chapter 2.

12-3 Optical microscopic analysis of seam cross section

The examination of a cross section of a welded HDPE sheets using an optical microscope can be performed on microtome slices. To obviate any difficulties encountered with the specimen preparation, a technique has been developed by Rollin (1989c) using optical fibres to light up surfaces of black polyethylene specimens obtained from simple die cuts. The specimen is therefore viewed with reflected light.

The specimens, approximately 10 mm wide and length equivalent or greater than the weld itself, are placed directly under the microscope in order to be observed and phoographed. Two optical fibers connected to the light source are placed at angles directly on top of the specimen to permit observation of a lighted surface. The angles of the lights can be infinitively varied to produce surface profile shadowing that will entrance the features of interest.

Using minimum resolution of 100 to 200 nm, macroscopic examination can be performed at magnifications ranging from 100 to 2000 on smooth surfaces. Similarly to the previous technique, carbon black dispersion, interface determination, localization of unbonded areas, displacement of a mass of polymer toward the edge of the seam, formation of a longitudinal channel across the seam, and identification of crazes and macro stress cracks can be observed with this technique.

Although these examinations can easily be performed with optical magnification, it is less applicable to microfractography because of limited resolution and depth of field (cannot focus on rough surfaces). A stereo microscope can be used for low magnification (<x20) fractography using adjustable fibre optic light sources. For higher magnifications, a scanning electron microscope is recommended for fractographic analyses.

12-4 Scanning electron microscopic analysis

The scanning electron microscope (SEM) is one of the most appropriate instrument for the investigation of microstructural changes and identification of micro stress cracks within welds. Fracture face characteristics provide information on the factors responsible for cracks initiation and the mechanism of propagation.

The minimum resolution range of SEM is 4 to 5 nm with a maximum useful magnifications up to 6000 X. This technique requires specimen cross sections to be coated with a paladium-gold alloy to improve electrical conductivity and to prevent the build up of static charges on the specimen surface.

The SEM may be used to complement the other microscopic techniques since it is more costly and time consuming. It can be used successfully to identify brittle fracture initiation sites and propagation of failure of samples collected in situ. Also it can be used in the detection of micro-stress cracks within the polymeric geomembrane or seam.

12-5 Seam examination

Photographs of seam cross sections are presented in this section to support microscopic analysis as a necessary technique both to establish their quality and to explain seam failure and processes.

12-5.1 Unbonded areas

As presented in Figure 12.4 a cross section of a fillet extrusion weld can be observed at a magnification of 13X. The mixing of both the sheet and the extrudate polymer was performed correctly at the weld as no interfaces can be detected between the upper and lower sheets and the extrudate. On the contrary, a horizontal interface can be observed in the magnified cross section of an hot wedge seam (75X) presented in Figure 12.5 indicating an unbonded area between the upper sheet and extrudate of an overlap extrusion weld.

259

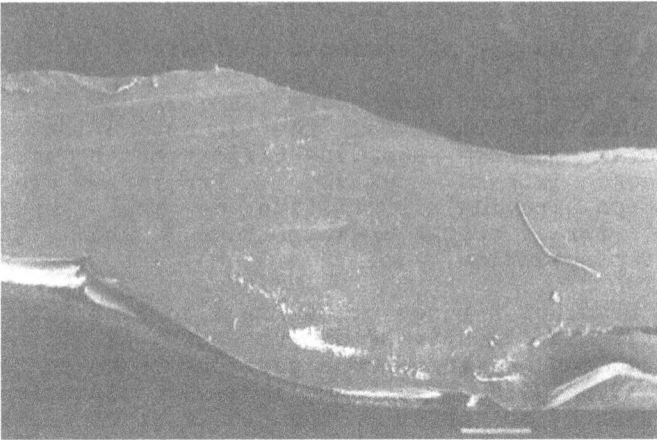

Figure 12.4   Cross section of a fillet extrusion
             HDPE weld

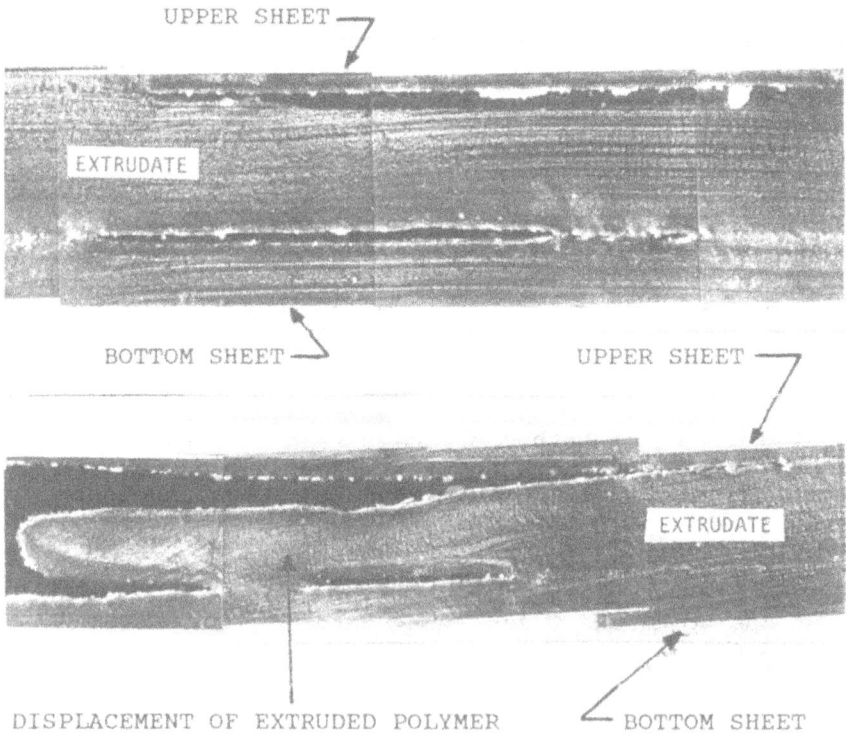

Figure 12.5   Photograph of a cross section of an
             overlap extrusion weld

12-5.2 Stress cracks

Residual stress cracks can be identified easily on microscopic photographs as shown in Figures 12.6 to 12.8. On Figure 12.6, a cross section magnified at 150X shows unbonded areas and the presence of channels and horizontal microcracks in the upper sheet and transverse microcracks in the extrudate of an overlap extrusion weld. These stress cracks are originating from the sheets interface and propagate perpendicularly through the extrudate thickness.

Similarly, microcracks are observed in Figure 12.7 at the interface of both sheets for a hot wedge weld magnified at 200X. In this case the microcracks are transverse in both the upper and bottom sheets and originated from the interface.

A larger magnification of these microcracks is presented in Figure 12.8 (350X). Many horizontal microcracks are associated with a large transverse stress crack of relatively important depth indicating a complex stress pattern.

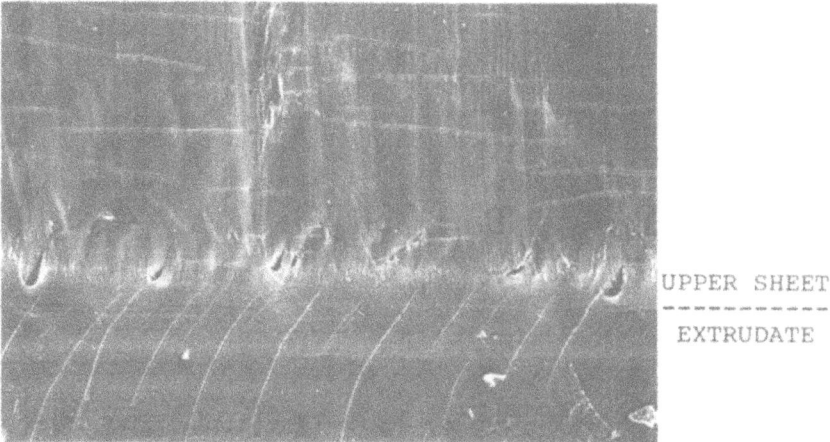

UPPER SHEET

EXTRUDATE

Figure 12.6   Photograph of an overlap extrusion weld

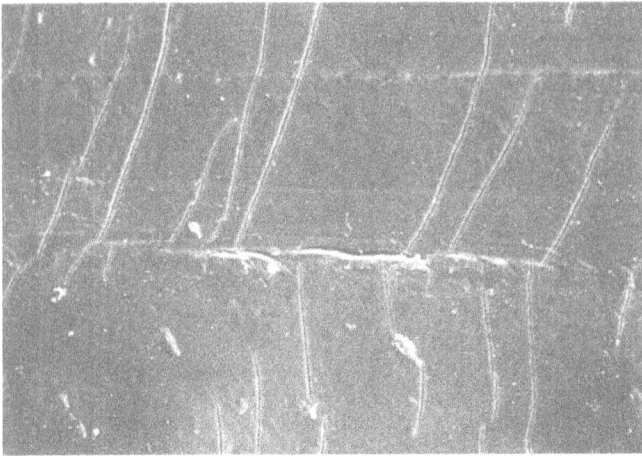

UPPER SHEET

BOTTOM SHEET

Figure 12.7   Photograph of a hot wedge seam

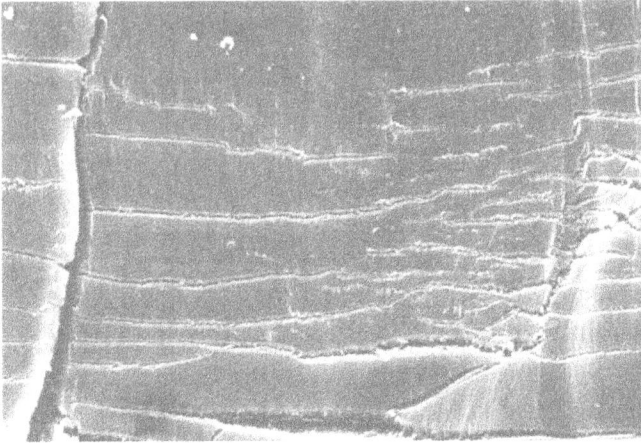

Figure 12.8   Photograph of magnified microstress cracks

12-5.3 Craze identification

When a seam peel test is performed, some peel separation may occur before the geomembrane itself breaks. The significance of such separation in the acceptance or rejection of a peel test is indicated in Figure 12.9 by the appearance of crazes that penetrate through 30% of the thickness of the geomembrane after separation has occurred. Since peel stresses can occur in the field, a seam which can be made to peel may be subject to crazing and ultimate premature failure as such crazes open up into cracks. Depending on the local stress situation, these cracks may propagate completely through the geomembrane sheet.

12-5.4 Residual stresses

The existence of residual stresses can be detected in microtome specimens by brilliant blue, yellow and orange colors which, in (Figure 12.10) can be observed at the root of the small extruded bead at the edge of a fused seam. When service stresses are applied to a seam, crazes are initiated at the residual stress area (Figure 12.11). These crazes eventually propagate upward through the geomembrane to form a brittle crack along the edge of the seam.

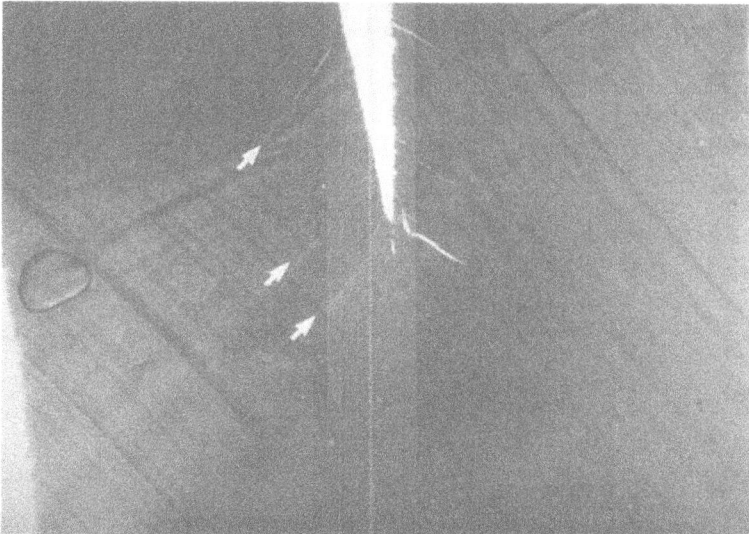

Figure 12.9   Microsection of a weld standing peel separation

Figure 12.12 shows part of a microsection through an extruded fillet seam at the bottom of the edge of the top sheet over which the extrudate has been deposited. There is an area of residual stress on the edge of the top sheet which, in service, subsequently initiated a crack up the edge of the top sheet. Some crazing that has been initiated at the notch stress concentrator where some of the extrudate has separated from the top of the bottom sheet can also be observed in this figure.

Figure 12.10   Microsection of a hot wedge weld

264

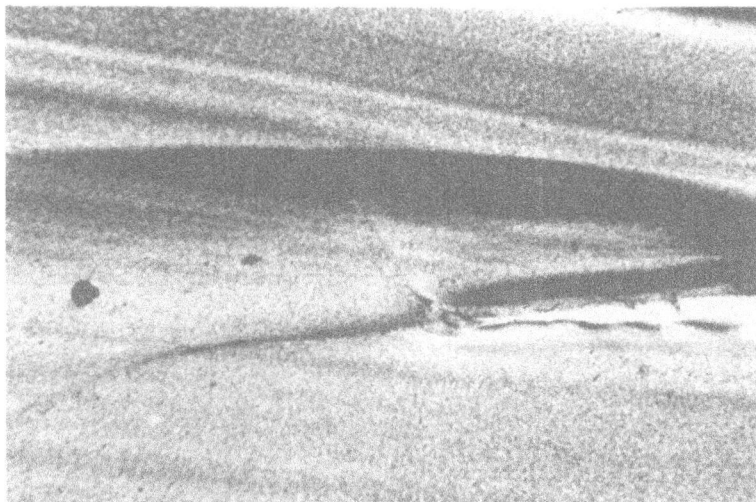

Figure 12.11   Microsection of a hot wedge weld

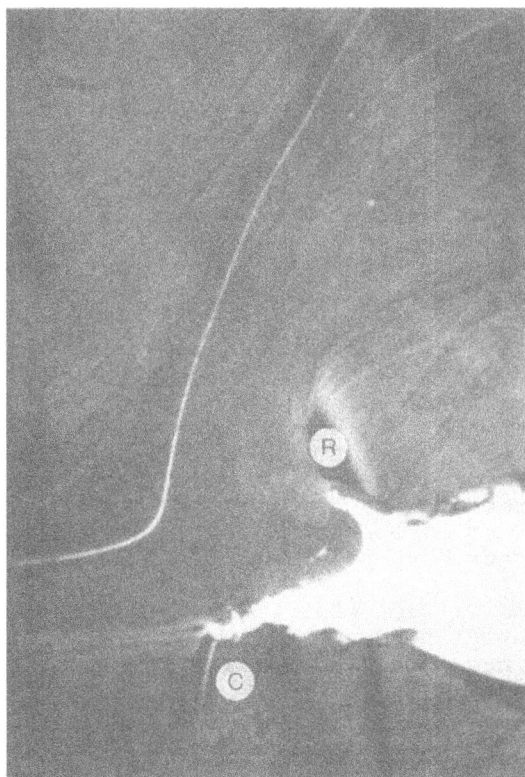

Figure 12.12   Microsection of an extruded fillet seam

12.6 Conclusions

Recently some cases have been reported of long term failure of HDPE seams (Halse (1989), Peggs (1988), (1989b) and Rollin (1989b)). These findings can be alarming if not well documented. As shown in this paper, microscopic analysis certainly advances useful information which helps to assess seam quality and to identify faulty seams not defined by etheir the shear nor the peel tests.

These methods are invaluable in detecting structural abnormalities likely to restrict the long term performance of seams and in determining the characteristics of fracture phenomena during failure analysis. These techniques can also easily be performed as customary test procedures on seam quality evaluation in an effort to improve proper welding guidelines for HDPE geomembranes.

12.7 References

Bell, G.R. and Cook, D.C., (1979), "Microtoming: An Emerging Tool for Analzing Polymer Structures", Plastic Engineering, August, pp 18-22

Halse, Y., Koerner, R. and Lord, A.E., (1989), "Laboratory Evaluation of Stress Cracking in HDPE Geomembrane Seams", Proceedings GRI-2, Durability and Aging of Geosynthetics, Journal Geotextiles and Geomembranes, Elsivier

Peggs, I.D., (1988), "Failure and Repair of Geomembrane Lining Systems", Geotechnical Fabrics Report, November, pp 13-16

Peggs, I.D. and Charron, R.M., (1989a), "Microtome Sections for Examining Polyethylene Geosynthetic Microstructures and Carbon Black Dispersion", proceedings of Geosynthetics'89 Conference, San Diego, February, pp 421-432

Peggs, I.D. and Carlson, D.S., (1989b) "Stress Cracking of Polyethylene Geomembrane: Field Experience", proceedings GRI-2, Durability and Aging of Geosynthetics, Journal of Geotextiles and Geomembranes, Elsevier, May

Rollin, A.L., Vidovic, A., Denis, R. and Marcotte, M., (1989a), "Microscopic Evaluation of HDPE Geomembrane Field Welding Techniques", proceedings ASTM symposium on Microstructure and the Performance of Geosynthetics, in press, Orlando, January

Rollin, A.L., Vidovic, A., Denis, R. and Marcotte, M., (1989b), "Evaluation of HDPE Geomembrane Field Techniques: Need to Improve Reliability of Quality Seams", Proceedings of Geosynthetics'89 Conference, San Diego, February, pp 443-455

Rollin, A.L., Vidovic, A. and Ciubotariu, V., (1989c), "Assessment of HDPE Geomembrane Seams", Proceedings of Second International Landfill Symposium, Sardinia, October

13 CHEMICAL COMPATIBILITY TESTING OF GEOMEMBRANES

L. G. TISINGER and I. D. PEGGS
GeoSyntec Inc., FL, USA
H. E. HAXO Jr
Matrecon Inc. Alameda, CA, USA

13.1 Introduction

Because of their low permeability characteristics and durability, geomembranes are finding increased use throughout the world in the lining
of waste storage and disposal facilities. The geomembranes that are
being used for waste containment range from plasticized polyvinyl
chloride (PVC) to polyethylene (PE) and laminated composites. The
selection of a geomembrane for a specific application is based on a
number of factors, including permeability and physical and chemical
properties. Critical factors for selecting a geomembrane include
durability in and chemical resistance to the actual environment in
which the material will be exposed. Geomembranes generally have good
resistance to many different environments; however, since the chemical
environments in storage and disposal facilities are site specific, it
is difficult to predict accurately the ability of a geomembrane to
maintain its initial properties for the lifetime of the installation
solely on the basis of the manufacturer's product literature. Therefore, chemical compatibility tests are performed using the actual
waste liquid to which the geomembrane will be exposed.

This chapter presents background on chemical compatibility testing
of polymeric compositions with particular reference to geomembranes.

The correct approach to assessing the chemical resistance of geomembranes is emphasized, and selected compatibility data on various geomembranes in different liquids are presented. These liquids include solvents, aqueous solutions, and actual wastes. This is followed by a discussion of the chemical compatibility test methods currently being used, such as the American Society for Testing and Materials (ASTM) test methods for polymeric compositions, National Sanitation Foundation methods for geomembranes, and the U.S. Environmental Protection Agency (EPA) Method 9090 for geomembranes (flexible membrane liners). Limitations of the latter method are presented, and recommendations for more accurately assessing the chemical compatibility of geomembranes are made.

13.2   Background

13.2.1 Chemical Compatibility Testing of Polymeric Materials

When in service, many rubber and plastic products, e.g., gaskets, O-rings, diaphragms, containers, and pipe and tank linings, are in contact with liquids. Such contact may, over long-term exposure, affect the performance of these products. A variety of chemical compatibility and simulation tests have been performed by polymer producers and manufacturers of polymeric products to aid in the development and characterization of polymers and the proper selection of a polymer for manufacturing specific products. Generally in these tests, the materials or products are measured for changes in weight, dimensions, and selected analytical and physical properties as a function of exposure time and temperature. Table 13.1 presents a list of selected ASTM standards often used in assessing the compatibility of a material or product with a specific liquid.

Of particular interest for compatibility testing of geomembranes with various liquids are such ASTM Test Methods as D471, which covers resistance of rubbers and elastomers to various liquids, and D543, which covers the resistance of plastics to chemical reagents. In both methods the changes in weight, dimensions, appearance, and strength properties are determined as a function of immersion time. The two tests generally involve the use of standard reagents of known composition as the immersion medium. For example, ASTM D543 recommends 50 standard reagents which cover a broad range of chemical types. The immersion time and temperature and the specific properties to be reported are generally arbitrarily set. Data from these types of tests are the basis for many of the compatibility tables published by the producers and manufacturers of rubber and plastic products (Harper, 1975, and Modern Plastics Encyclopedia, 1988).

Also of particular importance for geomembranes that are based upon semicrystalline polymers are two tests, i.e., ASTM D1693 and D2552, that can be used to assess the material's resistance to stress cracking, a brittle fracture phenomenon that compromises the durability, particularly of semicrystalline materials.

269

Table 13.1   ASTM standards[a] relating to chemical compatibility of polymeric materials and products.

C541-83	"Specification for Linings for Asbestos-Cement Pipe," Section 04.05.
C581-83	"Practice for Determining Chemical Resistance of Thermosetting Resins Used in Glass Fiber Reinforced Structures, Intended for Liquid Service," Sections 08.01 and 08.04.
C868-85	"Test Method for Chemical Resistance of Protective Linings," Section 04.05.
D471-79	"Test Method for Rubber Property--Effect of Liquids," Section 09.01.
D543-84	"Test Method for Resistance of Plastics to Chemical Reagents," Section 08.01 (ISO Method 175, 1981).
D896-84	"Test Method for Resistance of Adhesive Bonds to Chemical Reagents," Section 15.06.
D1239-55(1982)	"Test Method for Resistance of Plastic Films to Extraction by Chemicals," Section 08.01.
D1693-70(1980)	"Test Method for Environmental Stress-Cracking of Ethylene Plastics," Section 08.02.
D2552-69(1980)	"Test Method for Environmental Stress Rupture of Type III Polyethylenes Under Constant Tensile Load," Section 08.02, discontinued in 1988 (ISO Method 6252, 1981). A modified procedure is being drafted by ASTM Committee D35.
D3491-85	"Methods of Testing Vulcanizable Rubber Tank and Pipe Linings," Section 09.02.
G20-83	"Test Method for Chemical Resistance of Pipeline Coatings," Sections 06.01 and 14.02.

---

[a]Annual Book of ASTM Standards. Issued annually in several parts. American Society for Testing and Materials, Philadelphia, PA.

No specific ASTM method exists for assessing the chemical compatibility of geomembranes with waste liquids. Standard methods for the exposure of candidate geomembranes in waste liquids and for following the changes in properties are currently being drafted by ASTM Committee D35 on geosynthetics.

13.2.2 Compatibility Testing of Geomembranes

In the early 1970s, the EPA became concerned with effects on the environment of the landfilling of wastes, particularly with respect to possible pollution of the groundwater. In many sites the uncontrolled waste disposal on native soils had resulted in contamination of the ground, thus threatening the groundwater. Because of these concerns, the EPA became interested in the use of various man-made materials as possible linings for waste storage and disposal facilities. These materials were recognized for their low permeability to water and other species and had been successfully used for water storage, such as in the lining of reservoirs, and for water conveyance, such as in the lining of canals. Also, man-made materials had been used to line pits, ponds, and lagoons for the short-term storage of different waste waters, brines, and other liquids. However, the EPA was concerned about the ability of these materials to function adequately in extended service as liners for waste storage and disposal facilities because there was essentially no experience with these applications.

As a consequence, a series of research projects was undertaken by the EPA to assess the compatibility and durability of various potential lining materials, including geomembranes, on exposure to a range of waste liquids for durations up to several years and to assess the various factors that affect the compatibility of geomembranes and waste liquids. These wastes included municipal solid waste (MSW) leachate and a range of hazardous wastes. The projects also studied the effects of exposing geomembrane seams to various chemicals and attempted to develop acceptability criteria. A program was initiated to determine the relationship between laboratory studies and the field performance of geomembranes. The results of these studies were reported at 15 annual EPA research symposia, at various conferences, and in publications. Data from many of these projects have been summarized in EPA technical resource documents (Matrecon, 1983, and Matrecon, 1988).

13.2.2.1 Compatibility of Geomembranes with MSW Leachate

The initial project exposed 12 different potential lining materials, including six geomembranes, five asphaltic compositions, and a soil-cement composition, to MSW leachate generated in the landfill simulators (Haxo et al., 1985b, and Matrecon, 1988). The test samples were sealed as liners into the bases of the simulators, thus one side of each sample was exposed continuously to the leachate which was ponded at a 30-cm head above the liner. The effects of exposing the six geomembranes to leachate for 12 and 56 months are reported in Table 13.2. These results show significant absorption of leachate components during the course of the exposure, with low-density polyethylene (LDPE) having the least absorption and a potable-water grade chlorosulfonated polyethylene (CSPE-P) having the most. By the end of the exposure, there was a decrease in the extractables content of the geomembranes, indicating extraction by the leachate. Depending on the property, the physical properties appeared to reach a minimum or maximum value after 12 months, which may reflect the effect of maximum concentration of the leachate, as is shown in Table 13.3.

271

Table 13.2 Effects on properties of geomembrane liners of 12 and 56 months of exposure to leachate in MSW landfill simulators.

Item	Test method	Exposure time (mo.)	Base polymer					
			Butyl rubber	CPE	Potable-grade CSPE-R	EPDM	LDPE	PVC
Type of compound	---	---	XL	TP	TP	XL	CX	TP
Analytical properties								
Volatiles, 2 h at 105°C (%)	---	0	---	0.10	0.29	0.50	0.00	0.09
		12	2.02	6.84	12.78	5.54	0.02	3.55
		56	2.37	7.61	13.90	5.74	1.95	2.08
Extractables after removal of volatiles (%)	ASTM D3421, modified	0	11.0	7.5	3.8	31.8	---	37.3
		56	9.8	5.1	3.4	28.3	3.37	34.4
Solvent	---	---	MEK	n-Heptane	Acetone	MEK	MEK	CCl$_4$ + methanol
Physical and mechanical properties								
Thickness (mm)	---	0	1.60	0.81	0.91	1.30	0.30	0.53
		12	1.63	0.89	0.96	1.30	0.28	0.53
		56	1.63	0.94	0.94	1.24	0.25	0.56
Tensile strength (MPa)	ASTM D412	0	10.0	15.9	12.3	10.3	15.0	18.0
		12	9.75	12.6	11.5	10.2	17.2	16.4
		56	10.2	13.7	14.8	10.2	18.1	19.2
Elongation at break (%)	ASTM D412	0	400	410	250	415	505	280
		12	410	400	300	435	505	330
		56	405	385	235	375	540	340

continued . . .

Table 13.2   Continued.

Item	Test method	Exposure time (mo.)	Base polymer					
			Butyl rubber	CPE	Potable-grade CSPE-R	EPDM	LDPE	PVC
Stress at 200% elongation (MPa)	ASTM D412	0	4.86	9.31	10.7	5.28	8.82	13.8
		12	4.80	7.63	8.72	5.18	8.44	10.8
		56	5.25	7.98	12.8	5.60	9.28	12.7
Tear strength, Die C (kN/m)	ASTM D624	0	30.6	44.6	a	31.5	68.3	58.6
		12	35.0	56.0	a	34.1	86.7	78.8
		56	32.4	29.8	a	22.8	70.9	49.9
Hardness (Duro A points)	ASTM D2240	0	51.0	77.0	79.0	54.0	---	76.0
		12	50.5	65.5	64.0	51.5	---	64.0
		56	51.0	70.0	70.0	51.0	---	70.0
Puncture resistance	FTMS 101C, Method 2065							
Force (kg)		0	19.5	20.5	14.3	17.2	6.05	11.2
		12	21.6	21.7	24.8	17.5	6.45	13.1
		56	21.8	22.6	25.4	18.1	7.45	13.6
Travel of probe (mm)		0	31	26	15	37	19	18
		12	30	25	22	30	20	18
		56	32	25	22	30	32	21

a ASTM D624 is not appropriate for fabric-reinforced materials.

After the first year of exposure, the leachate decreased in concentration due to its ultimate exhaustion of extractable components from the waste in the simulators.

In addition to the samples exposed as liners in the simulator bases, 31 strip samples of various geomembranes and other polymeric compositions were buried vertically (on edge) in the sand above the liners. Incorporated in each of the samples was a seam, which was tested along with the geomembrane. The results of testing the strip samples were similar to those from testing the liners which had been exposed on one side only, indicating a correlation between immersion and one-sided exposure. The results of testing the seam portion of the strip samples varied greatly depending on the method of seaming.

Twenty-eight geomembrane samples were also immersed in a series of tanks through which a constant flow of leachate from the simulators was pumped. This group of 28 included samples of butyl rubber, chlorinated polyethylene (CPE), CSPE-P, ethylene propylene rubber (EPDM), elasticized polyolefin (ELPO), polybutylene (PB), polyester elastomer (PEL), LDPE, neoprene, PVC, and PVC with pitch so that different manufacturers and different compositions of a given polymeric type were represented. The results of testing these samples after 8 and 31 months of immersion are summarized in Table 13.4; for the types of geomembranes that were represented by more than one sample, a range of values is reported. The data vary from one polymeric type of geomembrane to another. The data also show that, even for a given polymeric type of geomembrane, there can be a substantial range of effects on properties.

Table 13.3 Analysis of MSW leachate[a] produced in simulators.

Test	Value
Total solids, %	3.31
Volatile solids, %	1.95
Nonvolatile solids, %	1.36
Chemical oxygen demand, g/L	45.9
pH	5.05
Total volatile acids, g/L	24.33
Organic acids, g/L	
Acetic	11.25
Propionic	2.87
Isobutyric	0.81
Butyric	6.93

[a]At the end of the first year of operation when the first set of liner specimens was recovered.

Table 13.4 Summary of the effects on geomembranes of immersion in MSW leachate for 8 and 31 months[a].

| Geomembranes | | Weight increase (%) | | Retained property | | | | Change in hardness (Duro A, points) | |
| Base polymer | Number in test | | | Tensile strength (%) | | Elongation (%) | | | |
		8 mo.	31 mo.[b]	8 mo.	31 mo.	8 mo.	31 mo.	8 mo.	31 mo.
Butyl rubber	1	1.8	25	90–97	92	104–106	90–92	0	−1
CPE	3	8–10	25–28	80–115	78–106	64–135	71–103	−5 to 1	−11 to −1
CSPE–P	3	13–19	19–32	82–124	103–138	97–107	69–86	−20 to −4	−21 to −3
ELPO	1	0.1	8	86–94	98–106	91–92	96–98	0	−1
EPDM	5	1–21	8–24	64–107	94–113	76–138	88–138	−1 to +2	−3 to +5
Neoprene	4	1–19	5–88	69–100	68–105	82–103	78–146	−11 to +5	−18 to +4
PB	1	0.1	——	96–99	84–97	96–97	86–89	−3	−3
PEL	1	2.0	16	99–115	81–90	101–108	80–96	−4	−3
LDPE	1	0.6	3	110–180	118–161	96–181	100–168	——	——
PVC	7	1–3	4–24	91–110	87–117	98–139	79–120	−2 to +1	−6 to +3
PVC + pitch	1	6	14	92	101–104	109–133	80–103	−2	+1

[a]Ranges of retention values for tensile strength and elongation are lowest and highest averaged values obtained for either machine or transverse directions of all tensile specimens within the group of slab specimens of a given polymer type.

[b]Some samples were inadequately cleaned, so some values are high.

13.2.2.2 Compatibility of Geomembranes with Hazardous Waste Liquids

In a second study of the compatibility of geomembranes with waste liquids, a series of exposure tests was conducted for periods of up to seven years (Haxo et al., 1985a; Haxo et al., 1986; Matrecon, 1988). In these tests a wide range of lining materials was exposed to approximately 10 different hazardous waste liquids of different composition, including two acidic wastes, two alkaline wastes, three oily wastes, a gasoline washwater containing lead, a pesticide, and a briny industrial waste. Data on these waste liquids are presented in Table 13.5. In addition, three liquids of known composition, i.e., deionized water, a 5% aqueous solution of salt, and a saturated solution of tributyl phosphate (0.1%), were included. Thirty-two geomembranes were tested; eight were installed in one-sided exposure cells in which the geomembranes functioned as liners at the bottom of the cells with wastes placed above them. Other tests included immersion tests, pouch tests, and tests in small tubs lined with geomembranes, filled with the wastes, and exposed outdoors to the coastal California weather.

Table 13.5   Composition of hazardous wastes[a] in exposure tests.

Type of waste	Name	Organic phase (%)	Water phase Amount (%)	Water phase pH	Solids phase (%)
Acidic	"HFL"	0	100	4.8	0
	"HNO$_3$–HF–HOAc"	0	100	1.5	0
Alkaline	"Slop Water"	0	100	12.0	0
	"Spent Caustic"	0	95.1	11.3	4.9
Lead	"Low Lead Gas Washing"	10.4	86.2	7.6	3.4
	"Gasoline Washwater"	1.5	98.1	7.9	---
Oily	"Aromatic Oil"	100	0	---	0
	"Oil Pond 104"	89.0	0	---	11.0
	"Weed Oil"	20.6	78.4	7.5	11.0
Pesticide	"Weed Killer"	0	99.5	2.7	0.5

[a]See Matrecon, 1988, for detailed data on the compositions.

The effects of the one-sided exposure of the eight geomembranes to the waste liquids are summarized in Table 13.6 for two exposure periods of approximately 500 and 1250 days. This table, which presents data on the tensile stress at 100% elongation of the respective geomembranes, illustrates the variation in the effects of the exposure. The changes in stiffness with exposure reflect either softening

276

caused by the absorption of components of the waste liquids or stiffening caused by the extraction of constituents of geomembrane compounds. No data are reported for the fabric-reinforced butyl rubber geomembrane because it broke at less than 100% elongation. The butyl rubber geomembrane showed modest absorption of the waste liquids, ranging from 1.4% for spent caustic to 11.5% for the highly acidic waste "HNO_3-HF-HOAc," as measured by percent volatiles. The magnitude of the changes in a property such as tensile stress at 100% elongation may be reflected in the overall performance of the geomembrane. It is believed that this is true in cases where the property values for exposed geomembranes are significantly higher or lower than the original values, e.g., 50% lower or 100% higher; however, no specific criteria have been established.

Table 13.6   One-sided exposure of geomembrane liners in exposure cells[a]--percent retention of original tensile stress at 100% elongation.

Base polymer	Original value[c] (MPa)	Retention in different wastes[b] (%)							
		Acidic		Alkaline		Oily			Pesticide
		"HFL"	"HNO₃-HF-HOAc"	"Slop Water"	"Spent Caustic"	"Lead Waste"	"Slurry Oil"	"Oil Pond 104"	"Weed Killer"
Butyl rubber (R)	d	---	---	---	---	---	---	---	---
CPE	6.29	---	97	---	82	56	---	53	94
		---	113	---	129	71	---	62	113
CSPE-P (R)	6.55	---	106	---	165	85	---	63	90
		---	112	---	200	118	---	96	118
ELPO	6.45	98	92	97	87	95	70	61	104
		---	98	---	87	104	60	71	91
EPDM	2.50	---	81	---	88	84	---	---	89
		---	70	---	108	80	---	---	87
Neoprene	3.22	---	---	---	90	53	---	50	62
		---	---	---	95	61	---	42	54
PEL	18.08	---	---	---	101	88	77	94	95
		---	---	---	109	88	---	85	96
PVC	6.96	153	200	---	99	82	113	152	100
		---	249	---	115	103	---	174	173

[a]Each pair of values represents two different exposure times, the first about 500 days and the second about 1250 days.

[b]See Table 13.5 for information regarding composition of wastes.

[c]Average of values in machine and transverse directions of specimens tested in accordance with ASTM D412 or D638 at 500 mm per minute.

[d]Broke at less than 100% elongation.

The effects of immersing various geomembranes in the wastes are reported in Tables 13.7 and 13.8. Table 13.7 presents data on the percent increase in weight of the immersed specimens, and Table 13.8 presents data on the percent retention of the tensile stress at 100% elongation.

The data are presented for each waste and geomembrane by pairs representing two exposure periods. The first exposure period was relatively short, about 50 to 250 days, and the second period was about 750 to 1500 days. In most cases, equilibrium had been reached by the end of the longer exposure period.

The responses of the geomembranes varied greatly from geomembrane to geomembrane and from waste liquid to waste liquid. The effects varied from essentially no change during the exposures to essential failure of the geomembrane. The maximum changes in weight ranged from essential dissolution of the geomembrane, e.g., the CPE geomembrane in "Slurry Oil," to the loss of weight of the PVC geomembrane in the "Slop Water," a highly alkaline waste. In most of the wastes, the changes in properties of the PE geomembranes were insignificant, i.e., in tenths of a percent. Overall, the oily wastes appeared to have the greatest effect on the geomembranes. The differences between geomembranes based on a given polymer are indicated by the data for the two PVC geomembranes.

The changes in the tensile stress at 100% elongation were often related to changes in weight. For example, the highly swollen geomembrane specimens tended to exhibit greater changes in tensile stress at 100% elongation than those that swelled less. This was particularly true of the specimens immersed in the oily waste liquids. The two PVC geomembranes showed substantial differences, particularly for the samples exposed in the acidic and alkaline solutions. These differences were probably due primarily to the differences in the amounts of plasticizer lost during the exposure period.

Overall, the results presented in Tables 13.7 and 13.8 indicate only limited changes in weight and changes in stress at 100% elongation of the PE samples after immersion in the various liquid media.

13.2.2.3 Need for a Compatibility Test

Based on the results of the studies described above and field experience with geomembranes used in lining ponds, it was apparent that many of the wastes of interest could significantly affect the geomembranes used as liners. These results indicate the need for selecting geomembranes for specific applications and for determining and ensuring compatibility between the geomembrane and the waste during the design and selection processes.

The EPA has developed EPA Method 9090 to assess the compatibility of geomembranes to be used to line hazardous waste storage and disposal facilities (EPA, 1986). In this test, samples of the geomembrane are immersed in a representative sample of the specific leachate or waste liquid to be contained, and the effect of the exposure on selected properties is measured. More recently, it was recognized that all materials used in the construction of a liner system,

Table 13.7 Effects on geomembranes of immersion in various hazardous wastes—percent increase in weight at the end of two exposure periods.

Increase in weight of samples on immersion in different wastes and liquids[a] (%)

Base polymer	Acidic "HFL"	Acidic "HNO3-HF-HOAc"	Alkaline "Slop Water"	Alkaline "Spent Caustic"	Brine 5% NaCl	Oily "Lead Waste"[b]	Oily "Slurry Oil"	Oily "Oil Pond 104"	Oily "Weed Oil"	Organic Pesticide trace Sat'd[c] TBP	Organic Pesticide cide "Weed Killer"	Water Deionized
Butyl rubber	2.74	1.39	2.04	0.37	0.87	20.1	32.3	97.5	70.8	18.1	0.76	1.34
	3.71	3.77	1.81	0.74	1.4	28.7	31.18	104	64.2	23.1	1.57	4.4
CPE	9.43	9.31	1.50	0.64	2.54	70.9	59.5	31.6	117	117	9.62	5.66
	12.9	19.9	1.89	1.11	1.3	119	d	36.9	----	121	12.7	12.4
CSPE-P (R)	6.75	10.3	3.79	3.32	5.75	83.0	51.1	75.10	202	31.5	13.07	7.99
	8.95	10.0	7.65	4.30	4.06	121	105	49.5	368	30.1	17.26	15.8
CSPE-P	5.41	7.46	3.84	2.17	5.06	69.6	53.2	58.5	211	38.3	12.3	7.73
	7.74	10.9	5.66	3.28	5.6	116	111	55.0	348	31.7	15.7	18.9
EPDM	16.7	18.3	3.13	0.23	1.19	29.3	35.3	80.1	79.4	5.2	8.09	1.93
	23.9	50.9	3.34	1.30	1.0	34.7	34.2	84.7	76.2	5.9	20.4	3.6
Neoprene	9.60	10.8	0.38	0.82	3.54	45.6	60.7	25.8	94.8	49.4	8.54	7.10
	12.0	17.4	2.66	1.53	----	59.1	142.6	26.3	89.3	41.1	11.4	11.4
HDPE	0.05	0.1	0.2	0.2	0.03	5.0	4.4	3.3	6.4	0.33	0.5	0.03
	0.16	0.2	0.52	0.01	0.09	4.5	8.0	6.6	7.3	0.5	0.2	0.6

continued. . .

Table 13.7   Continued.

Base polymer	Increase in weight of samples on immersion in different wastes and liquids[a] (%)											
	Acidic		Alkaline		Brine	Oily				Organic Pesticide		Water
	"HFL"	"HNO3-HF-HOAc"	"Slop Water"	"Spent Caustic"	5% NaCl	"Lead Waste"[b]	"Slurry Oil"	"Oil Pond 104"	"Weed Oil"	trace Sat'd[c] TBP	cide "Weed Killer"	Deionized
LDPE	-3.3	-0.3	3.5	0.1	0.07	3.1	8.5	8.4	10.7	0.44	0.2	0.03
	0.08	0.3	1.07	0.1	0.09	5.3	12.0	10.3	14.0	0.5	0.2	0.2
PVC	2.76	-2.82	-6.35	-3.00	-4.82	8.81	11.3	-1.54	33.4	39.7	0.46	1.18
	0.86	-6.12	-15.7	-0.89	-7.8	7.39	28.9	-0.54	24.7	40.7	0.95	-0.5
	7.60	19.8	-13.5	0.04	-1.51	2.22	7.2	-10.3	18.1	47.6	2.89	0.65
	14.3	28.2	-12.1	1.08	-1.8	-5.15	14.1	-9.9	25.2	47.5	1.62	-0.1

[a] See Table 13.5 for data on composition and Matrecon, 1988, for details regarding immersion times.

[b] Blend of three waste streams.

[c] Saturated solution of tributyl phosphate (TBP) in deionized water.

[d] Not measured because immersed specimen had become very "gooey" and seemed partially dissolved.

Table 13.8 Effects on geomembranes of immersion in various hazardous wastes—percent retention of tensile stress at 100% elongation at the end of two exposure periods.

Base Polymer	Original value[b] (MPa)	Acidic "HFL"	Acidic "HNO_3-HF-HOAc"	Alkaline "Slop Water"	Alkaline "Spent Caustic"	Brine 5% NaCl	Oily "Lead Waste"[c]	Oily "Slurry Oil"	Oily "Oil Pond 104"	Oily "Weed Oil"	trace Sat'd[dd] TBP	cide "Weed Killer"	Water Deion-ized
Butyl rubber	2.15	84 / 93	70 / 88	85 / 89	81 / 91	103 / 89	59 / 57	66 / 55	44 / 45	36-47[e] / 38	79 / 60	78 / 89	90 / 89
CPE	6.29	98 / 117	92 / 129	113 / 130	115 / 126	124 / 152	37 / 18	44 / f	44 / 48	8 / ---	6 / 9	102 / 103	129 / 123
CSPE-P	6.15	91 / 110	100 / 71	119 / 169	130 / 164	104 / 134	79 / 91	95 / 85	58 / 85	34 / f	80 / 106	102 / 120	143 / 106
EPDM	2.36	86 / 100	75 / 58	88 / 107	93 / 100	85 / 99	83 / 73	65 / 60	66 / 58	59-64[e] / 67	82 / 89	85 / 89	109 / 93
Neoprene	3.90	82 / 104	83 / 62	96 / 115	97 / 121	106 / 144	50 / 38	58 / 40	46 / 70	25-26[e] / 27	37 / 38	84 / 109	126 / 109
HDPE	17.55 / 18.06[i]	101 / 89[h]	103 / 79[h]	g / 90[h]	101 / 95[h]	g / 104[h]	100 / 82[h]	99 / 86[h]	102 / 88[h]	98 / 83[h]	92[h] / 89[h]	100 / 83[h]	g / 98[h]
LDPE	9.23 / 8.46[i]	97 / 101[h]	99 / 105[h]	97 / 98[h]	100 / 99[h]	104 / 103[h]	95 / 93[h]	91 / 100[h]	92 / 90[h]	88 / 89[h]	104[j] / 103[h]	94 / 100[h]	111 / 105[h]

continued . . . .

Table 13.8 Continued.

Base polymer	Original value[b] (MPa)	Acidic		Alkaline		Brine	Oily				Organic Pesticide		Water
		"HFL"	"HNO$_3$-HF-HOAc"	"Slop Water"	"Spent Caustic"	5% NaCl	"Lead Waste"[c]	"Slurry Oil"	"Oil Pond 104"	"Weed Oil"	trace Sat'd[d] TBP	cide "Weed Killer"	Deionized
PVC-1	6.96	99	199	161	103	136	89	114	118	32–35[e]	27	118	113
		124	252	239	123	185	87	107	145	45	28	130	116
PVC-2	12.13	70	68	183	86	96	83	99	145	37–41[e]	23[j]	83	100
		83	70	k	99	125	95	122	172	45	25[j]	99	104

[a]See Table 13.5 for data on composition and Matrecon (1988) for details regarding composition of wastes and immersion times.

[b]Average of values in machine and transverse directions of specimens tested in accordance with ASTM D412 or D638 at 500 mm/min. unless otherwise noted.

[c]Blend of three waste streams of gasoline tank wash waters.

[d]Saturated solution of tributyl phosphate (TBP) in deionized water.

[e]Waste stratified into an aqueous and an oily phase. Reported value ranges indicate the different effects of the two phases.

[f]Not measured because of condition of immersed specimen.

[g]Immersed sample failed at less than 100% elongation.

[h]Retention of values obtained for specimens tested at 50 mm/min.

[i]Average of values in machine and transverse directions of specimens tested at 50 mm/min.

[j]Original value of geomembrane = 11.80 MPa.

[k]Transverse direction failed at 80% elongation.

including the components of the leak detection and leachate collection systems, should also be tested for compatibility with the specific waste liquid to be contained. These materials can include geotextiles, geonets, geocomposites, and pipe. EPA Method 9090, which has become part of permit applications for the construction and operation of waste containment facilities in the United States, is discussed in more detail in Section 13.3.

13.2.2.4 Compatibility of Geomembrane Seams

Because the long-term durability of seams is important for liner systems, Morrison and Parkhill (1987) studied the effect of simulated service conditions on geomembrane seams. In this research program, factory and field seam samples of representative geomembranes available at the time were prepared by the method appropriate to the specific geomembrane and immersed in nine solutions. A total of 36 combinations of geomembranes and seaming methods were tested. The geomembranes included CPE, CSPE, ethylene interpolymer alloy (EIA), EPDM, linear-low density polyethylene (LLDPE), HDPE, PVC, and PVC/CPE. Samples of both the geomembranes and the geomembrane seams were immersed. The test liquids included 10% aqueous solutions of phenol, hydrochloric acid, sodium hydroxide, and methyl ethyl ketone (MEK), a 5% solution of furfural, methylene chloride (100%), saturated salt solutions at 23 and 50°C, and tap water.

The results of immersing the geomembranes in the test liquids are presented in Table 13.9. As with the data presented in the previous section, the changes in weight of the geomembrane specimens varied considerably from test liquid to test liquid, from geomembrane to geomembrane, and among the geomembranes of a given base polymer. The semicrystalline thermoplastic polymeric compositions dissolved in methylene chloride. Each of the PVC geomembranes showed significant loss in weight on immersion in the phenol and sodium hydroxide solutions. Some of the CPE geomembranes and the EIA geomembrane also lost weight on immersion in sodium hydroxide. The PE geomembranes exhibited the smallest changes in weight, although they gained weight on immersion in methylene chloride. Overall, these results again demonstrate the need for specific compatibility tests for geomembranes to be used in contact with different waste liquids.

After 52 weeks of immersion in the different test liquids, the seams were tested in shear, peel, and constant load tests. Selected results are presented in Table 13.10. The results indicate that some of the organics have a severe effect on the geomembranes and the seams of the geomembranes. The results also indicate the aggressiveness of the sodium hydroxide solution on the seam strength. The seams of the HDPE geomembranes made by three standard methods, i.e., lap welding, thermal wedge welding, and fillet welding, were satisfactory after the 52 weeks of exposure to all of the test media. One set of LLDPE seams made by the hot wedge technique was satisfactory after 52 weeks of exposure in the various liquids, whereas the seams of a second set made by the same method were questionable.

13.2.2.5 Developing Chemical Compatibility Criteria

Since there were no established or accepted benchmarks for geomembrane performance based on immersion tests, Bellen et al. (1987) conducted a

Table 13.9   Change in weight of geomembranes exposed to various test liquids for 52 weeks[a]

Geomembrane			Change (% by weight)									
Base polymer	Type of compound[b]	Nominal thickness (mm)	Tap water	Saturated NaCl[c]		10% Phenol	10% HCl	10% NaOH	10% MEK	100% CH$_2$Cl$_2$	5% Furfural
				23°C	50°C						
CPE	TP	0.76	10.19	1.41	1.10	25.61	1.41	−2.07	29.89	−100.00	67.60
	TP	0.76	9.78	1.27	1.99	25.88	1.24	−2.22	38.63	−100.00	80.00
	TP(R)	0.41	22.10	2.37	3.09	37.11	9.47	8.25	18.25	−100.00	47.81
	TP(R)	0.91	9.27	1.20	0.48	6.10	−0.22	−5.37	12.72	−100.00	24.92
CSPE	TP(R)	0.76	4.06	0.65	1.74	14.53	0.60	1.20	4.60	−100.00	14.99
	TP(R)	0.91	8.10	2.23	4.45	16.54	2.90	2.25	16.01	−100.00	23.89
	TP(R)	0.91	5.81	3.00	4.46	17.54	2.68	2.15	7.25	−100.00	18.09
	TP(R)	0.91	6.77	1.27	3.29	38.04	6.04	3.60	10.19	−100.00	38.55
	TP(R)	0.91	4.92	2.46	2.96	16.64	9.15	14.05	6.30	−100.00	17.43
	TP(R)	0.91	5.12	2.46	2.73	16.68	3.68	14.59	6.47	−100.00	17.76
	TP(R)	0.91	11.72	2.96	4.15	19.03	20.15	11.94	14.82	−100.00	22.01
EIA	TP(R)	0.97	4.03	1.81	2.38	−100.00	7.12	−5.22	5.77	−100.00	32.30
EPDM	XL(R)	0.76	3.55	1.55	2.94	8.61	3.76	1.29	4.74	4.03	11.80
HDPE	CX	0.76	−0.01	−0.01	0.14	−0.48	−0.28	0.32	0.39	6.74	0.83
	CX	2.00	0.05	0.06	0.13	0.12	−0.29	0.19	0.55	4.07	0.59
	CX	2.00	0.01	0.01	· · ·	−0.01	−0.31	0.12	0.18	3.17	0.34
	CX	2.00	0.06	0.02	0.00	−0.41	−0.26	0.14	0.28	4.78	0.51

continued · · ·

Table 13.9 Continued.

Geomembrane			Change (% by weight)								
				Saturated							
Base polymer	Type of compound[b]	Nominal thickness (mm)	Tap water	NaCl[c]		10% Phenol	10% HCl	10% NaOH	10% MEK	100% CH_2Cl_2	5% Furfural
				23°C	50°C						
LLDPE	CX	0.76	0.00	0.07	0.27	-0.50	-0.79	0.21	0.63	7.52	0.72
PVC	TP	0.75	1.57	-0.97	-0.54	-16.38	6.41	-18.11	2.91	-100.00	5.55
	TP	0.76	2.42	-0.81	-0.57	-15.90	3.69	-19.65	5.51	-100.00	13.12
	TP	0.76	1.53	-0.18	-0.50	-12.78	7.04	-14.97	11.37	-100.00	15.35

[a] All solutions are aqueous; exposure was at room temperature (23°C), except where otherwise indicated. A "5%" or "10%" solution means 5 or 10 g of solvent per 100 g H_2O, respectively. Methylene chloride was neat; MEK = methyl ethyl ketone.

[b] TP = thermoplastic; XL = crosslinked; CX = semicrystalline thermoplastic; R = fabric-reinforced.

[c] Saturated solution at 23°C is 26.5% by weight (36.1 g per 100 g H_2O); saturated solution at 50°C is 27.0% by weight (37.0 g per 100 g H_2O).

Source: Morrison and Parkhill, 1987, pp 81–85.

Table 13.10 Performance of selected geomembrane seam samples exposed to various test liquids[a].

Base polymer	Sample[b]	Nominal thickness (mm)	Type of seam[d]	Tap water	Test liquid[c] Saturated NaCl[e] 23°C	50°C	10% Phenol	10% HCl	10% NaOH	10% MEK	100% CH$_2$Cl$_2$	5% Furfural
CPE	L	0.76	SA	U	S	S	U	S	U	U	U	U
	M	0.76	TDI	U	S	S	U	S	U	U	U	U
	A(R)	0.91	BSA	U	U	U	U	U	U	U	U	U
	B(R)	0.91	SA	S	S	U	U	U	U	U	U	U
CSPE	D(R)	0.91	THA	U	U	U	U	S	U	U	U	U
	E(R)	0.91	THA	S	S	Q	U	Q	U	U	U	U
	E(R)	0.91	AD	S	S	S	U	Q	U	S	U	S
	F(R)	0.91	BSA	S	S	S	U	S	U	S	U	U
	H(R)	0.91	TDI	S	S	S	U	S	U	S	U	U
	H(R)	0.91	SA	S	S	S	U	S	U	S	U	Q
EIA	J(R)	0.97	THA	S	S	S	U	S	U	U	U	U
EPDM	K(R)	1.00	VZ	S	S	S	S	S	U	S	S	S
	K(R)	1.00	GTC	U	U	S	U	U	U	U	Q	U
HDPE	O	2.00	EFW	S	S	S	S	S	S	S	S	S
	P	2.00	ELW	S	S	S	S	S	S	S	S	S
	Q	2.00	TDW	S	S	S	S	S	S	S	S	S
LLDPE	R	0.76	THW	S	S	S	S	S	S	S	S	S
	R	0.76	THW	Q	Q	Q	Q	Q	Q	Q	Q	U

continued . . .

Table 13.10 Continued.

Base polymer	Sample[b]	Nominal thickness (mm)	Type of seam[d]	Tap water	Saturated NaCl[e] 23°C	Saturated NaCl[e] 50°C	10% Phenol	10% HCl	10% NaOH	10% MEK	100% CH$_2$Cl$_2$	5% Furfural
PVC	S	0.76	SA	S	S	S	U	S	U	U	U	U
	T	0.76	TDI	S	S	S	U	S	U	U	U	U
	U	0.76	SA	S	S	S	U	S	U	U	U	U
PVC– CPE	---	0.76	TDI	U	U	U	U	U	U	U	U	U

(header spanning columns: Test liquid[c])

aSummary of the results of testing samples after 3, 6, and 12 months of exposure. U = unsatisfactory; Q = questionable; S = satisfactory. Performance was considered unsatisfactory if there was more than a 20% loss in either shear or peel strength after exposure.

bIdentification for geomembrane as tested in immersion; R = fabric-reinforced.

cAll solutions are aqueous; exposure was at room temperature (23°C), except where otherwise indicated. A "5%" or "10%" solution means 5 or 10 g of solvent per 100 g H$_2$O. MEK = methyl ethyl ketone; methylene chloride was neat.

dAD = adhesive; BSA = bodied solvent adhesive; EFW = extrusion fillet weld; ELW = extrusion lap weld; GTC = gum tape and cement; SA = solvent adhesive; TDI = thermal--dielectric; TDW = thermal--hot dual wedge; THA = thermal--hot air; THW = thermal--hot air; VZ = vulcanized plus 19.0-in. cap strip.

eSaturated solution at 23°C is 26.5% by weight (36.1 g per 100 g H$_2$O); saturated solution at 50°C is 27.0% by weight (37.0 g per 100 g H$_2$O).

Source: Morrison and Parkhill, 1987, pp 10–15.

test program to generate data on the chemical resistance of commercial geomembranes. It was anticipated that the data would be useful in assessing the results of compatibility tests, such as those performed in accordance with EPA Method 9090, and could be used to develop general criteria to assess the chemical compatibility of a specific geomembrane proposed for use in lining specific waste storage and disposal facilities.

In this program, six commercial geomembranes were immersed in 20 different chemical solutions or liquids, including acids and bases, polar and nonpolar organics, organic and inorganic solutions, at different concentrations. The geomembranes were immersed at 23 and 50°C in the solutions for 1, 7, 14, 28, and 56 days (a short-term test) and for four-month increments up to two years (a long-term test). The immersed samples were measured for changes in appearance, weight, dimensions, and tensile and tear properties. The six geomembranes that were selected for the test program were unreinforced CPE, CSPE (industrial-grade), ECO, EPDM, HDPE, and PVC. They represent a range of different polymer types, including variations in chemical composition, polarity, crystallinity, and crosslink density. Of the six geomembranes, five were 0.76 mm in nominal thickness and the sixth, an ECO geomembrane, was 1.52 mm in thickness.

The 20 chemical liquids used in the study were: water, hydrochloric acid, sodium hydroxide, sodium chloride, potassium dichromate, phenol, furfural, methyl ethyl ketone, 1,2-dichloroethane, and ASTM #2 oil. Four of the chemicals, all organics, were each used at three concentrations and the NaCl was used at two concentrations to give information on the effect of concentration.

The authors observed five basic types of response of the materials to chemical immersion:

- Changes in physical properties and weight.
- Swelling by changes in dimensions.
- Swelling and softening with loss of strength.
- Shrinking and stiffening with loss of elongation.
- Combination of swelling and shrinking depending on immersion conditions.

The response of geomembranes to increased immersion temperature indicated that higher temperatures (at least up to 50°C) can be used to accelerate the material response for some geomembranes but not others. With caution, the effects of temperature could be distinguished from a chemical response.

Bellen et al. (1987) proposed a mathematical curve fitting method for evaluating immersion data as a function of time. The method assumes the liner approaches a limit of physical property change asymptotically. The method can be used to predict the ultimate end point of physical property change and sampling time intervals for continued immersion testing in the specific chemical solution or liquid.

Some of the general conclusions of this study of geomembranes
in simple chemical solutions or liquids are:

- Immersing a geomembrane in the waste it is intended to contain
  and determining changes in physical and analytical properties
  is essential because each waste liquid is unique.
- In general, the magnitude of a geomembrane's response to an
  aqueous solution containing an organic solvent is a function
  of its concentration. However, solutions containing low concentrations of some chemicals can have a more significant
  effect than solutions containing higher concentrations of other
  chemicals.
- In evaluating the chemical compatibility between a geomembrane
  and a given waste, the ability of the geomembrane to come to an
  equilibrium in the chemical environment as well as the magnitude of changes in properties needs to be considered.
- For some geomembrane-waste combinations, increasing the immersion temperature can be used to accelerate testing. However, for
  others, increasing the temperature to 50°C produced a different
  rather than an accelerated response.

13.2.2.6 Factors Affecting Compatibility of Geomembranes and Waste Liquids

A further study was undertaken by Haxo et al. (1988) to investigate
various factors that affect the compatibility of geomembranes with
waste liquids. The purpose of the study was to ultimately develop
methods to predict such compatibility. The major factors were:

- Solubility parameters of the polymer with respect to those of
  constituents of the liquid.
- Crosslinking of the polymer.
- Crystallinity content of the polymer.
- Filler content of the compound.
- Plasticizer content of the compound.
- Soluble constituents in the compound.

This investigation was conducted in three basic areas:

- Swelling of geomembranes and other geomembrane-related
  compositions in organics, and calculation of the solubility
  parameters of these compositions.
- Distribution of organics between aqueous solutions, such as
  leachates, and geomembranes.
- Variables in EPA Method 9090 compatibility testing of
  geomembranes and waste liquids.

Equilibrium swelling of 28 geomembrane-related compositions was
determined in 30 organics and deionized water. These 28 polymeric
materials included thermoplastic, crosslinked, and semicrystalline
compositions, of which 22 were commercial geomembranes or sheetings
and 6 were known compositions prepared in the laboratory for this
study. Basic polymer and compound variations, e.g., differences in
polymer type, level of crystallinity, crosslink density, filler
level, and amount of and type of plasticizer, were assessed.

The specific organics were selected on the basis of:

- Representation of a full range of solubility parameter
  values.
- Presence in actual waste streams.
- Solubility in water.
- Density, melting point, and boiling point.
- Vapor pressure.

Table 13.11 presents values for equilibrium percent volume changes obtained on eight of the geomembranes immersed in the various solvents. The table also presents the extractables contents and the solubility parameters of the respective geomembranes. The swelling data, which are arranged by increasing solubility parameter of immersion liquid, show the overall importance of the solubility parameters on the swelling that can take place. The data also show the loss in weight of many of the geo membranes resulting from the extraction of the plasticizers.

The most significant results of this investigation were:

- The degree of crystallinity of the polymer appeared to be the
  dominant factor in reducing the swelling of a polymeric
  composition in all of the organics and appeared to override
  both the crosslinking and the solubility parameters.
- Crosslinking in an amorphous polymer reduced swelling in all of
  the organics.
    Note: The crystallinity and crosslinking factors are not
    additive. The introduction of crosslinking in semicrys-
    talline polymeric compositions tends to reduce the degree
    of crystallinity.
- Although the magnitude of swelling of amorphous polymers
  could, in many instances, be estimated from the proximity of
  the values of the component solubility parameters of the
  polymer and those of the organic, the swelling of a geomembrane in many combinations could only be roughly estimated
  based upon the type of organic. The matching of the Hildebrand
  solubility parameter values remains a necessary but not
  sufficient condition for swelling. Swelling tests should be
  performed to ensure that an amorphous geomembrane will not
  swell in a particular organic. Thus, empirically derived
  data are still needed for untested combinations of organics
  and geomembranes.
- With waste liquids that contained dissolved organics, the
  organics were absorbed by the geomembrane. The amounts
  absorbed depended on: (1) the relationship of the solubility parameters of the organic and the geomembrane; and
  (2) the solubility of the organic in water.

Table 13.11 Equilibrium percent volume change of selected geomembranes on immersion in organic solvents of increasing solubility parameter[a].

Item	Hildebrand solubility parameter (δ_o)	Base polymer (%)							
		CPE	CSPE	EPDM	HDPE	HDPE-A	PEL	PVC	PVC
Extractables		14.85[b]	7.15[c]	22.78[d]	0.73[d]	2.09[d]	1.09[b]	34.57[e]	40.12[e]
Hildebrand solubility parameter (o)	—	9.3	9.4	8.9	7.4	7.7	10.6	10.1	9.6
Solvent									
Isooctane (Ref. Fuel A)	7.0	4.24	8.71	68.6	7.06	15.9	0.86	-21.7	-22.1
n-Octane	7.6	6.08	12.9	86.8	8.49	18.6	2.88	-19.9	-20.3
Cyclohexane	8.2	26.6	99.7	126	11.8	34.7	4.46	-19.9	-16.4
Methyl isobutyl ketone	8.3	D[f]	40.8	-5.20	3.62	4.36	10.5	D	D
Isoamyl acetate	8.4	D	45.4	5.70	4.24	7.72	10.5	245	D
o-Xylene	8.8	41.2	153	103	12.6	28.8	17.1	7.92	-17.3
Di(2-ethylhexyl) phthalate	8.9	362	29.2	-4.16	0.46	3.12	0.72	176	143
Ethyl acetate	8.9	D	16.5	-4.96	2.63	3.27	11.6	147	D
Methyl ethyl ketone	9.3	D	26.8	-8.33	2.53	3.42	12.9	D	D
Trichloroethylene	9.3	D	D	135	10.5	32.1	25.1	17.0	276
Cyclohexanone	9.6	D	101	-3.00	3.03	5.40	15.8	D	D
Acetone	9.8	103	13.3	0.88	2.42	1.41	11.2	172	D
Tetralin	9.8	D	181	68.6	7.67	16.9	24.7	111	D
Tetrachloroethylene	9.9	123	161	146	13.7	41.4	14.5	-2.64	0.19
2-Ethyl-1-hexanol	9.9	4.57	3.12	-4.83	0.52	1.44	1.51	-12.8	3.0

continued . . .

Table 13.11 Continued.

Item	Hildebrand solubility parameter (δ_o)	Base polymer (%)							
		CPE	CSPE	EPDM	HDPE	HDPE-A	PEL	PVC	PVC
Solvent (cont'd)									
Cyclohexanol	10.9	2.14	5.66	-7.47	0.86	1.82	2.41	-11.2	0.37
N,N-dimethylacetamide	11.1	D	20.9	3.93	1.67	2.02	16.5	D	D
Benzyl alcohol	11.6	35.2	8.79	-0.38	0.94	0.61	17.1	-11.8	-9.5
1-Propanol	12.0	2.13	3.13	-6.40	0.77	1.34	7.28	-19.0	-19.4
Propylene-1,2-carbonate	13.3	23.0	5.63	-0.08	0.28	0.09	3.30	11.9	27.3
2-Pyrrolidone	13.9	112	11.2	1.29	0.87	0.54	4.68	278	D
Methanol	14.5	3.56	4.58	1.59	1.40	2.86	4.72	-17.7	-11.7
Ethylene glycol	16.1	2.21	3.22	0.49	0.41	0.43	1.67	3.43	4.9
Water	23.4	3.99	4.44	1.18	0.62	1.60	2.13	1.58	4.2

aTaken from Table 4-18 and Tables 5-49 through 5-51 (Matrecon, 1988).

bWith n-heptane.

cWith acetone.

dWith methyl ethyl ketone.

eWith 2:1 mixture of CCl4 and CH3OH.

fD = dissolved or disintegrated.

13.3 EPA Method 9090

EPA Method 9090, "Compatibility Test for Wastes and Membrane Liners," which is currently used for chemical compatibility testing of geomembranes with leachates and other liquids that might be contained, is described in this section. EPA Method 9090 requires that physical and mechanical properties of the materials be measured after 30, 60, 90, and 120 days of exposure to a waste liquid at temperatures of 23 and 50°C. These properties and the associated test methods are shown in Table 13.12, which shows that the specific monitoring tests differ with the types of geomembranes that are tested.

The containers, or cells, used for material exposure to leachate must be non-interactive with the geomembrane and the leachate. Exposure cells made of polymeric materials are typically not used because they may compete with the geomembrane for the organic constituents, and thus reduce the concentration of available constituents that might be absorbed by the geomembrane. Therefore, leachates that contain appreciable concentrations of organic constituents are placed in exposure cells either of glass or of stainless steel. Conversely, inorganic leachates are tested in synthetic or glass exposure cells. Figure 13.1 presents a schematic drawing of a cell that has been used in performing the EPA Method 9090 exposure.

To simulate landfill conditions, which in some situations are anaerobic, the geomembranes are exposed to the leachate in sealed exposure cells. To preclude stagnation and stratification of the leachate around the geomembrane, the leachate is continuously circulated through pumping, stirring, or continuous agitation of the cell (mechanical vibration).

Leachates may contain significant quantities of volatile organic constituents that may either escape through evaporation during sample retrieval or produce significant pressure buildup within the cell. In order to avoid pressure buildup, reflux condensers or liquid seals are installed on the lids of the exposure cells. To maintain the initial level of components throughout the test duration, leachate in the cells is replaced after each exposure period. Alternatively, the most active components may be added to the leachate (spiking) as levels decrease during the progression of the test. This process requires periodic monitoring of the concentrations of volatile leachate constituents through the use of gas chromatography. Gas chromatography is used for the separation and measurement of concentration of volatile organic mixtures.

The geomembrane is cut into appropriately sized samples to fit into the exposure cells. The samples are positioned separately from one another in the exposure cells to ensure that the entire surface area of each sample is in contact with leachate.

Table 13.12 Properties and test methods used in performing EPA Method 9090 to determine analytical and physical properties of geomembranes.

Property	Geomembranes without fabric reinforcement			Fabric reinforced
	Thermoplastic	Crosslinked	Semicrystalline	
Analytical properties				
Volatiles	a	a	a	a
Extractables	b	b	b	b
Physical properties				
Mass	EPA Method 9090	EPA Method 9090	EPA Method 9090	EPA Method 9090
Thickness (average)	ASTM D638[c]	ASTM D412[c]	ASTM D638[c]	ASTM D751, Section 6[c]
Dimensions	EPA Method 9090	EPA Method 9090	EPA Method 9090	EPA Method 9090
Tensile properties[d]	ASTM D638	ASTM D412	ASTM D638 (modified)[d]	ASTM D751, Method B
Tear resistance	ASTM D1004 (modified)	ASTM D624, Die C	ASTM D1004	ASTM D751, Tongue Method (200 x 200-in. test specimen)[d]
Modulus of elasticity	na	na	ASTM D882, Method A (modified)[e]	na
Hardness (Duro A or D)	ASTM D2240	ASTM D2240	ASTM D2240,	ASTM D2240
Puncture resistance	FTMS 101C, Method 2065	FTMS 101C, Method 2065	FTMS 101C, Method 2065	FTMS 101C, Method 2065

[a]EPA SW-870, Appendix III-D (Matrecon, 1983) or ASTM D3030.

[b]EPA SW-870, Appendix III-E (Matrecon, 1983) or ASTM 3421; solvents vary with type of geomembrane.

[c]Reported thickness values are averages of all values measured on test specimens used in the physical property testing.

[d]NSF Standard 54 (NSF, 1985).

[e]Measured using 12.7 x 152-mm strip specimens with an initial jaw separation of 50.0 mm at the standard strain rate of 0.1 mm/mm·min. Using a specimen size large enough so that specimens would be tested with an initial jaw separation of 254 mm as specified by ASTM D882-83 would result in higher values.

na = not applicable.

THERMOMETER

GEAR-MOTOR

SEPTUM FOR
SAMPLING
LIQUID

STANDPIPE

TEFLON
GASKET

VENT/DRAIN
TUBE

LID

TEFLON
GASKET

LINK LOCK

TEMPERATURE
CONTROLLER

316 STAINLESS
STEEL TANK

SPECIMEN
HOLDER

HEATER

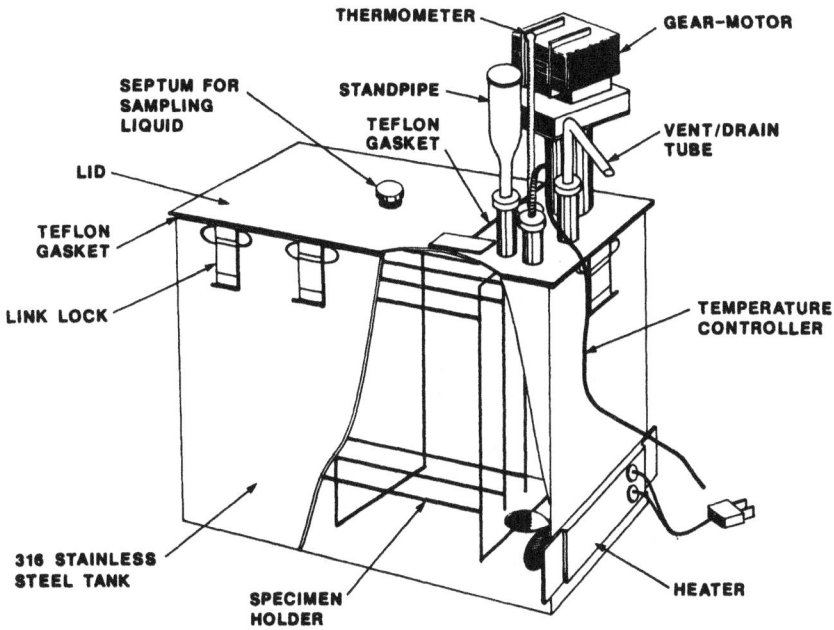

Figure 13.1   Schematic of an exposure tank suitable for use in
performing the EPA Method 9090 compatibility test.

295

At the conclusion of each 30-day exposure period, leachate is pumped from the cells into a waste drum. The exposure cells are then opened and the geomembrane samples are retrieved. The samples taken from the 50°C cells are transferred to leachate kept at 23°C for 3 hours in order to effect proper conditioning to the laboratory atmosphere. After being retrieved, the samples are quickly dipped in water to remove excess leachate and are sealed in PE bags until the commencement of testing. Testing is performed within 24 hours of removal of the samples from leachate, in accordance with the procedures prescribed in EPA Method 9090.

13.3.1 Purpose
EPA Method 9090 is currently the most comprehensive test method accepted by regulatory agencies for chemical compatibility testing of geomembranes. It is designed to assess the effect of a special leachate on the properties of a geomembrane. It combines analytical and physical measurements to provide an indication of the extent of the effects of the leachate on the geomembrane. The changes in mechanical properties give an indication of the magnitude of potential degradative changes that may affect overall performance. Two temperatures (23 and 50°C) are used in order to observe the behavior of the geomembrane under both typical landfill conditions (23°C) and accelerated aging conditions (50°C). The purpose of exposure at the elevated temperature is to give some indication of the ability of the geomembrane to withstand the potentially aggressive landfill environment for the projected lifetime of the landfill.

13.3.2 Test Results
Table 13.13 presents typical results for physical property values obtained in an EPA Method 9090 compatibility test of a PE geomembrane exposed to a hazardous waste leachate.

The results show essentially no changes in mass, thickness, and dimensions of the PE geomembrane over the entire 120-day test period and thus indicate that interaction of the geomembrane with the leachate was negligible. In the event of appreciable interaction, such as absorption of organic leachate components, increases in mass would be observed with associated changes in thickness and dimensions. Such changes would be indicative of a partial solvation process (when solvent molecules are soluble in the polymer). Solvation causes the material to soften, thereby reducing hardness and modulus and probably tensile strength. Geomembranes that contain significant amounts of plasticizers and other soluble additives may lose those components of the compound. Such a process would be reflected in various changes in their mechanical properties, e.g., increases in modulus, hardness, and possible tensile strength, but with a loss in elongation and flexibility.

Table 13.13   Percentage change in physical
property versus exposure period.

Property	Exposure Period (days)			
	30	60	90	120
Mass (%)				
23°C	0.19	-0.10	0.19	0.00
50°C	0.05	1.09	0.05	0.05
Thickness (%)				
23°C	-1.19	1.19	1.10	1.16
50°C	-1.18	0.00	2.33	1.22
Dimensions (%)				
MD				
23°C	0.02	-0.33	0.00	0.17
50°C	-0.84	-0.18	-0.10	0.03
XD				
23°C	0.00	-0.73	-0.23	-1.12
50°C	-0.93	0.00	-0.86	-0.80

MD = Machine Direction
XD = Cross-Machine (Transverse) Direction

The mass and dimensional properties are measured nondestructively,
therefore the same specimen is measured before, during, and after
exposure for 120 days.   The results presented in Table 13.13 reveal
stability in these properties and minimal variability.
    Table 13.14 presents the mechanical properties of a PE geomembrane
in the same compatibility test that generated the data shown in
Table 13.13.

Table 13.14   Percentage change in mechanical property
versus exposure period

Property	Exposure Period (days)			
	30	60	90	120
Hardness (%)				
23°C	2.3	1.0	8.4	5.3
50°C	-1.9	-3.1	10.0	8.7
Stress at Yield (%)				
MD				
23°C	-8.1	10.0	-8.5	-11.8
50°C	-5.3	7.7	-6.3	-13.4
XD				
23°C	3.0	16.4	-5.5	-5.5
50°C	2.5	10.0	-6.3	-8.6
Elongation at Yield (%)				
MD				
23°C	-1.2	-3.1	-0.1	2.9
50°C	-11.1	-11.8	-1.9	-6.7
XD				
23°C	7.3	-1.4	5.8	6.2
50°C	0.2	-7.9	-4.3	-7.4
Stress at Break (%)				
MD				
23°C	17.3	30.4	18.7	19.9
50°C	-9.0	3.1	-8.4	-4.4
XD				
23°C	4.7	32.8	-16.4	10.6
50°C	-15.7	-35.8	-34.6	-25.0
Elongation at Break (%)				
MD				
23°C	14.8	8.9	20.4	22.2
50°C	-9.0	-8.2	-4.4	2.3
XD				
23°C	1.3	18.6	-0.5	-25.0
50°C	-14.6	-28.1	-15.5	-14.2

continued . . .

Table 13.14   Continued.

Property	Exposure Period (days)			
	30	60	90	120
Modulus of Elasticity (%)				
MD				
23°C	-7.0	-9.9	-4.6	-26.5
50°C	-22.4	-19.4	-17.5	-38.9
XD				
23°C	4.0	1.4	-4.1	-26.5
50°C	-23.9	-1.0	-17.2	-30.3
Initial Tear Resistance (%)				
MD				
23°C	1.0	-6.3	-10.1	1.7
50°C	-1.1	-11.0	-11.2	-2.1
XD				
23°C	5.3	10.4	3.5	13.9
50°C	1.0	8.0	-1.2	11.3
Puncture Strength (%)				
23°C	-10.5	0.1	-4.2	12.3
50°C	-4.5	11.7	-2.3	-4.2

MD = Machine Direction
XD = Cross-Machine (Transverse) Direction

The mechanical properties show significant changes. However, the percentage changes are erratic, and as a result, trends cannot be established. The scatter observed appears to be attributable to inherent material variability. Furthermore, fundamental material relationships are not observed. For instance, increases in stress values should be accompanied by reductions in elongation values, but such a relationship is not observed in these test results. The analysis of the physical properties (e.g., Table 13.13) displayed little change to indicate significant interaction with the leachate, thus significant changes in the mechanical properties are not expected. The mechanical properties should show changes that are related to one another. For instance, reductions in stress at yield should be accompanied by related reductions in stress at break, puncture strength, initial tear resistance, and modulus of elasticity. Such behavior is not observed, therefore the effect of material variability on mechanical property test results is significant.

Modulus of elasticity is typically measured by drawing a line tangent to the initial, apparently linear section of the stress/strain curve. This measurement is subjective since a linear portion of a stress/strain curve of a viscoelastic material such as PE is not likely to exist.   The most consistent method of calculating modulus of elasticity is to expand the stress/strain curve and employ a computer to perform the calculation at the steepest part of the curve. This method will preclude operator bias that is typically observed in manual calculation.   The computer-based procedure produces lower values than the manufacturer's specifications; however, since relative changes are the important factor in chemical compatibility test programs, consistency in the test method is needed in order to assess the effect of the chemical environment on the geomembrane as a function of time.

As presented in Table 13.13, much stability is observed in the nondestructive test results since material variability is not a factor; the same specimens are used for measurement after each exposure period.   Conversely, the mechanical properties display the effects of material variability since different specimens are tested after each exposure period.   Furthermore, since the tests are conducted over a period of four months, the effect of potential operational error may be significant. Increasing the number of specimens required for each test should minimize the effect of material variability.   In EPA Method 9090, five specimens in each direction are required to measure stress at yield and stress at break; three specimens in each direction are required for initial tear resistance; and two specimens are required for puncture strength.   The number of required specimens is low, especially for materials that possess localized variations in thickness, crystallinity, and other properties.

Interpretation of EPA Method 9090 test results requires careful assessment of the mass, thickness, and dimensional values (Table 13.13) since they provide a fundamental indication of the mechanism of degradation.   In order to assess trends in mechanical properties, an understanding of the physical properties is essential. For instance, an increase in mass indicates that leachate components are being absorbed by the geomembrane.   Such an interaction should be accompanied by increases in both elongation at yield and at break and reductions in stress (at yield and at break), puncture strength, initial tear resistance, modulus of elasticity, and secant modulus (for PVC, vulcanized, and fabric-reinforced geomembranes).   These interactions are not always observed, as presented in Table 13.13, therefore a greater emphasis is placed on the mechanical property values when an interpretation of EPA Method 9090 test results is involved.   The EPA has developed a computer program, "FLEX," based upon an expert system to aid in the interpretation of the data generated in EPA Method 9090 tests of geomembrane (Landreth, 1989).

Knowledge of the concentrations of various leachate components is also important for EPA Method 9090 data interpretation.   When leachates contain high levels of organic constituents, increases of weight with corresponding reductions in strength and modulus of elasticity are expected.   Conversely, some leachates may promote migration of plasticizers and other additives out of the geomembranes, which may occur with organic-based leachates.   Loss of plasticizers leads to loss of both weight and flexibility. Inorganic-based leachates may cause oxidative reactions on the surface of the geomembrane (when the leachate contains appreciable quantities of potentially oxidizing species such as sulfates and nitrates). Changes in mass, dimensions, and thickness may not occur. However, increases in hardness values may be observed since oxidative by-products may cause crosslinking of the polymer (Tisinger, 1989). Oxidation may also produce a reduction in tensile strength and elongation at break.

13.4 Modified Approach to Conventional Chemical Compatibility Testing

The interactions that may occur between leachates and geomembranes are complex. The bulk property testing that is used to monitor the behavior of a geomembrane in leachate may not be sensitive enough to detect the onset of degradative processes over the relatively short time period of typical geomembrane chemical compatibility test programs. The selection of a geomembrane for a specific application depends on the geomembrane's ability to retain its initial properties in such a test program. Since the duration of typical chemical compatibility test programs is relatively short compared with the expected service life, more sensitive techniques that can detect microstructural changes in the geomembrane should be incorporated in the testing of exposed geomembranes. Furthermore, exposure conditions need to be reevaluated for possible modification of both test duration and the method by which geomembrane samples are exposed to leachate.

13.4.1 Sump Exposures
Geomembrane samples could be exposed to a leachate in either a leachate collection or monitoring sump at a landfill. The geomembrane samples could be placed in perforated, inert containers and immersed in leachate. The perforated cell would allow fresh leachate to percolate through the container and continuously contact the geomembrane samples. The geomembrane samples would then be monitored for the same properties as those of the laboratory-exposed samples. Comparison of the property monitoring tests for both the laboratory-exposed samples and sump samples could then be carried out. Samples exposed in the sump could also be left in their perforated containers for extended periods of time long after the laboratory exposures have ended. This procedure would allow the operator to monitor the long-term effect of the leachate on the geomembrane under landfill conditions. In addition, as the waste stream changes, exposed geomembrane samples would be available for assessing changes in properties.

13.4.2 Constant Laboratory Exposures
An extension of the sump exposures is the constant laboratory exposure of geomembrane samples to leachate for any real time period of interest. Samples are periodically retrieved and tested for predetermined properties. The leachate is changed as the character of the waste stream changes or as plans to change the waste stream constitution are proposed. In the latter, case any incompatibility of the geomembrane with the new waste stream can be pre-determined. This type of test program will provide an early warning of possible

loss in the values of important properties as caused by both long-term exposure and waste stream change. Remedial actions can be prepared in advance.

13.4.3 Microstructural Analysis

The use of techniques that are typically employed for polymer characterization may be used to monitor the microstructure of geomembranes exposed to leachate. These techniques have significant advantages over measuring the bulk properties. For example, the amount of geomembrane required for these measurements is small, thus a smaller volume of leachate is required for exposures. These techniques also detect the onset of degradation and are sensitive enough to allow for accurate elucidation of the degradative mechanism. The following techniques may be employed:

- Differential Scanning Calorimetry (DSC),
- Thermal Gravimetric Analysis (TGA), and
- Infrared Spectrophotometry (IR).

DSC, TGA, and IR are used to measure thermal, compositional, and structural properties, respectively. These techniques are discussed in the following sections.

13.4.3.1 Differential Scanning Calorimetry

DSC monitors the thermal energy required to maintain a test specimen at the same temperature as a reference specimen while being heated at a constant rate of increasing temperature. This energy is exhibited as a function of the reference temperature in a thermogram. The thermogram may display endotherms corresponding to (net) energy absorbed in the specimen and exotherms corresponding to (net) energy emitted. From endotherms, melting ranges and degrees of crystallinity may be derived. Exotherms provide data for the assessment of the oxidative stability of the material based on either the oxidative induction time or oxidative induction temperature. Both oxidative induction time and oxidative induction temperature are performed in a labile atmosphere (air or oxygen). The time or temperature at which the onset of the exotherm is observed is the oxidative induction time or oxidative induction temperature, respectively. The degree of crystallinity tests are performed in a nitrogen atmosphere. From the resulting endotherm, the heat of fusion of a semicrystalline material can be determined.

13.4.3.2 Thermal Gravimetric Analysis

TGA is used for the determination of both the composition and temperatures at the onset of decomposition of polymeric materials. In a TGA, the weight of a specimen is monitored as the temperature is increased at a constant rate. The analyses are typically performed in an inert atmosphere initially to preclude weight gain due to oxidation. The various components that comprise a material have different thermal stabilities, thus decomposition takes place at different temperatures. When the weight becomes constant the residue consists of char, carbon black, and ash. Oxygen is then introduced into the instrument to burn off the char and carbon black, thus leaving the ash.

13.4.3.3 Infrared Spectrophotometry

IR spectra provide information on the structural characteristics of a material. Infrared energy causes vibrational motion of molecules and their components. This analytical technique involves exposing the material to infrared radiation at decreasing frequencies. When the IR frequency matches the vibrational frequency of the molecular component, the molecular component will vibrate (or absorbs radiation at that specific IR frequency). This frequency scan generates a spectrum of bands, each band corresponding to a specific vibrational frequency or range of frequencies where infrared radiation is absorbed by the specimen. The molecular components of any given molecule display a characteristic "spectrum of bands," representative of the particular chemical and physical environments of these moieties. Changes due to oxidation are observed by increased absorbance of those moieties that contain oxygen.

13.4.3.4 Stressed Specimens

Geomembranes and geomembrane seams may experience significant stresses while in use on the side slopes of landfills due to the in-plane components of the applied stress while also being exposed to leachates. The geomembranes may also experience contraction and expansion due to temperature fluctuations, and shrinkage at constant elevated temperatures. The effect of constant stress on the geomembranes and geomembrane seams needs to be assessed when the geomembrane is exposed to leachate.

The stress crack growth rate of semicrystalline geomembranes and geomembrane seams may be monitored while the samples are exposed to leachate. In these tests, ASTM D638 type IV (dogbone) specimens are notched to induce plane strain test conditions. The specimens are then installed in a constant load test rig and subjected to a constant load which is a fraction of its load at yield. The specimens are exposed to leachate in individual chambers. Reference values are obtained from notched specimens tested on the same rig but not exposed to leachate. The crack growth rate at the root of the notch is measured periodically or the test is allowed to run until failure occurs. The test data provide qualitative information on the mechanical durability of the geomembrane and geomembrane seams when under constant stress in leachate while in service.

The specimens may be subjected to cyclic loading while being exposed to leachate. The procedure for these tests is the same as that for the constant load specimens except that a cyclic load is applied at a predetermined rate. Crack growth rates are determined as a function of leachate exposure time. Durability data are generated in a shorter time period.

303

13.5 Summary

Not only must geomembranes used as liners for waste storage and disposal facilities have a low permeability to waste liquid, but they
must also maintain their properties over the designed service life of
the installation. This requires compatibility of the geomembrane with
the waste liquid to be contained. A large number of different types
of geomembranes varying in base polymer and in compounding ingredients
may initially be considered for each specific installation. Because
waste liquids are highly complex mixtures containing organic and inorganic constituents, it is necessary to assess and ensure compatibility between the proposed geomembranes and the site specific waste
liquid. Since geomembranes are polymeric materials, they can absorb
many constituents from a waste liquid, resulting in swelling and other
chemical effects, such as crosslinking and chain scission. These, in
turn, can cause changes in many of the bulk properties of the geomembrane, some of which could drastically compromise the performance of a
liner.

Much has been learned with respect to the basic factors involved
with the chemical compatibility of polymeric materials with various
liquids. The importance of crystallinity, solubility parameters,
crosslinking, molecular weight, and other such parameters as they
apply to relatively simple solvents and solutions are well recognized.

A wide range of tests has been developed by the polymer industry
to assess the compatibility of polymeric materials for various applications for containing simple liquids or mixtures consisting of relatively few species. However, because waste liquids can contain a
great number of constituents at very low concentrations, the absorption of selected species can be very slow--on the order of months or
years. These species will partition between the waste liquid and the
geomembrane at ultimate equilibrium concentrations. Because of the
complexity of waste streams and the synergism between low concentrations of many constituents, empirical data are required for untested
combinations of waste liquids and geomembranes.

At present, the EPA requires that geomembranes proposed for use in
constructing a hazardous waste disposal facility be tested for compatibility with the waste liquid in accordance with EPA Method 9090 as
part of the permit application process. The method is divided into
two basic parts:

- The exposure portion of the test, which attempts to simulate
  some of the conditions of service and is basically an
  immersion-type test run for 120 days.
- Determination of the properties of the geomembrane before and
  after exposure.

The tests include analytical procedures for indicating the amount of
liquid that has been absorbed and the geomembrane constituents that
have been lost to the waste liquid, as well as mechanical tests that
measure changes in properties such as tear resistance and tensile
strength.

Changes in properties during the exposure result principally from the absorption of waste constituents or the loss of components of the geomembrane, depending on the compatibility of the geomembrane with the waste liquid.

Only EPA Method 9090 has a related set of acceptance/rejection criteria, and these are in the form of an experimental expert system (FLEX). There are still many aspects of the test data which FLEX is not able to satisfactorily handle.

As presently performed, EPA Method 9090 has a number of major deficiencies:

- There are no effective criteria for defining acceptable changes in properties during the exposure.
- Changes in the properties as determined in the laboratory have not been correlated with those that occur in service and with long-term performance of liner systems.
- The length of the exposure period is comparatively short in contrast to the length of service that is anticipated for most of the geomembranes.
- The test program does not reflect the changing composition of the leachate or waste liquid in service.
- The monitoring tests may not be sensitive to the onset of degradation during the course of the exposure, i.e., the bulk property measurements are not sufficiently sensitive to detect the changes that cause degradation on a microstructural or local (surface) level.
- The specimens are not subject to any stress during the exposure.

All of these deficiencies lead to difficulties in interpreting results of a Method 9090-type test.

Because of the importance of chemical compatibility testing in assessing the potential performance of geomembranes that are in contact with waste liquids, some of the measures presented in this chapter should be introduced into the tests that are performed, e.g., sump exposures, real-time laboratory exposures, microstructure monitoring tests.

REFERENCES AND BIBLIOGRAPHY

Annual Book of ASTM Standards. Issued annually in several parts.
American Society for Testing and Materials, Philadelphia, PA.

Bellen, G., Corry, R., and Thomas, M.L. (1987). Development of Chemi-
cal Compatibility Criteria for Assessing Flexible Membrane Liners.
EPA 600/2-87/067 (NTIS No. PB 87-227 310). U.S. Environmental
Protection Agency, Cincinnati, OH. 492 pp.

Tisinger, L.G., and Carraher, C.E. (1989). "Nitric Acid Induced Degradation of High Density and Linear Medium Density Polyethylene:
Physical Property Sensing Techniques," Proceedings of the Symposium on Microstructure and the Peformance of Geosynthetics, ASTM,
Orlando, FL, January 1989.

Harper, C.A., ed. (1975). Handbook of Plastics and Elastomers.
McGraw-Hill Book Company, New York, NY. 1,014 pp.

EPA. (1986). EPA Method 9090. Compatibility Test for Wastes and
Membrane Liners. Test Methods for Evaluating Solid Waste. Vol.
1A: Laboratory Manual, Physical/Chemical Methods. 3rd ed.
SW-846. U.S. EPA, Washington, DC. September 30, 1986.

EPRI. (1989a). Study of Lining Materials for Use in the Management of
Wastes Generated by Coal-Fired Power Plants. Second Interim
Report. EPRI GS-6381. Electric Power Research Institute, Palo
Alto, CA. 181 pp.

EPRI. (1989b). Study of Lining Materials for Use in the Management
of Wastes Generated by Coal-Fired Power Plants. Final Report.
Electric Power Research Institute, Palo Alto, CA. (In preparation).

Haxo, H.E. (1983). Chemical Compatibility of Lining Materials with
Different Waste Fluids. Proceedings of the Colloque Sur L'
Etancheite Superficielle des Bassins, Barrages et Canaux. Volume
II. Sponsored by the Centre National du Machinisme Agricole, du
Genie Rural, des Eaux etdes Forets. Paris, France. pp 13-19.

Haxo, H.E. (1985). Polymeric Membrane Lining Materials for Waste
Management Applications. Kautschuk + Gummi · Kunststoffe. May.
pp 485-493.

Haxo, H.E. (1989a). Compatibility of Flexible Membrane Liners and
Municipal Solid Waste Leachate. Proceedings of the 15th Annual
Research Symposium: Remedial Action, Treatment and Disposal of
Hazardous Waste, April 10-12, 1989. U.S. EPA, Cincinnati, OH.
18 pp. (In press).

Haxo, H.E. (1989b). Synthetic Lining Systems for Land Waste Disposal
Facilities. Proceedings of SARDINIA '89 - Second International
Landfill Symposium, Vol. 1, October 9-13, 1989. Universita di
Cagliari, Porto Conte (Alghero) - Italy. pp XVII-1--XVII-13.

Haxo, H.E., and Haxo, P.D. (1989). Environmental Conditions Encoun-
tered by Geosynthetics in Waste-Containment Applications. Pro-
ceedings of GRI 2: Aging and Durability of Geosynthetics,
December 8-9, 1988, Sponsored by the Geosynthetic Research
Institute, Philadelphia, PA. Edited by R.M. Koerner. Elsevier
Applied Science Publishers, Ltd. 20 pp. (In press).

Haxo, H.E., Haxo, R.S., Nelson, N.A., Haxo, P.D., White, R.M., and Dakessian, S. (1985a). Liner Materials Exposed to Hazardous and Toxic Wastes. EPA-600/2-84-169 (NTIS No. PB 85-121-333). U.S. EPA, Cincinnati, OH. 256 pp.

Haxo, H.E., Haxo, N.A. Nelson, Haxo, P.D., White, and Dakessian, S. (1986). Liner Materials Exposed to Toxic and Hazardous Wastes. Waste Management and Research 4. pp 247-264.

Haxo, H.E., Lahey, T.P., and Rosenberg, M.L. (1988). Factors in Assessing the Compatibility of FMLs and Waste Liquids. EPA/600/2-88/017 (NTIS No. PB 88-173-372/AS). U.S. EPA, Cincinnati, OH. 143 pp.

Haxo, H.E., and Nelson, N.A. (1984). Factors in the Durability of Polymeric Membrane Liners. In: Proceedings of the International Conference on Geomembranes, Denver, CO. Volume II. Sponsored by Industrial Fabrics Association International. St. Paul, MN. pp 287-292.

Haxo, H.E., White, R.M., Haxo, P.D., and Fong, M.A. (1985b). Liner Materials Exposed to Municipal Solid Waste Leachate. Waste Management and Research 3. pp 41-54.

Landreth, R.E. (1989). FLEX — An Expert System to Assess Flexible Membrane Liner Materials. Proceedings of Geosynthetics '89, San Diego, CA. February 21-23, 1989. IFAI, St. Paul, MN. pp 397-407.

Modern Plastics Encyclopedia. (1988). McGraw-Hill, Inc., pp 510-562.

Matrecon, Inc. (1983). Lining of Waste Impoundment and Disposal Facilities. SW-870 Revised. U.S. EPA, Washington, DC. 448 pp. GPO #055-00000231-2.

Matrecon, Inc. (1988). Lining of Waste Containment and Other Impoundment Facilities. 2nd Revised Edition. EPA/600/2-88/052. U.S. EPA, Cincinnati, OH. 991 pp.

Morrison, W.R., and Parkhill, L.O. (1987). Evaluation of Flexible Membrane Liner Seams After Chemical Exposure and Simulated Weathering. EPA-600/2-87-015. U.S. EPA, Cincinnati, OH. 296 pp.

National Sanitation Foundation. (1985). Standard Number 54: Flexible Membrane Liners. Rev. Standard. National Sanitation Foundation, Ann Arbor, MI.

U.S. General Services Administration. (1980). Method 2065: Puncture Resistance and Elongation Test (1/8-Inch Probe Method), and Method 2031: Tetrahedral-Tip Probe Method. Federal Test Method Standard 101C. U.S. GSA, Washington, DC.

14 BIOLOGICAL RESISTANCE TESTING

R. DEGEIMBRE, J-M. RIGO and J. WIERTZ
Geotextiles and Geomembranes Research Centre, University of
Liege, Belgium

14.1 Introduction

Geomembranes are subjected to various stress originated by
environment.
The surrounding environment can be namely characterized
physical, chemical and biological parameters.
These latters are not fully independant one from each other
interelations do exist. Synergies may occur for the activation
passivation of the reactions.
This contribution will examine the biological paramete
their mode of actions and the test procedures used to simul
their effect in laboratory.

14.2 The biological parameters

14.2.1 Characterization
The biological stresses result from the action of micro-organi
living in contact or near the geomembranes. These organisms
be plants, animals or protist.
These three categories can be subdivided in sub categories li
for plants : bacterias, actinomyectes, fungi, algae
macroscopic plants.

14.2.2 Mode of action
The mode of action of the biological parameters and the
kinetics depend upon the organisms activity (nutriti
reproduction,...).
Different schemes are possible :
- the geomembrane or some of its components are assimilated by
  the biological agent;
- the products of the biological agent metabolism act on the
  geomembrane or some of its components.
It must be pointed out that the resulting actions may
direct (assimilation or chemical attack by the metaboli
products) or indirect.
On the other hand, it is important to take into account not o
the effects on the geomembrane but also on the waterproof
system : geomembrane, geotextiles, protection, ballast drain
systems....
Rollin demonstrated earlier that drainage system may be affect
by biological action.

At this level, it is important to observe that the geomembranes made of polymeric materials are not specially sensitive to biological attack by assimilation. In fact, this process supposes the assimilation of molecules from the material by the cells of the biological agent. In this case particularly one has macromolecules that cannot be directly assimilated by the cells.

14.3 Biological attack testing methods

14.3.1 Introduction
Simulations of the biological attacks can be made in laboratory. These are accelerated tests where an artificial attack is performed in an activated medium where some parameters of the expected environment are present and in general under artificially increased importance. For example : the activity level can be increased by higher temperature or the daily action duration is increased or both.
The acceleration possibilities are not unlimited because the biological agents do not react systematically and proportionnally to the accelerators.
Generally, experiments are run in optimized conditions : temperature, humidity, quantity and shape of the exposed samples. Practically, to run an accelerated biological attack test is not easy to be done. The effective activity of the biological agent is to be controlled on reference materials. This latter must be sensible and adequate to the biological agent.

14.3.2 Bacteriological resistance
Plastic resistance to bacteria attack may be determined by procedures described in ASTM G 22.
The procedure described in ASTM G 22 consists of the following steps :
- selection of suitable specimens for determination of pertinent properties;
- inoculation of specimens with suitable organisms;
- exposure of inoculated specimens under conditions favorable to growth;
- examination and rating for visual growth and
- removal, sterilization and evaluation of specimens.

The resin portion of plastic materials is usually resistant to bacteria, in that it does not serve as a carbon source for the growth of bacteria. It is generally the other components, such as plasticizers, lubricants, stabilizers and colorants that are responsible for bacterial attack on plastic materials. It is important to establish the resistance of plastics to microbial attack when plastics are used under conditions of high temperature and humidity favorable for such attack.

The effects to be expected are :
- surface attack, discoloration and loss of transmission (optical);
- removal of susceptible plasticizers, modifiers and lubrican
resulting in increased modulus (stiffness), changes in weig:
dimensions and other physical properties.

Other effects include preferential growths caused nonuniform dispersion of plasticizers, lubricants and ot: processing additives. Pronounced physical changes may be obser on products in film form or as coatings where the ratio surface to volume is high, and where nutrient materials such plasticizers and lubricants continue to diffuse to the surface they are utilized by the organisms.

Since attack by organisms involves a large element of cha: due to local accelerations and inhibitions, the order reproducibility may be rather low. To assure that estimates behavior are not too optimistic, the greatest observed degree deterioration should be reported.

Conditioning of specimens such as exposure to leachi: weathering, heat treatment, etc.. may have significant effects the resistance of plastics to bacteria. Determination of the effects is not covered in ASTM G 22.

Tests are realized in thermal and humidity control incubators : 36 ± 1°C and more than 85 % R.H.

The simplest specimen may be a 50 by 50 mm piece, a 50 diameter piece or a piece (rod, tubing) at least 75 mm (3 i: long cut from the material to be tested. Completely fabrica' parts or sections cut from fabricated parts may be used as t: specimens. On such specimens, observation of effect is limited appearance, density of growth, optical reflection transmission, or manual evaluation of change in physi: properties such as stiffness.

Film-forming materials such as coatings may be tested in form of films, at least 50 by 50 mm in size. Such films may prepared by casting on glass and stripping after cure, or impregnating ignited glass fabric.

For visual evaluation, three specimens shall be inoculated. the specimen is different on two sides, three specimens of ea: face up and face down, shall be tested.

14.3.3 Resistance to microorganisms
The evaluation of the geomembranes microorganisms resistance to be estimated following two main procedures : fungi and s: exposure.

14.3.3.1 Resistance to fungi
Fungi include yeasts, molds and mushrooms. They depend on orga: matter for carbon, nitrogen and other elements. Their numbers : be very large as much as 10 to 20 million per gram of dry s: and their population is constantly changing. Placing geomembra: in decomposing organic residue often causes concern.

However, the high-molecular-weight polymers generally used for geomembranes seem very insensitive to such degradation. ASTM G 21 deals with resistance of plastics to fungi.

Of very real concern, however, is the possibility of fungal deposits clogging and blinding flow through or within the geotextiles, geonets and drainage geocomposites that are often associated with geomembranes. These items will be discussed subsequently.

The ASTM G 21 procedure consists to describe :
- selection of suitable specimens for determination of pertinent properties;
- inoculation of the specimens with suitable organisms;
- exposure of inoculated specimens under conditions favorable to growth;
- examination and rating for visual growth and
- removal of the specimens and observations or testing, either before cleaning or after cleaning and reconditioning.

The resin portion of these materials is usually fungus-resistant in that it does not serve as a carbon source for the growth of fungi. It is generally the other components, such as plasticizers, cellulosics, lubricants, stabilizers and colorants, that are responsible for fungus attack on plastic materials. It is important to establish the resistance to microbial attack under conditions favorable for such attack, namely, a temperature of 2 to 38°C and a relative humiditiy of 60 to 100 %.

The effects to be expected are as follows :
- surface attack, discoloration, loss of transmission (optical);
- removal of susceptible plasticizers, modifiers and lubricants, resulting in increased modulus (stiffness), changes in weight, dimensions and other physical properties and deterioration of electrical properties such as insulation resistance, dielectric constant, power factor and dieletric strength.

The simplest specimen may be a 51 by 51 mm piece, a 51 mm diameter piece or a piece (rod or tubing) at least 76 mm long cut from the material to be tested. Completely fabricated parts or sections cut from fabricated parts may be used as test specimens. On such specimens, observation of effect is limited to appearance, density of growth, optical reflection or transmission, or manual evaluation of change in physical properties such as stiffness.

Film-forming materials such as coatings may be tested in the form of films at least 51 by 25,4 mm in size. Such films may be prepared by casting on glass and stripping after cure, or by impregnating (completely covering) filter paper or ignited glass fabric.

For visual evaluation, three specimens shall be inoculated. If the specimen is different on two sides, three specimens of each, face up and face down, shall be tested.

Such as in AFNOR NFX 41.514, samples are placed in conta
simultaneously with different fungi organisms (6 or 12)
complete media (for invasion tests) or uncomplete media (f
comestibility tests). The first one includes carbon hydrate su
as glucose and the latter doesn't in this case; the fungi has
find its food in the specimen of geomembrane under analysis.

The all medium (sample, biological agent, artificial mediu
is placed in a box. This is installed in a temperature (30°C) a
relative humidity (95 %) controlled oven. The attack is do
without light.

As presented here above, after this period, usual observatio
are made on the attacked geomembrane specimen.

A quotation of the resistance is established from 0 to 4 (0 wh
no growth of fungi is observed and 4 in case of comple
invasion).

In order to control the activity of the fungi agents, tests a
done in the same conditions as described before but where t
specimen is replaced by cotton.

After such an exposure, the geomembrane specimens are washed
immersion in a mercuric chloride solution and than in water.

The specimen are ready for various controls : weight los
mechanical, chemical or physico-chemical controls.

Degeimbre (1980) submitted a (polyester polypropylen
geotextile to fungi attack following the NF X 41.514 Te
procedure. After 6 weeks of exposures to envasion conditions a
comestibility conditions visual observations were done.

The geotextile was completely envased (degree 4) and partial
covered by mushroom (degree 2) (degree 0 corresponds to
developement of fungi and degree 4 to an important development
when the specimen is not visible anymore).

After observations, the specimen are cleaned with a 1 % mercur
chloride solution and the remaining tensile characteristics we
determined.

Table 14.1 gives the obtained test results.

Table 14.1   Tensile characteristics modification after fungi
              exposure from Degeimbre (1980)

Specimen	Tensile resistance resistance (kN/m)	Elongation at maximum strength (%
Reference	19,0	17,0
After fungi attack test	14,4	17,0
After comestibility test	13,0	17,5

After the exposures, thickness, infra-red spectroscopy and wat
permeability were also controlled but the remaining presence
cleaning liquids and fungi rendered the obtained resul
difficult to be analysed.

In fact, the complete removal of fungi request a great amount of energy that destroy partially the specimen.

14.3.3.2 Soil exposure

The artificial soil is a mould with the next physical and chemical characteristics :
- 6 < pH < 7,5;
- carbon/nitrogen ratio between 8 and 15;
- humidity content 30 ± 2 %.

The medium must have an activity such as a cotton fabric will loose its tensile resistance after an exposure of 8 days in a 30°C and 95 % relative humidity oven.
Samples to be tested are submitted to the same treatment during 2 or 4 weeks or more.

In parallel, samples of the geomembrane to be tested previously impregnated with mercuric chloride are placed in a sterelised mould (3 days at 120°C). These samples that are thus protected against the fungi attack will be submitted to the thermal and humidity ageing in the soil. This helps to establish the distinction between the fungi and the thermal ageing.

14.3.4 Roots resistance

The swiss standard SIA 280 describes a procedure for testing the roots resistance of materials.
Following this procedure, three containers (20 cm diameter approximatively) are filled up to mid-height with a earth constitued of a mix of vegetable earth and peat. The specimen to be tested covers this first layer (see photo 1). The junction with the container is waterproofed. The sample is than covered by a second earth layer (about 100 mm). Lupinus albus seeds are planted. This operation is done during winter; the containers are placed in a green-house and an artificial light is added.

Photo 1 : root resistance of geomembranes

313

After 8 weeks of exposure, the plants have reached th
maturity. The sample is removed and examined. The observat
consists to see if roots have crossed the sample.
The activity of the lupinus albus plants is checked on referei
membranes made of a 20 mm thick 85/40 bitumen layer. This lat'
must be crossed by the roots.
Other tests can also be done where the plants are not
aggresive but their duration is longer (up to 4 years).

14.3.5 Resistance to animals
The determination of the geomembrane resistance to animals is i
easy to be realised because the test conditions are difficult
be reproduced.
Animals types are numerous and different : insects, millipede
woodlice, mites, rats...
They differ by their "food" interest, appetit, biological cycl
live-time...
It must be checked that the membrane doesn't constitute a fe
for the considered species. This is rarely the case taking
molecular weight of the exposed plastics into account.
If the membrane is an obstacle to the normal growth or
progression of the animal in search of foods, it may be loca
destroyed.
     It is finally important to check the resistance of
geomembrane to the metabolism products of the animals.

14.4 Conclusions

Such as it happens for all the accelerated ageing test,
simulation of microorganisms attack is a difficult task.
Synergies effects are to be expected and in general, for on s
exposure, microorganisms, plant and animal agression
happening at the same time.
     All the hereabove described testing procedures are trying
simulate biological attacks one by one. For some cases
biological factors are limited to one subcomponent (see ro
resistance with lupinus albus only).
     A second problem is that the results obtained by these te
are not specially linked with the performances characteristics
the geomembrane.
As it is the case for all the durability problems of structu
or complex systems, it will be necessary to have an idea of
behavior of all the components versus the biological agents :
instance drainage systems may be affected by biological activi
This will be easier if the engineer has informations on
behavior of each component of the system.
     These informations can be originated by laboratory te
methods but also by field observations and cases analysis.
The latter must help to define correctly the global and lo
environments supported by the geomembranes and the waterproof
systems. They must also indicate clearly the obtained degradat
versus the nature and shape of the materials.

14.5 Bibliography

ASTM G 22 (1976). **Standard recommended practice for determining resistance of plastics to bacteria.**

ASTM G 21 (1975). **Standard recommended practice for determining the resistance of synthetic polymeric materials to fungi.**

DEGEIMBRE (1980). **Determination of the geotextiles resistance to fungi attack.** Internal report. LMC, University of Liege.

KOERNER (1989). **Designing with geosynthetics.** 2d edition, Drexel University.

NF X 41-514 (1961). **Plastic materials protection – part 2 : test methods for determining the microorganism resistance of manufactured products.**

SIA 280 (1975). **Test methods on materials, waterproofing membranes in plastic materials.**

15 CHEMICAL IDENTIFICATION METHODS USED TO CHARACTERIZE POLYMERIC GEOMEMBRANES

Y. H. HALSE
Geosynthetic Research Institute, Drexel University,
Philadelphia, PA, USA
J. WIERTZ and J-M. RIGO
Geotextiles and Geomembranes Research Centre, University of
Liege, Belgium

15.1 Introduction

It must be clearly recognized by geomembranes manufacturers, users and designers that each type of polymer used in geomembrane manufacturing is actually a compounded material which is given a generic name, e.g., polyethylene, polyvinyl chloride, etc. Within each of these types, the compound (consisting of resin, fillers, antidegradants, etc.) can vary considerably. Haxo (1988) gives the following information, where the basis of each classification is 100 parts of the resin and thereafter ingredients are added depending upon the particular type and use of the geomembrane (Table 15.1).

Table 15.1   Major components in various types of polymeric geomembranes, after Haxo (1988)

Component	Composition in Parts by Weight		
	Elastomers	Thermoplastic Amorphous	Thermoplastic Semicrystalline
Polymer or Alloy	100	100	100
Oil or Plasticizer	5-40	5-55	0-10
Fillers			
carbon black	5-40	5-40	2-5
inorganics	5-40	5-40	--
Antidegradants	1-2	1-2	1
Crosslinking			
inorganic	5-9	0-5	--
sulfur	5-9	--	--

In light of the above variability it is often necessary to identify the specific ingredients after the geomembrane is manufactured. Some of these reasons are as follows:

- To check on quality control during the manufacturing process
- To verify that the installer is furnishing the specified compound, i.e., chemical fingerprinting

316

- To investigate unusually poor field performance, i.e., forensic analysis
- To assess and keep track of the aging process
- To predict long-term performance
- To be able to formulate better compounds knowing the precise ingredients of those currently in use

It will be noted that this concluding chapter of the book forms somewhat of a departure from the previous chapters, in that focus is on the microscopic (or molecular) level rather than on macroscopic (or large scale) behavior of the geomembrane. This being the case, it bears heavily on chemical analysis methods of which there is a tremendous body of available information, as for example Skoog and West (1980), Turi (1981), Haslam (1972) and Krause et al. (1983). Furthermore, in the polymer identification area this information is changing and growing rapidly. Thus a good deal of selectivity in choosing the methods to be presented is required. The methods we will focus upon are listed in below, and brief section will be written on each method:

- thermogravimetric analysis (TGA)
- differential scanning calorimetry (DSC)
- thermomechanical analysis (TMA)
- infrared spectroscopy (IR)
- chromatography (GC or HPLC)
- density determination (ϱ)
- melt index (MI)
- gel permeation chromatography (GPC)

As a preview of the discussion to follow, Table 15.2 summarizes the advantages and disadvantages of each of the methods.

Since the number of methods is large, we have been arbitrarily selective in limiting the number of geomembranes to analyze. Among the polymeric geomembranes, only those in widespread current use will be illustrated. They are HDPE (high density polyethylene), VLDPE (very low density polyethylene), PVC (polyvinil chloride) and CSPE (chlorosulfonated polyethylene). Please note that HDPE is the common terminology used in the geomembrane community. However, according to a strict classifications, e.g. ASTM D-1248 (1989), it actually is made from a MDPE resin. Only the addition of carbon black raises the density of the resulting compound to a value within the high density range. The analysis methods, however, are equally well suited for other polymeric geomembranes, such as elastomeric geomembranes, as butyl rubber or ethylene propylene dyene monomer (EPDM), as for example presented by Vanni et al. (1989).

15.2 Thermogravimetric analysis (TGA)

There are a number of thermal analysis techniques that are useful in determining the chemical nature of polymeric materials used for manufacturing geomembranes. The first which we review is that of thermogravimetric analysis, or simply "TGA".

Table 15.2 Chemical identification methods described in this chapter.

Method	Information Obtained	Advantages*	Disadvantages
TGA	.polymer, additives and ash contents .carbon black amount .decomposition temperatures	.straightforward measurement .high accuracy .for all polymers	.qualitative results .high cost
DSC	.melting point .crystallinity .oxidation time .glass transition	.straightforward measurement .high accuracy .for all polymers	.qualitative results .limited for chlorinated polymers .high cost
TMA	.coefficient of linear thermal expansion .softening point .glass transition	.straightforward measurement .high accuracy .for all polymers	.high cost
IR	.identifies additives .identifies fillers .identifies plasticizers .rate of oxidation reaction	.for all polymers	.specimen preparation .no resin information .high cost
GC and HPLC	.identifies additives .identifies plasticizers	.straightforward measurement	.specimen prep. .no resin information .high cost
ϱ	.density .crystallinity	.straightforward measurement .accurate values .low cost	.only selected polymers .no direct resin information
MI	.melt index .flow rate ratio	.related to molecular weight .straightforward measurement .low cost	.limited for chlorinated polymers .empirical values
GPC	.molecular weight distribution	.accurate values .only valid technique for molecular weight .for all polymers	.not very common .tedious test .specimen preparation .very high cost

* An advantage common to all methods is the extremely small sample size required for testing in comparison to traditional physical and mechanical test specimen sizes.

318

TGA is an instrument that permits the continuous weighing of a specimen as a function of increasing temperature. The test is subjected to heat at a constant rate, or it may be isothermally maintained at a fixed temperature under inert atmosphere or air. Any reaction which involves changing of the specimen's mass can be detected by the instrument's recording balance. The technique provides information on the following:

- Determination of the amount of moisture, volatiles, carbon black and ash.
- Decomposition temperatures of the various components in the sample.
- Determination of chemical compounds in the specimen and their quality.

An example of the TGA procedure for analyzing polymeric geomembranes liners is described. A 20 mg specimen is placed in the holder, and then the apparatus is saturated with nitrogen gas at the flow rate of 50 cm^3/min. The instrument is adjusted to give a 100% full-scale deflection for the weight of the specimen. Thus the result will be read directly in percent weight lost. The specimen is heated from room temperature up to 650°C at a rate of 20°C/min for polyethilene and 10°C/min for other polymers. It then maintains this upper temperature for at least 10 minutes until no weight loss occurs. At this time the residue is free of all polymers and now includes only carbon black, filler, and/or carbonaceous polymer residue. Air is then introduced to oxidize the residue for determining the ash content.

Typical thermograms for HDPE and VLDPE are illustrated in Figure 15.1. Both curves show only a single decomposition response, corresponding to the break-down of the polyethylene polymer main chains. The onset temperature of this break-down process for VLDPE (465°C) is only a few degrees lower than the HDPE (471°C). This suggests that the variation in the density between various PE resins does not significantly affect the decomposition temperature. The major component in both polymers is polyethylene (97.3%). Most of the remainder is carbon black (2.4%). When air was introduced into the apparatus, it oxidized the carbon and the amount of ash content was found to be 0.3%.

In contrast to the response of PE shown in Figure 15.1, note the TGA thermogram of PVC which is shown in Figure 15.2. There are two decomposition processes occurring at onset temperatures of 270°C and 352°C. The first process includes the dehydrohalogenation to form polyenes (given out HCl) and the degradation of plasticizers. The second process is the degradation phases of polyenes. After these are complete, air is introduced to oxidize the carbon black and carbonaceous polymer residue. The final residue is ash, which amounts to 6.7%.

Figure 15.3 shows the TGA thermogram of CSPE. Two major thermal degradation processes occur at onset temperatures of 304°C and 439°C. The loss of SO_2Cl groups and plasticizers, and dehydrohalogenation is the first process. The second process is polymer main-chain breakdown,

319

described by Smith (1966). However, the amount of plasticizer content in both PVC and CSPE polymers cannot be deduced using TGA. This is due to the dehydrohalogenation process which overlaps the degradation of the plasticizer.

The amount of plasticizer in the polymer can be determinated using some type of extraction technique. To do so the polymer is first dissolved in acetone, and then petroleum ether is added to precipitate out the polymer. Then by simple weighing the amount of plasticizer can be obtained. Additionally, the type and amount of plasticizers can also be determined using gas chromatrophy and infrared spectroscopy. Both techniques will be described in subsequent sections.

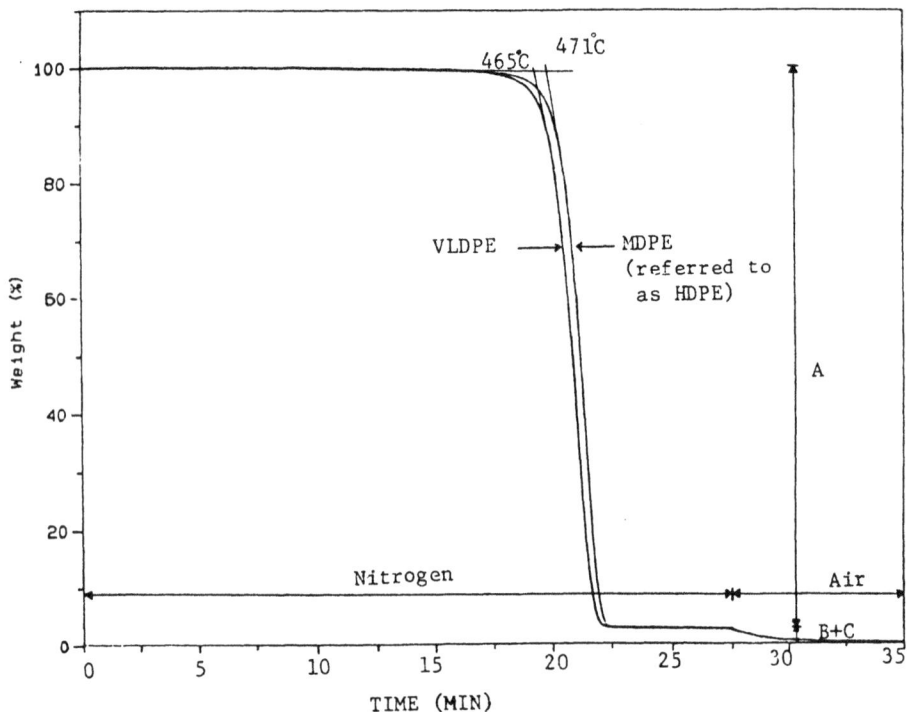

Heating Rate = 20°C/min

A = 97.3% = polymer
B = 2.4% = carbon black
C = 0.3% = ash

Fig. 15.1 TGA thermograms of HDPE and VLDPE geomembranes.

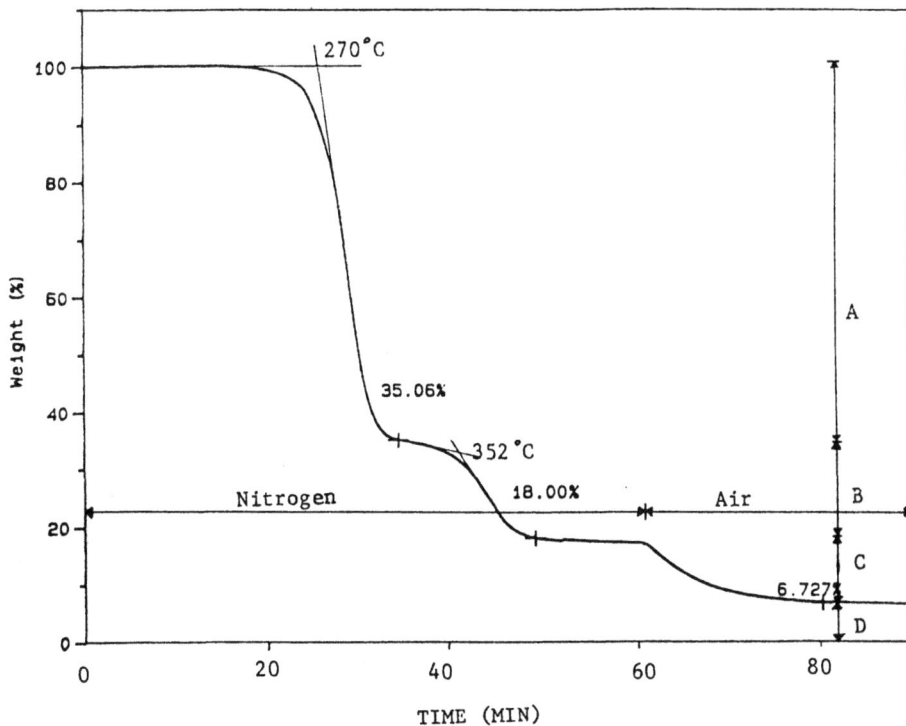

Heating Rate = 10°C/min

A = 65% = plasticizers + HCl from the polymer
B = 17% = residual polymer
C = 11% = carbonaceous polymer residue + carbon black
D = 7% = ash

Fig. 15.2 TGA thermogram of a plasticized PVC geomembrane.

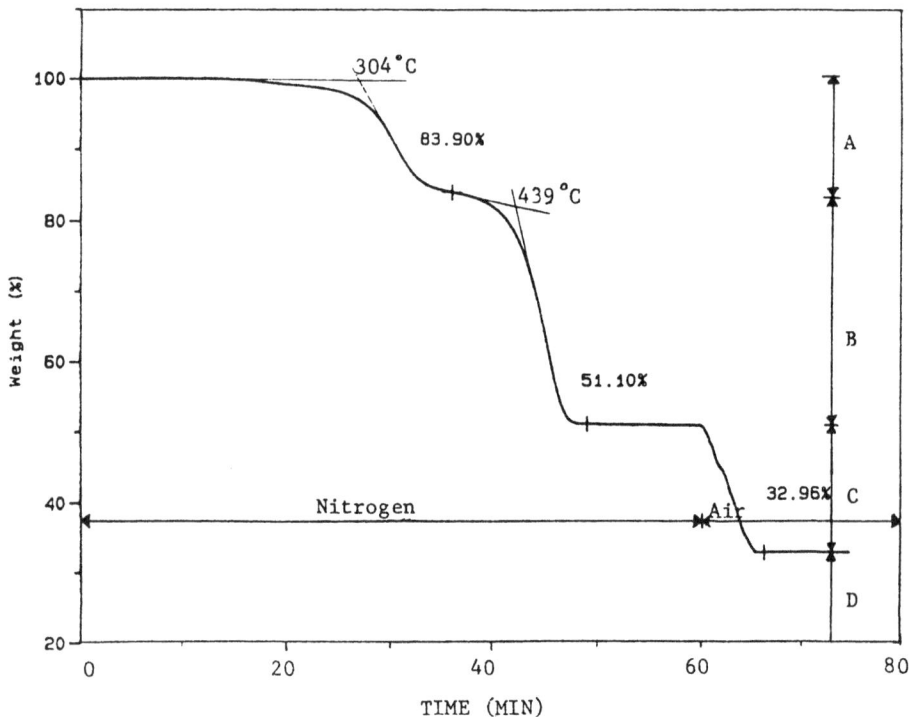

Heating rate = 10°C/min

A = 16% = plasticizers + HCl and SO_2Cl from the polymer
B = 33% = residual polymer
C = 18% = carbonaceous polymer residue + carbon black + filler
D = 33% = ash

Fig. 15.3   TGA thermogram of a CSPE geomembrane.

15.3 Differential scanning calorimetry (DSC)

Differential scanning calorimetry, or "DSC", is another thermal method used to characterize polymers. It is an instrument that permits the continuous comparing of the power difference between a test specimen and a blank reference during heating. The specimen is placed in an aluminum pan with a cover, and is referenced to an identical, but empty, aluminum pan with a cover. The differential power to maintain constant temperature between the specimen and the reference is monitored. Any reactions involving heat transition can be detected. Endothermal (such as melting) and exothermal (such as crystallization) reactions will appear as peaks in opposite directions to each other. The glass transition temperature will be noted by a step in the thermocurve. Information about the polymer which are determined using DSC are melting point (T_m), crystallinity, oxidation stability, glass-transition temperature (T_g), and vaporization.

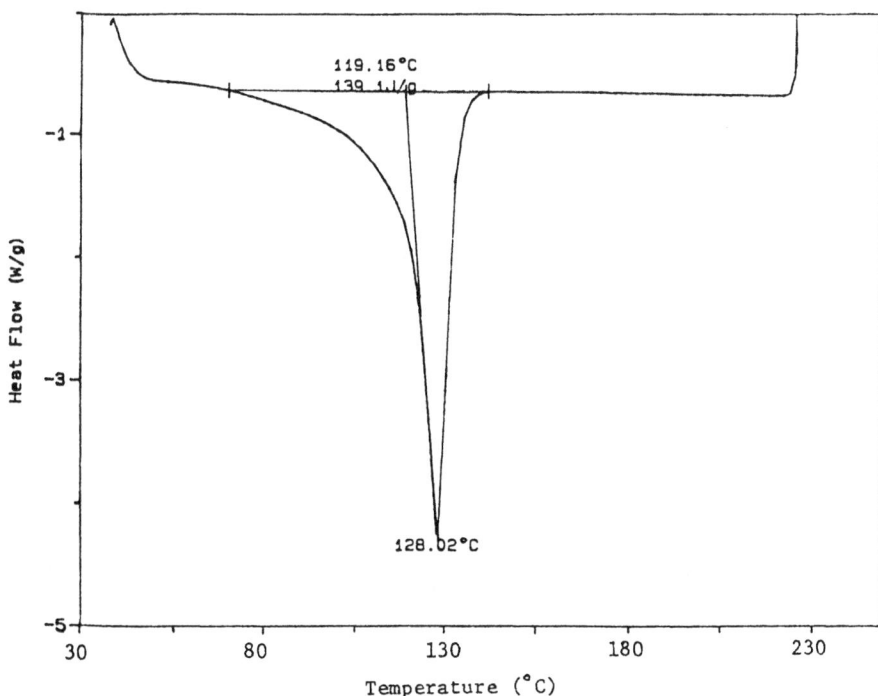

Fig. 15.4   DSC thermogram of HDPE geomembrane.

The procedure for operation of a DSC is very simple. A specimen weighing 2 to 3 mg is placed in the sample holder. The apparatus is then saturated with nitrogen gas using a flow rate of 50 cm^3/min. For PE geomembranes, the sample is heated from room temperature to 200°C at the rate of 20°C/min to observe the melting. HDPE resins shows a single melting peak at around 128°C (as shown in Figure 15.4), whereas VLDPE shows a double melting peak at temperatures of 102°C and 118°C (as shown in Figure 15.5). The double melting peaks in VLDPE can be caused by two factors. One is the chemical nature of the polymer, such as two polymers being blended together or two distinguished molecular weights. The other factor is recrystallization of polymer during thermal analysis. The second melting peak is the melting of more perfect crystals which are formed by recrystallization after the initial melting. The recrystallization process generated an exothermic peak which is strongly dependent on the heating rate. At a slow heating rate, there is higher degrees of recrystallization, resulting a more pronounced peak, as described by Turi (1981).

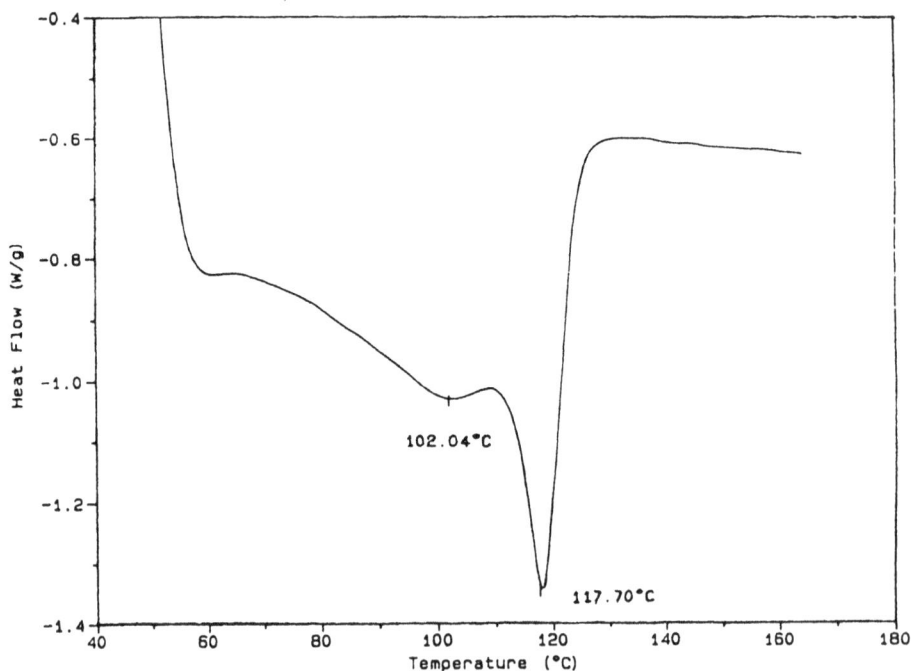

Fig. 15.5   DSC thermogram of VLDPE geomembrane.

324

In this particular VLDPE specimen, the multiple melting peaks are probably caused by recrystallization rather than chemical nature of the polymer. This is because that its TGA thermograph (Figure 15.1) only shows one decomposition temperature, reflecting a homogeneous material.

The heat of fusion (ΔH) is equal to the area under the melting peak. It represents the amount of crystallinity in the polymer. The percent crystallinity is deduced by comparing the ΔH of the specimen to the theoretical ΔH value of the 100% crystalline polymer which is 285 J/g for HDPE. In Figure 15.4, the heat of fusion of this HDPE geomembrane is 139 J/g, so the percent crystalline of the polymer is 49%.

The oxidation stability of PE can be obtained by measuring the oxidation induction time (OIT). It is the length of time for oxidation to occur under an isothermal condition. Figure 15.6 shows a thermograph of HDPE which is heated to 225°C under nitrogen atmosphere and then it is held at such temperature with air flowing until oxidation takes place. The OIT for this specimen is 10.7 minutes.

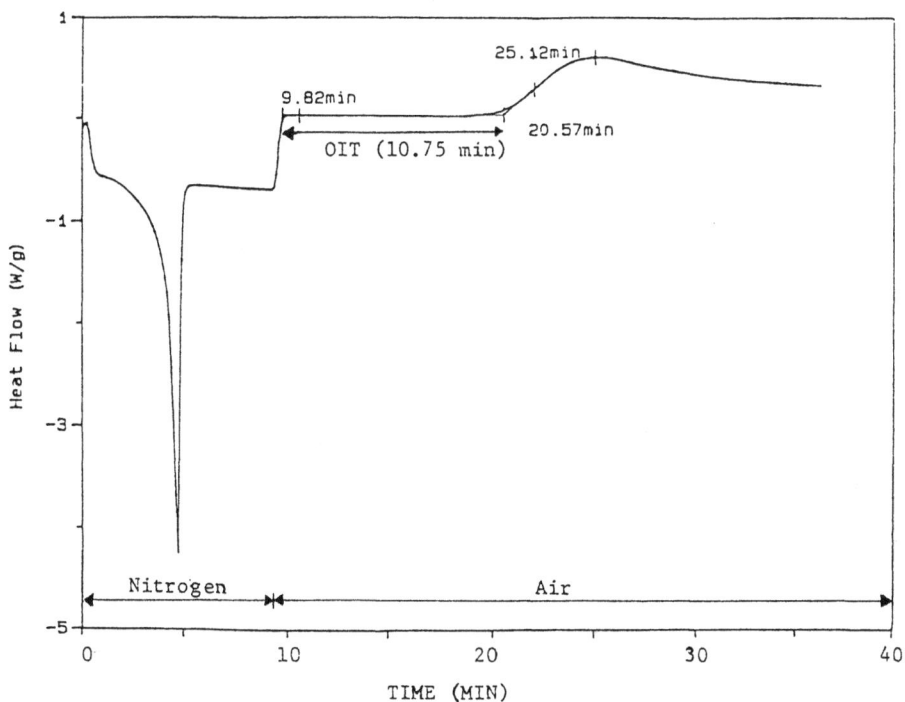

Isothermal temperature = 225°C

Fig. 15.6   DSC thermogram of HDPE geomembrane showing OIT (Oxidation Induction Time).

According to ASTM D-3895 (1988) "Copper-Induced Oxidative Induction Time of Polyolefins by Thermal Analysis", the isothermal temperature is 200 ± 2°C, with oxygen flowing at the rate of 50 ± 5 ml/min.

It must be noted, however, that the OIT value is very sensitive to the type of antioxidants that are in the compound. Hence caution must be taken when comparing OIT values of different PE geomembranes. These geomembranes must contain the same type of antioxidant, if not the OIT comparison will be questionable.

The evaluation of the glass transition temperature (T_g) of PE requires a cooling chamber, since the T_g of PE is below zero (-125°C). The specimen is cooled down to liquid nitrogen temperature (-178°C) and then heated up. The T_g value is indicated when the curve shows an abruptly shifted. For HDPE material, the T_g is not very pronounced due to the limited amount of amorphous phase material. For evaluation of T_g, the HDPE specimen should be heated above the melting point and then quenched down to -178°C as soon as possible so that more amorphous phase material can be generated.

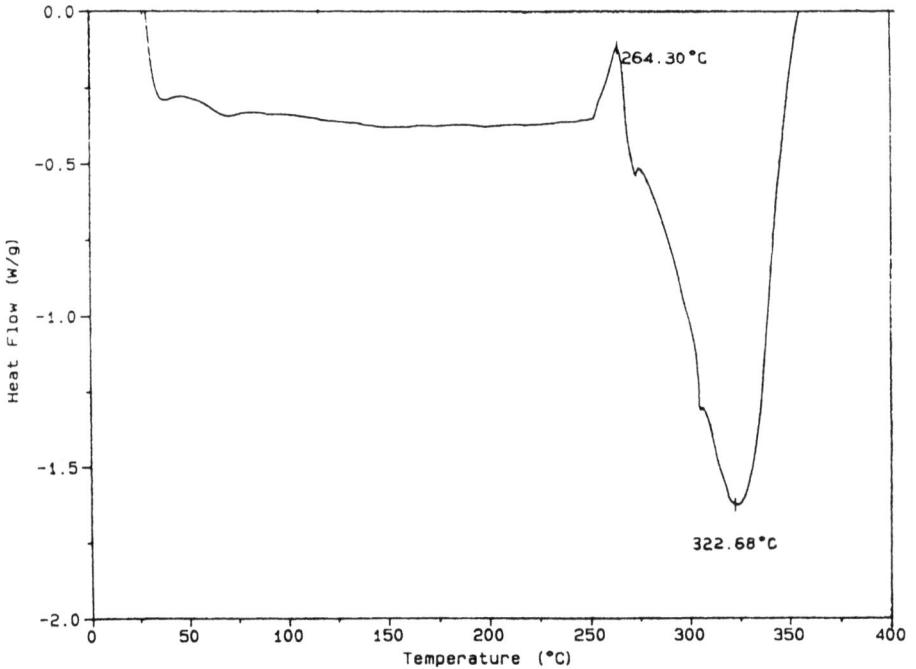

Fig. 15.7   DSC thermogram of PVC geomembrane.

For other polymeric geomembranes, such as CSPE, CPE and PVC, their melting behavior is much more difficult to be detected using DSC. These polymers are very sensitive to thermal degradation, as shown by the TGA results. In addition, they have relatively low crystallinity when compared to PE. Figure 15.7 shows a DSC thermogram of a PVC geomembrane. There is an exothermic peak at 264°C which is due to degradation of the stabilizers. Its maximum value will shift to higher temperatures as the amount of stabilizer is increased, as described by Turi (1981). The large endothermal peak at 323°C is due to dehydrohalogenation, which is the beginning of its thermal degradation. A DSC thermograph of CSPE polymer is given in Figure 15.8. The first transition is an endothermal reaction at 253°C and is caused by the loss of SO_2Cl groups, according to Smith (1966). The second large endothermal reaction at 324°C is due to dehydrohalogenation, which is the beginning of its thermal degradation.

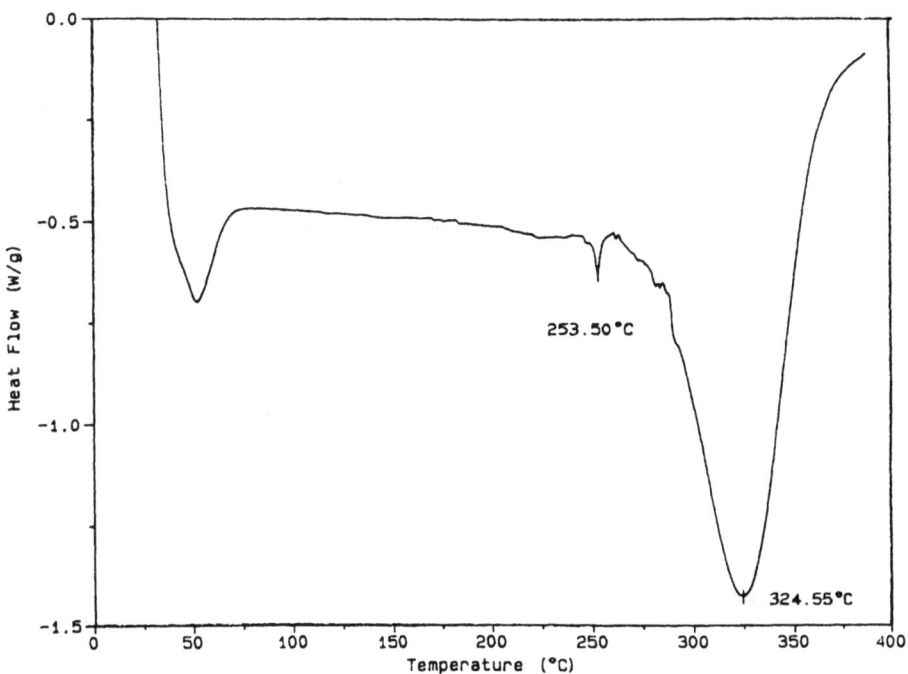

Fig. 15.8 DSC thermogram of CSPE geomembrane.

15.4 Thermomechanical analysis (TMA)

Thermomechanical analysis, or "TMA", is a technique in which the deformation of a material is measured under static load, as the material is subjected to a controlled temperature gradient. Hence the linear coefficient of thermal expansion of the material can be deduced by using this method. In addition, the glass transition temperature and softening point can also be detected.

In a TMA test the sample is placed on the platform of a quartz sample tube. A quartz probe is connected to the armature of a linear variable differential transformer and any change in the position of the armature results in an output voltage from the transformer which is then recorded. The probe assembly includes a weight tray, which permits a choice of loadings on the sample surface. In general, most polymers are tested at a heating rate of 5 to 10°C/min under a nitrogen atmosphere. The load acting on the specimen is 5 grams. Table 15.3 (after Harper, 1975) gives the coefficient of linear thermal expansion of various types of polymeric geomembranes. More details about the application of thermomechanical analysis to geosynthetics are illustrated by Thomas and Verschoor (1988).

The use of thermomechanical analysis for the evaluation of the durability of a PVC geomembrane 24 years old was recently presented by Cazzuffi and Venesia (1990).

Table 15.3 The coefficient of linear thermal expansion of polymeric geomembranes, after Harper (1975).

Geomembrane Material	Coefficient of Linear Expansion $(m/m/°C \times 10^{-5})$
HDPE	11.0 - 13.0
MDPE	14.0 - 16.0
LDPE	10.0 - 22.0
PVC	7.0 - 25.0

15.5 Infrared absorption spectroscopy (IR)

Infrared spectroscopy, or "IR", is a technique used to qualitatively and quantitatively analyze of polymers. It can identify the specific functional groups in the polymer, and determine the concentration of such groups.

The functional groups in the molecule are always in vibrational or rotational motion. If the frequency of the incident radiation matches a natural vibrational frequency of the group, a net energy transfer will occur, resulting in a change in the amplitude of the molecular vibration. Therefore, the absorption bands in an infrared spectrum are at frequencies corresponding to the frequencies of vibration of the functional group concerned.

Certain functional groups in the polymer, such as $-CH_2-$ and $-C=N-$, exhibit discrete absorption spectra. Their bands are virtually unchanged in any compound in which they occur. On the other hand, for some other groups, such as $C=O$ and $C-O$, their absorption peak position are affected by the adjacent groups in the molecule. For example, the $C-O$ peak in methanol is 1034 cm^{-1}; in ethanol it is 1053 cm^{-1}. These variations result from a coupling of the $C-O$ stretching with adjacent $C-C$ or $C-H$ vibrations. Therefore, IR spectroscopy not only can identify the group but also can provide information on the nature of adjacent groups in the molecule. Generally, there are many different functional groups in the repeat unit of the polymer. The absorption spectrum of the polymer is basically the sum of the individual spectrum of these groups. For example, the spectrum of polystyrene consists the bands of $-CH_2-$, benzene, and $C-H$.

One of the applications of this technique in geomembranes is to study thermal and photodegradation of the polymers. For example, the oxidation reaction of polyethylene forms a carbonyl group ($C=O$), which has an absorption band from 1710 to 1735 cm^{-1}. The degrees of oxidation of the polymer can be obtained by comparing the concentration of carbonyl groups in the unexposed material to those of the exposed material. In addition, it can be used to evaluate the suitable stabilisation in PE. Figure 15.9 shows the extent of oxidation in different formulated PE. In this case, formulation C9 would be better than C6. Another common application is to identify the type of plasticizers used in the polymer after solvent extraction.

IR has also been applied to analyze bitumen materials. The acid index of the bitumen correlated to the concentration of carboxylic polar groups at 1705 cm^{-1}. The concentration of paraffinic groups by absorption is measured at 720 cm^{-1}. For modified bitumen materials, the nature of substitutes used in modification can be determined using IR, e.g. atactic polypropylene and SBS.

Specimen preparation is an important step to obtain a good and complete spectrum. For PE geomembranes, the specimen can be in the form of a thin film (0.03-0.005 mm), which is obtained by using a microtome. As an alternative, the specimen can be in the form of powder, which is obtained by grinding the polymer under liquid nitrogen temperature. For other types of geomembranes the polymer can be dissolved in a solvent and then cast into a thin film.

15.6 Chromatography (GC and HPLC)

Chromatography is a method that permits one to separate, isolate and identify closely related components of complex mixtures. All of the various chromatography methods employ a stationary phase and a mobile phase. Components of a mixture are carried through the stationary phase by the flow of the mobile phase. Separations are based on differences in migration rates among the specimen components. The test specimens must be soluble in the mobile phase. They must also be capable of interacting with the stationary phase either by dissolving in it, being adsorbed by it, or reacting with it chemically.

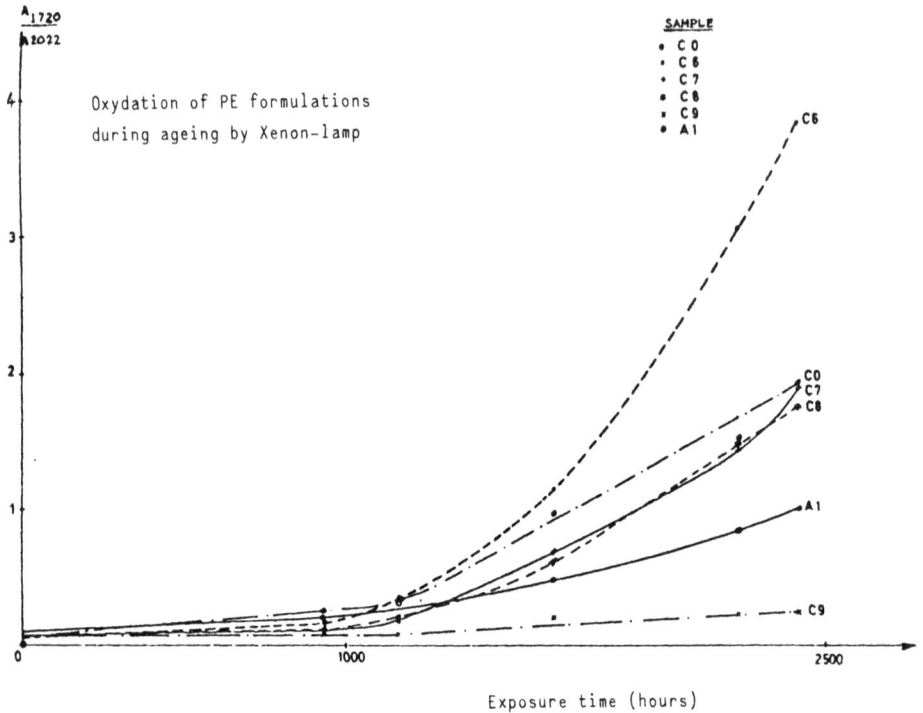

Fig. 15.9 Degree of oxidation in different formulated PE resins.

In gas chromatography, or "GC", the components of a vaporized specimen are fractionated. The mobile phase is gas (nitrogen or helium), and the stationary phase is a liquid supported on an inert solid matric of the column. The rates at which the various components move along the column depends upon their tendency to dissolve in the liquid phase. The components with low solubility in the liquid phase will rapidly flow through the column.

The specimen can be prepared in the liquid or gaseous phase, and is injected into apparatus. The sample port, located at the head of the column, is heated 50°C above the boiling point of the solution. It then is fractionated by the column. Finally the concentration of each component is obtained by the detector. The result, which is called a chromatogram, is expressed in a plot of concentration against time. A series of symmetric peaks is obtained. The position of the peaks serve to identify the components of the specimen; the areas under the peaks are related to the concentration.

This technique has been widely used to analyze plasticizers in geomembranes. It can determine both the amount of plasticizer in the material and can identify their types. However, the plasticizers must first be extracted from the geomembrane. The recommended solvents for various geomembranes are list in Table 15.4, modified from Haxo (1988).

Table 15.4 Extraction solvents for different polymeric geomembranes, modified from Haxo (1988).

Geomembrane Types	Extraction Solvent
PVC	2:1 blend of carbon tetrachloride and methyl alcohol Ethyl-ether
CPE	n-Heptane
CSPE	Acetone

In <u>liquid chromatography</u>, or "LC", the mobile phase is liquid and the stationary phase is liquid or solid. The classical method employs a glass tube with solid particles as the stationary phase. To assure reasonable flow rates, the solid particles are kept larger than 150 microns to 200 microns in diameter. However, the separation process is still very time consuming. Recently, high performance liquid chromatography, or "HPLC", enables one to obtain a high flow rate, using packings with particle diameters as small as 10 microns, and pumping at high pressures.

The advantage of HPLC is that the sample does not need to be transformed from the liquid into the gaseous phase. This is one of the techniques used to determine and identify the additives in PE geomembranes. However, a high temperature environment is required during the process.

15.7 Density determination (ϱ)

The crystallinity of a polymer is directly related to its density. For a homogeneous polymer (i.e., one without such additive as inorganic fillers or plasticizers), its crystallinity can be determined by measuring the density.

The most common method to measure the density of a geomembrane is ASTM D-792 (1988) "Specific Gravity (Relative Density) and Density of Plastics by Displacement". A geomembrane specimen of 1 to 50 g is weighed in air and then in water. The specific gravity of the specimen is the ratio of the weight in air to the weight difference in air and in water. For geomembranes which are lighter than water, a sinker must be applied. If a sinker is not applied, a low density liquid must be used instead of water. This particular test method is very straight forward and the major apparatus is a analytical balance, with a precision within 0.1 mg. The accuracy of the test is about 0.05%, but note that the specimen size is relatively large, particularly for geomembranes with low density.

An alternative density measuring method is ASTM D-1505 (1988) "Density of Plastics by the Density-Gradient Technique". This test method is designed to yield results which are accurate to better than 0.05%. In addition, the required size of the specimen is very small. A one meter long glass gradient tube is filled with solution consisting of a mixture of water and alcohol in the density a range of no more than 0.02 g/cm^3. Hence the sensitivity of the column is $2x10^{-5}$ $g/cm^3/mm$. For PE it would typically be 0.930 g/cm^3 at the top, to 0.950 g/cm^3 at the bottom. A minimum of five glass floats covering the desired density range are evenly distributed through the tube. Their position is plotted against density so that a calibration curve is obtained. A test specimen with size not less than 0.127 mm^3 is gently placed in the tube, and its equilibrium position is noted. The density of the specimen can then be deduced using the calibration curve.

This technique is often used to measure the density of PE geomembranes. However, the carbon black in the geomembrane increases the total density value. Therefore the true resin density must be calculated by knowing the amount and density of the carbon black. The different density ranges for various PE are as shown in Table 15.5.

Table 15.5 Density ranges for various polyethylene types, modified from Kurtz (1985).

Types Range	Density Range of Resin (g/cm^3)	Calculated Density of Compound (g/cm^3)
HDPE	0.940 to 0.970	0.951 to 0.981
MDPE	0.930 to 0.940	0.941 to 0.951
LDPE	0.915 to 0.930	0.926 to 0.941
VLDPE	0.900 to 0.906	0.911 to 0.917

* Calculations assume that the percent carbon black in the material is 2.5%, and the density of the carbon black is 1.8 g/cm^3.

The test method also has provisions for measuring the density of polymers greater than one. In such cases, the column is filled with a different mixture of two solutions, depending on the polymer. For example, water and calcium nitrate is used for measuring polyethylene terephthalate. However, the actual polymer's density in highly compounded materials, such as PVC, CSPE and CPE geomembranes is very difficult to deduce. This is because more than one additive has been compounded in these geomembranes, and each additive will affect the density of the product in some largely unknown manner.

15.8 Melt index (MI)

The melt index test, or "MI", is used to measure the flow rate of a molten polymer at a specified temperature and pressure. This is an important property for the processing of PE geomembranes and one that is routinely measured for quality control during manufacturing. In addition, the MI value is qualitatively related to the molecular weight of the polymer. A high MI value means a low molecular weight polymer and vice-versa.

The flow rate ratio (FRR) is the ratio of two MI values under different pressures at the same temperature. It indicates the width of the molecular weight distribution curve. A high value reflects a broad distribution curve and vice versa. Details of the test method and the various conditions are described in ASTM D-1238 (1988). For PE, conditions P (190°C and 5 kg) and E (190°C and 2.16 kg) are commonly used. The resulting MI value is expressed in grams per 10 minutes. Typical MI values of PE geomembranes under condition E are between 0.2 and 1.0 grams per 10 minutes, and the FRR (P/E) is around 2.5 grams per 10 minutes. The test, however, is not accurate for polymers which contain plasticizers and thus is not applicable to PVC or CSPE.

15.9 Gel permeation chromatography (GPC)

In general, the molecular weight of polymer is expressed in terms of one, or all, of the following defined values:

Weight Average Molecular Weight ($\bar{M}_w$): $$\bar{M}_w = \frac{\sum N_x M_x^2}{\sum N_x M_x}$$

Number Average Molecular Weight ($\bar{M}_n$): $$\bar{M}_n = \frac{\sum N_x M_x}{\sum N_x}$$

Solution Viscosity Molecular Weight ($\bar{M}_v$): $$\bar{M}_v = \left[\frac{\sum N_x M_x^{\alpha+1}}{\sum N_x M_x}\right]^{1/\alpha}$$

where:
M_x = molecules with weight x
N_x = number of molecules with weight x
α = a constant

All of these values reflect some aspects of the average value of the molecular weight of the polymer. However, all polymers consists of a large distribution of different molecular weights, as shown schematically in Figure 15.10. The positions of each of the molecular weight average values are also indicated in the same figure.

Small changes in the high molecular weight region can improve the properties of a polymer significantly. Such changes can only be detected by measuring the molecular weight distribution. Gel permeation chromatography, or "GPC", is the generally acceptable technique for this measurement, as described by Allcock (1981). This is essentially a process for fractionation of polymer according to their size.

Therefore, it has two functions: analyze the molecular weight distribution of the polymer and fractionate the polymer into various groups, according to their molecular weight for any further analysis.

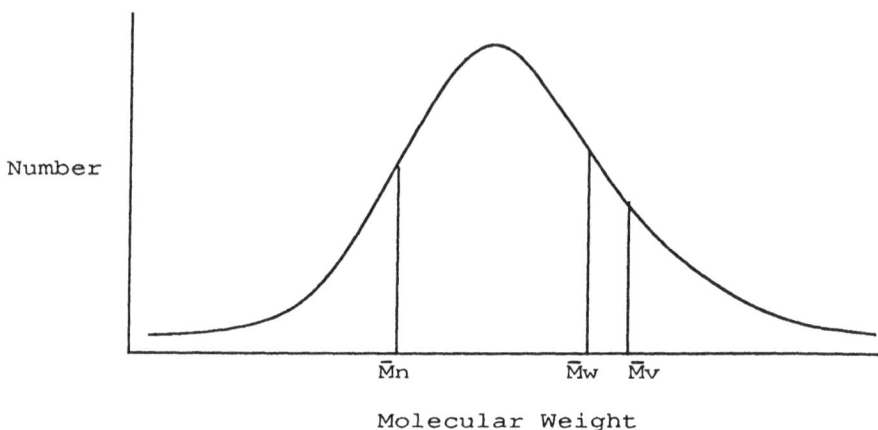

Fig. 15.10 Molecular weight distribution curve.

    A polymer is dissolved the same solvent used for transport and are
introduced into the system by means of a syringe. This solution flows
through a column which consists of a series of different pores sizes
or a single pore size. As the dissolved solute passes through a
column, the smaller molecules will "enter" and "explore" the openings
of the pores. This is dual action of diffusion and exclusion occurs
continuously throughout the separation process. This results that the
larger molecules eluted before the smaller ones. After passage through
the column system, the eluent passes through detector which can be a
differential refractive index (or absorbance) between the eluted
solution and the pure solvent is plotted as function of time,
resulting a plot of the molecular-weight distribution.
    The columns are commonly packed with polystyrene crosslinked with
divinyl benzene. The pore size is determined by the amount of
crosslinking, since the tunnels are formed by the solvent-swelled
cavities that exist between the crosslinks. The choice of the moving
phase, i.e. solvent, is very critical. It should be an effective
solvent for the specimen and be able to cause certain degree of
swelling of the polystyrene, thereby exposing pores of the packing to
the solvent.
    The cost of GPC equipment is very high and the specimen
preparation is very tedious. High temperature is required during the
entire process for testing PE materials, and a significant amount of
filtration is needed to separate the carbon black particles from the
polymer.

334

15.10 Conclusions

Knowledge of the specific compound from which a geomembrane is made is valuable information in many situations. It can be analogized to one inquiring as to what specific type of a clay soil is being used for a particular application, e.g., kaolinite, illite or montmorillonite, and furthermore what is the predominate ion, its concentration and species. In fact, the variation of polymer compounds is every bit as great, if not greater, than the variation of types of clay soils. Furthermore, polymeric materials from which geomembranes are made are specifically formulated from the outset leading to an endless variety of present and future formulations.

The chemical identification methods described and illustrated in this chapter are those listed in Table 15.2. Also included in this table are some comments as to the individual method's advantages and disadvantages for geomembrane identification studies.

It is felt that the type of information gained in the use of the chemical identification tests described in this chapter is paramount to both the short and long term behavior of polymeric geomembranes. We say short-term, for the stress-strain-time behavior of the geomembrane bears direct relationship to the type of polymer and its specific compounds. We say long-term, for an understanding of polymeric aging and durability lies within an understanding the geomembrane's molecular structure.

References

Allcock, H. (1981) "Contemporary Polymer Chemistry", published by Prentice and Hall.

ASTM D-792 (1988) "Specific Gravity (Relative Density) and Density of Plastics by Displacement", Vol. 08.01, 299-302.

ASTM D-1238 (1988) "Flow Rates of Thermoplastics by Extrusion Plastometer", Vol. 08.01, 402-410.

ASTM D-1248 (1989) "Polyethilene Plastics Molding and Extrusion Materials", Vol. 08.01, 423-428.

ASTM D-1505 (1988) "Density of Plastics by the Density-Gradient Technique", Vol. 08.01, 468-473.

ASTM D-3895 (1988) "Copper-Induced Oxidative Induction Time of Polyolefins by Thermal Analysis", Vol. 08.03, 251-255.

Cazzuffi. D.A. and Venesia, S. (1990) "The Use of Thermal Analysis for the Evaluation of the Durability of a PVC Geomembrane 24 years old", presented at ASTM Symposium on Microstructure of Geosynthetics, Orlando, January 1989, published by ASTM (STP 1076) in 1990.

Harper, A.C. (1975) "Handbook of Plastics and Elastomers", published by McGraw-Hill Book Company.

Haslam, J. "Identification and Analysis of Plastics", second edition, Butterworth and Co., London.

Haxo, H.E. Jr.,(1988) "Lining of Waste Containment and Other Impoundment Facilities", EPA/600/2-88/052, Matrecon Inc.

Krause, A., Lange, A. and Ezrin, M. (1983) "Plastics Analysis Guide: Chemical and Instrumental methods", published by Hanser Publisher, Munich.

Kurtz, S.J., "VLDPE Plugs a Gap in PE's Density Spectrum", Plastics Engineering, 59-62, September.

Skoog, D.A. and West, D.M., "Principles of Instrumental Analysis", second edition, published by Saunders College/Holt. Rinehart and Winston.

Smith, D.A. (1966) "Characterization on the Thermal Analysis", Polymer Letters, Vol. 4, 215-221.

Thomas, R.W. and Verschoor, K.L. (1988) "Thermal Analysis of Geosynthetics", Geotechnical Fabrics Report, May-June.

Turi, E.A. (1981) "Thermal Characterization of Polymer Materials", published by Academic Press.

Vanni, A., Saini, G. and Campi, E. (1989) "Indagini sulla durabilità di geomembrane a contatto con percolati di rifiuti industriali. Compatibility Tests of Geomembranes with Industrial Leachates" (in Italian, with an English abstract), presented at Second Italian Conference on Geosynthetics in Earth Structures, Bologna, October 1988, published in L'Ingegnere, 1-4, 1989.

APPENDIX

SYMBOLS

This list of symbols has been prepared by
J.P. Giroud

1   General symbols

1.1 *Dimension symbols*

Symbols used for the dimensions are:

    L : length
    M : mass
    T : time
    - : dimensionless

1.2 *Unit symbols*

m	meter
m^2	square meter
m^3	cubic meter
mm	millimeter
μm	micron
g	gram
mg	milligram
kg	kilogram

s	second
N	newton
kN	kilonewton
Pa	pascal
kPa	kilopascal
MPa	megapascal
J	joule
tex	tex
L	liter
o	degree
%	percent
-	pure number

The following relationships exist:

$$1 \text{ Pa} = 1 \text{ N/m}^2$$

$$1 \text{ tex} = 1 \text{ mg/m}$$

$$1 \text{ J} = 1 \text{ mN}$$

1.3 *Mathematical symbols*

ln x   natural logarithm of x

lg x   logarithm of x base 10

1.4 *Subscripts*

Subindex	Applies to
a	air or active (earth pressure)
d	dry state
f	failure or final
g	geotextile
H	horizontal
i	immediate or initial
p	passive (earth pressure)
r	radial
s	solid particles
sat	saturated

sec	secant
u	undrained conditions
V	vertical
w	water
x, y	two orthogonal horizontal axes
z	vertical axis
o	at rest or initial conditions
1,2,3	principal directions

1.5 Geometry and kinematics

L	L	(m)	length
B	L	(m)	breadth
H	L	(m)	height, thickness
D	L	(m)	depth
z	L	(m)	vertical coordinate
d	L	(m)	diameter
A	L^2	(m^2)	area
V	L^3	(m^3)	volume
t	T	(s)	time
v	$L\ T^{-1}$	(m/s)	velocity
g	$L\ T^{-2}$	(m/s^2)	acceleration due to gravity ($g = 9.81\ m/s^2$)

1.6 Stress and strain

u	$ML^{-1}T^{-2}$	(kPa)	pore pressure
			pressure of the fluid in the voids of a porous medium (soil, geotextile)
σ	$ML^{-1}T^{-2}$	(kPa)	normal stress
σ'	$ML^{-1}T^{-2}$	(kPa)	effective normal stress
			normal stress transmitted by intergranular contacts (σ' = σ - u for saturated soils)
τ	$ML^{-1}T^{-2}$	(kPa)	shear stress
$\sigma_1, \sigma_2, \sigma_3$	$ML^{-1}T^{-2}$	(kPa)	principal stresses

ε	$-$	(-, %)	strain, élongation

*change in length per unit
length in a given direction
(called elongation if
expressed in % and
applied to a geotextile)*

$\varepsilon_1, \varepsilon_2, \varepsilon_3$	$-$	(-,%)	principal strains

1.7 *Hydraulic parameters*

h	L	(m)	hydraulic head or potential

*sum of pressure height
(u / γ_w) and geometrical
height (z) above a given
reference level*

q	$L^3 T^{-1}$	(m³/s)	rate of discharge

*volume of water seeping
through a given area per
unit of time*

v	$L\ T^{-1}$	(m/s)	discharge velocity

*rate of discharge per total
unit area perpendicular to
direction of flow*

i	$-$	(-)	hydraulic gradient

*loss of hydraulic head per
unit length in direction
of flow*

j	$ML^{-2}T^{-2}$	(kN/m³)	seepage force per unit volume

*force per unit volume of
a porous medium generated
by action of the seeping
fluid upon the solid
elements of the porous
medium ($j = i\gamma_w$)*

2 Properties of fluids

ρ_w	ML^{-3}	(kg/m³)	density of water
γ_w	$ML^{-2}T^{-2}$	(kN/m³)	unit weight of water

η_w $ML^{-1}T^{-1}$ (kg/ms) dynamic viscosity of water

NOTE: Instead of w , use any other appropriate subscript of other fluids (e.g.: η_a, dynamic viscosity of air)

3   Properties of soils

3.1 *Solid particles and their distribution*

ρ_s ML^{-3} (kg/m^3) density of solid particles

          ratio between mass and volume of solid particles

γ_s $ML^{-2}T^{-2}$ (kN/m^3) unit weight of solid particles

          weight of solid particles per unit volume

d   L   (mm)   particule diameter

          particle size as determined by sieve analysis or wet mechanical analysis

d_n L   (mm)   n percent-diameter

          diameter corresponding to n percent by weight of finer particles

c_U -   (-)   uniformity coefficient

          defined as : d_{60}/d_{10}

3.2 *Physical properties of soils*

ρ ML^{-3} (kg/m^3) density of soil

          ratio between total mass and total volume of soil

γ $ML^{-2}T^{-2}$ (kN/m^3) unit weight of soil

          ratio between total weight and total volume of soil

341

ρ_d ML^{-3} (kg/m^3) density of dry soil

ratio between mass of solid particles and total volume of soil

γ_d $ML^{-2}T^{-2}$ (kN/m^3) unit weight of dry soil

ratio between weight of solid particles and volume of soil

ρ_{sat} ML^{-3} (kg/m^3) density of saturated soil

ratio between total mass and total volume of completely saturated soil

γ_{sat} $ML^{-2}T^{-2}$ (kN/m^3) unit weight of saturated soil

ratio between total weight and total volume of completely saturated soil

ρ' ML^{-3} (kg/m^3) density of submerged soil

difference between density of soil and density of water

γ' $ML^{-2}T^{-2}$ (kN/m^3) unit weight of submerged soil

difference between unit weight of soil and unit weight of water

e –   (–) void ratio

ratio between volume of voids and volume of solid particles

n –   (–. %) porosity

ratio between volume of voids and total volume of soil

w –   (–.%) water content

ratio between weight of pore water and weight of solid particles (expressed in percentage)

S_r -   (-. %)   degree of saturation

ratio between volume of pore water and volume of voids

3.3 *Permeability of soils*

k LT^{-1} (m/s)   coefficient of permeability (or hydraulic conductivity)

ratio between discharge velocity and corresponding hydraulic gradient (v / i)

3.4 *Mechanical properties of soils*

E $ML^{-1}T^{-2}$ (kPa)   deformation modulus

ratio between a given normal stress change and the strain change in the same direction (all other stresses being constant)

υ -   (-)   Poisson's ratio

ratio between strain changes perpendicular to and in the direction of a given uniaxial stress change

k_s $ML^{-2}T^{-2}$ (kN/m^3)   modulus of subgrade reaction

ratio between change of vertical stress on a rigid plate and corresponding change of vertical settlement of the plate

E_{oed} $ML^{-1}T^{-2}$ (kPa)   oedometric modulus

C_c -   (-)   compression index

slope of virgin compression curve in a semi-logarithmic plot "effective pressure-void ratio":Cc = -Δe/ Δlgσ'

c_v L^2T^{-1} (m^2/s)   coefficient of consolidation

T_v	–	(–)	time factor
			defined as $T_v = t\, c_v / d^2$, t being the time elapsed since application of a change in total normal stress
τ_f	$ML^{-1}T^{-2}$	(kPa)	shear strength
			shear stress at failure in rupture plane through a given point
c'	$ML^{-1}T^{-2}$	(kPa)	effective cohesion
ϕ'	–	(°)	effective angle of internal friction
			shear strength parameter with respect to effective stresses. Defined by the equation: $\tau_f = c' + \sigma' \tan \phi'$
c_u	$ML^{-1}T^{-2}$	(kPa)	apparent cohesion
ϕ_u	–	(°)	apparent angle of internal friction
			shear strength parameters with respect to total stresses. Defined by the equation: $\tau_f = c_u + \sigma \tan \phi_u$. In undrained situation, with saturated cohesion soils, c_u is also called undrained shear strength
CBR	–	(–)	California Bearing Ratio
q_c	$ML^{-1}T^{-2}$	(kPa)	static point resistance (or cone resistance)
			average pressure acting on the conical point in the standard static penetration test
f_s	$ML^{-1}T^{-2}$	(kPa)	local side friction
			average unit side friction acting on the friction sleeve in the standard static cone penetration test
N	–	(–)	SPT blow count
			standardized result of the Standard Penetration Test (number of blows for 30 cm)

P_l $ML^{-1}T^{-2}$ (kPa)   pressuremeter limit pressure

limit pressure defined in the standard Ménard pressuremeter test

E_M $ML^{-1}T^{-2}$ (kPa)   pressuremeter modulus

conventional modulus defined in the standard Ménard pressuremeter test.

3.5 *Consistency of soils*

w_L -   (-.%)   liquid limit

water content of a remolded soil at transition between liquid and plastic states (determined by a standard laboratory test)

w_p -   (-, %)   plastic limit

water content of a remolded soil at transition between plastic and semi-solid states (determined by a standard laboratory test)

w_s -   (-, %)   shrinkage limit

maximum water content at which a reduction of water content will not cause a decrease in volume of the soil mass

I_p -   (-,%)   plasticity index

difference between liquid and plastic limits

I_L -   (-,%)   liquidity index

defined as $(w - w_p) / I_p$

I_C -   (-,%)   consistency index

defined as $(w_L - w) / I_p$

e_{max} -   (-)   void ratio in loosest state

maximum void ratio obtainable by a standard laboratory procedure

e_{min} — (-) void ratio in densest state

maximum void ratio obtainable by a standard laboratory procedure

I_D — (-) density index

defined as $(e_{max} - e) / (e_{max} - e_{min})$

4 Properties of geotextiles

4.1 *Solid elements and their distribution*

ρ_f ML^{-3} (kg/m^3) density of fibers or filaments

d_f L (μm) diameter of fibers or filaments

λ ML^{-1} (tex) linear density of yarns, fibers or filaments

mass per unit length of yarns, fibers or filaments

O_n L $(mm, \mu m)$ n-percent opening size

opening size corresponding to n-percent of finer openings (as measured by a test to be defined, such as sieving glass beads) (e.g. O_{95}, opening size corresponding to 95% of finer openings, sometimes called "Equivalent Opening Size")

Note: Standard sieve numbers vary from one country to another. Opening sizes expressed by a standard sieve number would not be understood by most of the readers. Consequently, authors are strongly requested to express opening size in millimeters (mm) or microns (μm). Standard sieve numbers can be given in parenthesis after the value in mm or μm.

4.2 Physical properties of a geotextile

μ	ML^{-2}	$(kg/m^2, g/m^2)$	mass per unit area
			ratio between mass and area of geotextile
e	-	(-)	void ratio
			ratio between volume of voids and volume of solid elements of a geotextile
n	-	(-, %)	porosity
			ratio between volume of voids and total volume of geotextiles
T_g	L	(mm)	thickness of geotextile

4.3 Hydraulic properties of a geotextile

k_n	LT^{-1}	(m/s)	coefficient of normal permeability of a geotextile
k_p	LT^{-1}	(m/s)	coefficient of permeability in the plane of a geotextile
Ψ	T^{-1}	(s^{-1})	permittivity of a geotextile $(\Psi = k_n/H_g)$
θ	$L^2 T^{-1}$	(m^2/s)	transmissivity of a geotextile $(\theta = k_p/H_g)$

4.4 Mechanical properties of a geotextile

α_ϵ	MT^{-2}	(kN/m)	force per unit width of the geotextile at a given elongation ϵ (e.g. α_{30} is the force per unit width of the geotextile at 30% elongation)
α_f	MT^{-2}	(kN/m)	force per unit width of the geotextile at failure
J	MT^{-2}	(kN/m)	"modulus" of the geotextile
		Note:	The "modulus" of a geotextile is defined as a force per unit width while modulus is usually defined as a force per unit area.

347

J_i	MT^{-2}	(kN/m)	initial (tangent) "modulus" of the geotextile
$J_{t\epsilon}$	MT^{-2}	(kN/m)	tangent "modulus" of the geotextile at elongation ϵ (e.g. J_{t30} is the tangent "modulus" of the geotextile at 30% elongation)
$J_{sec\ \epsilon}$	MT^{-2}	(kN/m)	secant "modulus" of the geotextile between 0 and elongation ϵ (e.g. $J_{sec\ 30}$ is the secant "modulus" of geotextile between 0 and 30% elongation)
υ_g	-	(-)	Poisson's ratio of a geotextile
ζ_t	M^2T^{-2}	(N/tex)	tenacity of a thread
			quotient of the tensile force at failure of a thread to its linear density
ζ_g	M^2T^{-2}	(J/g)	tenacity of a geotextile
			quotient of the force per unit width at failure of a geotextile to its mass per unit area

(Note: 1 J/g = 1000 N/tex)

F_G	MLT^{-2}	(N,kN)	breaking force of geotextile as measured in grab test
F_P	MLT^{-2}	(N, kN)	breaking force of geotextile in a puncture test (to be defined)
F_T	MLT^{-2}	(N, kN)	breaking force of geotextile in a tear test (to be defined)
P_B	$ML^{-1}T^{-2}$	(kPa, MPa)	bursting pressure of a geotextile.

5   Practical problems

5.1 *General*

FS	-	(-)	Factor of safety

5.2 *Earth pressure*

δ	–	(°)	angle of wall friction
			angle of friction between wall and adjacent soil
a	$ML^{-1}T^{-2}$	(kPa)	wall adhesion
			adhesion between wall and adjacent soil
K_a, K_p	–	(–)	active and passive earth pressure coefficients
			dimensionless coefficients used in expressions for active and passive earth pressure
K_o	–	(–)	coefficient of earth pressure at rest
			ratio of lateral to vertical effective principal stress in the case of no lateral strain and a horizontal ground surface

5.3 *Foundations and embankments*

B	L	(m)	breadth of foundation or embankment
L	L	(m)	length of foundation or embankment
D	L	(m)	depth of foundation or base of embankment beneath ground
s	L	(m)	settlement
U	–	(–,%)	degree of consolidation
			ratio of settlement at a given time to final settlement

5.4 *Slopes*

H	L	(m)	vertical height of slope
D	L	(m)	depth below toe of slope to hard stratum
β	–	(°)	angle of slope to horizontal

Conversion factors

Conversion factors between the English system and SI are as follows:

Length, area, volume

1 mil	= 25.4 microns (μm)
1 inch	= 25.4 millimeters (mm)
1 foot	= 0.305 meter (m)
1 yard	= 0.914 meter (m)
1 square foot	= 9.29×10^{-2} square meters (m^2)
1 acre	= 4047 square meters (m^2)
1 cubic foot	= 2.83×10^{-2} cubic meters (m^3)
1 gallon	= 3.79×10^{-3} cubic meters (m^3)
1 gallon	= 3.79 liters (L)

Mass

1 ounce	= 28.3 grams (g)
1 pound	= 0.454 kilogram (kg)
1 ton (US)	= 907 kilograms (kg)
1 long ton (GB)	= 1016 kilograms (kg)

Linear Density

1 denier	= 0.111 milligram per meter (mg/m)
1 denier	= 0.111 tex (tex)

Mass per unit area

1 ounce per square yard	= 33.9 gram per square meter (g/m^2)
1 ounce per square yard	= 3.39×10^{-2} kilogram per square meter (kg/m^2)

Density

1 pound (mass) per cubic foot	= 16.0 kilograms per cubic meter (kg/m^3)

Force

1 pound	= 4.45 newtons (N)
1 kip	= 4.45 kilonewtons (kN)

Force per unit width or per unit length

1 pound per inch	= 175 newtons per meter (N/m)
1 pound per inch	= 0.175 kilonewton per meter (kN/m)

Stress, pressure

1 pound per square inch	= 6895 pascals (Pa)
1 pound per square inch	= 6.89 kilopascals (kPa)
1 pound per square foot	= 47.9 pascals (Pa)
1 kip per square foot	= 47.9 kilopascals (kPa)
1 ton (US) per square foot	= 95.8 kilopascals (kPa)

Unit weight or reaction modulus

1 pound (force) per cubic foot = 157 newtons per cubic meter (N/m^3)

1 pound (force) per cubic foot = 0.157 kilonewton per cubic meter (kN/m^3)

Velocity, Coefficient of Permeability

1 centimeter per second = 0.010 meter per second (m/s)

Flow rate

1 gallon per minute = 6.31×10^{-2} liter per second (L/s)

1 gallon per minute = 6.31×10^{-5} cubic meter per second (m^3/s)

1 cubic foot per second = 2.83×10^{-2} cubic meter per second (m^3/s)

1 cubic foot per minute = 0.472 liter per second (L/s)

1 cubic foot per minute = 4.72×10^{-4} cubic meter per second (m^3/s)

1 gallon per day = 4.38×10^{-5} liter per second (L/s)

Dynamic viscosity

1 poise = 0.100 kilogram per meter-second (kg/sm)

1 pound (force) second per square foot = 0.478 kilogram per meter-second (kg/sm)

(Note: 1 kg/sm = 1 Pa.s)

Gravity

g = 32.2 feet per square second = 9.81 meters per square second (9.81 m/s^2)

U. S. Sieve Designation	Opening Size	
	mm	μm
200	0.075	75
170	0.090	90
140	0.106	106
120	0.125	125
100	0.150	150
80	0.180	180
70	0.212	212
60	0.250	250
50	0.300	300
45	0.355	355
40	0.425	425
35	0.500	500
30	0.600	600
25	0.710	710
20	0.850	850
18	1.000	
16	1.18	
14	1.40	
12	1.70	
10	2.00	
8	2.36	
7	2.80	
6	3.35	
5	4.00	
4	4.75	

INDEX

For Product Safety Concerns and Information please contact our EU
representative GPSR@taylorandfrancis.com
Taylor & Francis Verlag GmbH, Kaufingerstraße 24, 80331 München, Germany